EMC

IN MEMORIAM

Paul Chatterton

B.Sc., Ph.D., C.Eng., F.I.E.E., S.M.I.E.E.E., F.Inst.P.

formerly
Reader in Electrical Engineering and Electronics, University of Liverpool

Paul Chatterton died on the night of 26 May 1991, only 10 days after the manuscript of this book was delivered to the publishers. Although often in pain and totally exhausted, Paul worked on the manuscript to the end, constantly striving for precision, clarity and economy of expression. He was characteristically very self-critical and was strongly motivated by a desire to encapsulate in this book a view-point, moulded by considerable experience in teaching, industry and consultancy work, that would help young people in particular to start working in EMC. There is in this book a lot of the soul of a kind and gentle man.

MAH
29 May 1991

EMC

Electromagnetic Theory to Practical Design

Paul A. Chatterton

and

Michael A. Houlden

University of Liverpool
UK

JOHN WILEY & SONS
Chichester · New York · Brisbane · Toronto · Singapore

Other Wiley Editorial Offices

John Wiley & Sons, Inc., 605 Third Avenue,
New York, NY 10158-0012, USA

Jacaranda Wiley Ltd, G.P.O. Box 859, Brisbane,
Queensland 4001, Australia

John Wiley & Sons (Canada) Ltd, 5353 Dundas Street West, Fourth Floor
Etobicoke, Ontario M9B 6H8, Canada

John Wiley & Sons (SEA) Pte Ltd, 37 Jalan Pemimpin 05-04,
Block B, Union Industrial Building, Singapore 2057

Library of Congress Cataloging-in-Publication Data:

Chatterton, Paul A.
EMC : electromagnetic theory to practical design / Paul A.
Chatterton, Michael A. Houlden.
p. cm.
Includes bibliographical references and index.
ISBN 0 471 92878 X
1. Electromagnetic compatibility. 2. Electronic apparatus and
appliances—Design and construction. I. Houlden, M. A. (Michael
A.) II. Title.
TK6553.C43 1991
621.382′24—dc20 91-28311
CIP

A catalogue record for this book is available from the British Library.

Typeset by Mathematical Composition Setters Ltd, Salisbury, Wiltshire
Printed in Great Britain by Courier International, East Kilbride, Lanarkshire

Dedicated to Amnesty International
on the 30th anniversary of its foundation by one caring man,
Peter Benenson

PAC
MAH
May 1991

'...The first process therefore in the effectual study of the science, must be one of simplification and reduction of the results of previous investigation to a form in which the mind can grasp them. The results of this simplification may take the form of a purely mathematical formula or of a physical hypothesis. In the first case we entirely lose sight of the phenomena to be explained; and though we may trace out the consequences of given laws, we never obtain more extended views of the connexions of the subject. If, on the other hand, we adopt a physical hypothesis, we see the phenomena only through a medium, and are liable to that blindness to facts and rashness in assumption which a partial explanation encourages'

James Clerk Maxwell (1856)

'My neighbor has had a new heart pacemaker fitted. Every time he makes love my garage doors open'

Bob Hope, the American comedian, (circa 1975)

Contents

Preface	**xiii**
1 An Overview of Electromagnetic Compatibility	**1**
1.1 Historical Perspective and Basic Definitions	1
1.2 Examples of EMC Problems	4
1.3 Formal Definition of EMC	6
1.4 Noise Sources	8
1.4.1 Natural sources	8
1.4.2 Man-made sources	9
1.5 General Methods of Solving Interference Problems and Meeting EMC Requirements	15
1.6 EMC Regulations and Testing	16
1.7 Organisation of the Book	22
References	22
2 The EMC Implications of Basic Electromagnetic Ideas	**25**
2.1 Introduction	25
2.2 The Basic Fields of Electromagnetism	25
2.2.1 Ferromagnetic material	28
2.2.2 Summary	29
2.3 Maxwell's Equations from an EMC Perspective	30
2.3.1 Maxwell's equations	30
2.3.2 Poisson's and Laplace's equations	32
2.4 Near and Far Field Approximations and Energy Flow	33
2.4.1 Short wire antenna	34
2.4.2 Small loop antenna	34
2.4.3 The near and far field	35
2.4.4 Energy flow around a short wire antenna	37
2.4.5 High and low impedance fields	38
2.4.5.1 Fields around a short wire antenna	38
2.4.5.2 Fields around a loop antenna	40
2.4.6 Reactive fields	40
2.5 Waves in Different Media	41
2.5.1 Waves in a dielectric medium	41
2.5.1.1 Refractive index	43
2.5.1.2 Characteristic impedance of a dielectric	43
2.5.2 The impedance of the near field	44
2.5.3 The importance of the concept of impedance	46
2.5.4 Impedance in front of a boundary	47
2.5.4.1 Half-wave dielectric windows	49
2.5.4.2 Quarter-wave coatings	50
2.5.5 Summary of the impedance concept	51

2.5.6 Plane waves in an arbitrary medium 51
2.5.6.1 The propagation constant γ 51
2.5.6.2 Skin depth 53
2.5.6.3 Current distributions in conductors 55
2.5.7 Wave propagation in a good conductor 56
2.5.8 Intrinsic impedance of conductors 56
2.5.9 Diffusion 58
2.6 Integral Forms of Maxwell's Equations 60
2.6.1 Faraday's and Ampère's laws 60
2.6.2 Electric fields in conductors 62
2.7 Some Illustrative EMC Examples 63
2.7.1 Interference in a small loop 63
2.7.2 Interpreting measurements at different distances 64
2.7.3 Capacitive and inductive coupling 65
2.7.4 Switching transients 66
2.7.4.1 Switching a transformer on 67
2.7.4.2 Switching a transformer off 68
2.7.4.3 Very early time transients 69
2.8 Summary 71
References 71

3 Waves in Guiding Structures and Layered Materials 73

3.1 Impedance in Lossy Materials 73
3.1.1 TEM waves incident on boundaries 75
3.2 Transmission of a TEM Wave 78
3.2.1 A first approximation to the transmission coefficient 78
3.2.2 Re-reflection effects 79
3.2.3 Decibels, shielding effectiveness and nepers 82
3.2.4 Transmission through conductors 83
3.3 The Reflection Coefficient from Multi-layered Media 86
3.4 Design of Absorbers 87
3.4.1 Introduction 87
3.4.2 Some factors in absorber design 88
3.4.2.1 A hypothetical absorber 88
3.4.3 Performance of absorbers at different frequencies 90
3.4.4 Examples of real absorbers 91
3.5 Overview of Transmission Line Theory 92
3.5.1 Basic concepts 93
3.5.2 Impedance and phase shift of an ideal line 94
3.5.3 The characteristic impedance of a lossy transmission line 95
3.5.4 Wave velocity down a lossy line 96
3.5.5 Voltage and current reflection coefficients 97
3.5.6 Input impedance of short transmission lines 99
3.5.7 Concluding remarks 100
3.6 Coupling between transmission lines 101
3.6.1 Inductively coupled directional couplers 101
3.6.2 Coupling over short line lengths 103
3.7 Coupling to transmission lines 105
3.7.1 Mathematical framework 106
3.7.2 Coupling of shield currents to the signal wires 108
3.8 Waveguides and Cavities 110
3.8.1 Waveguides 110

3.8.2 Cut-off frequency and attenuation constant 111
3.8.3 Shielding effectiveness of apertures 112
3.8.4 Cavities and cavity resonance 114
References 116

4 Shielding Theory and Practice **119**
4.1 Basic Physical Ideas 119
4.2 Static (or Quasi-static) Field Shielding 121
4.2.1 Magnetostatic shielding 121
4.2.2 Superconducting shields 124
4.2.3 Electrostatic shielding 124
4.3 Equivalent Circuit Models of Shielding 128
4.3.1 Electric field screening 129
4.3.2 Quasi-static magnetic field screening 132
4.3.3 Conclusions 135
4.4 Plane Wave or Transmission Line Models of Shielding 136
4.5 Extensions of Plane Wave Theory into Non-ideal Situations 139
4.6 The Relationships of Shielding Theories to Practical Shields 145
4.6.1 Apertures 150
4.6.2 Windows and thin conducting films 155
4.6.3 Alternative ways of describing shield quality 157
4.6.4 Cables and connectors 161
4.7 Some Conclusions and a Comment about Grounding 163
References 164

5 Spectral Analysis and Antenna Theory in EMC **167**
5.1 Basic Concepts 167
5.1.1 Harmonic distortion 168
5.1.2 Intermodulation distortion or mixing 168
5.2 Spectral Analysis 169
5.2.1 The Fourier series 170
5.2.2 An example of a Fourier series 171
5.2.3 Fourier series of trains of pulses 172
5.2.4 Fourier transforms 175
5.2.5 Spectrum analysers 176
5.2.5.1 The fast Fourier transform 177
5.2.6 Effect of finite rise times 178
5.2.7 Noise voltage in a coil 179
5.2.8 An approximation for the Fourier spectrum 181
5.2.9 Interference bandwidth 184
5.3 Antennas and Radiation 185
5.3.1 Differential-mode and common-mode radiation 185
5.3.2 General properties of antennas 186
5.3.2.1 Power pattern, directivity, and gain 186
5.3.2.2 Radiation resistance 187
5.3.2.3 Effective aperture 188
5.3.3 Slot antennas and apertures 188
5.4 Prediction and Measurement of Radiated Fields 189
5.4.1 Introduction 189
5.4.2 Mathematical basis 190
5.4.2.1 Units 191

5.4.3 Radiation from a loop 192
5.4.3.1 Loops with impedance $Z < Z_0$ 192
5.4.3.2 Loops with impedance $Z > Z_0$ 194
5.4.4 Prediction of radiated fields 196
5.4.4.1 The basic calculation 196
5.4.4.2 A spreadsheet to calculate radiated field strengths 196
5.4.5 Common mode radiation from cables 198
5.4.6 Computer codes to predict radiation 200
5.4.7 Broadband antennas 200
5.5 Generation of Electromagnetic Fields for EMC Testing 203
5.5.1 Introduction 203
5.5.2 Crawford cells 206
5.5.3 The GTEM cell 208
5.5.4 Reverberating chambers 210
References 210

6 Important Factors in Practical EMC Design **213**
6.1 Introduction 213
6.2 Some Simple Coupling Estimates 214
6.3 Grounding 223
6.3.1 Introduction 223
6.3.2 Safety and signal grounds 225
6.3.3 Cable grounds and 'pigtails' 230
6.3.4 Grounding for single and nested shielded enclosures 233
6.4 Passive Components and Filters 238
6.4.1 Passive components 238
6.4.1.1 Introduction 238
6.4.1.2 Conductors 239
6.4.1.3 Resistors 239
6.4.1.4 Capacitors 240
6.4.1.5 Inductors 241
6.4.2 Filters 242
6.5 Isolation and Suppression 247
6.5.1 Isolation techniques 247
6.5.1.1 Balanced circuits 249
6.5.1.2 Transformers and common mode chokes 250
6.5.1.3 Opto-isolators and optical fibres 252
6.5.2 Suppression techniques 253
6.6 Circuit Design for EMC 257
6.7 EMC System Design 265
References 267

Appendix A Some Useful Contact Addresses for EMC Standards **269**
Appendix B The Curl Vector and Stokes' Theorem **273**
Reference 277
Appendix C Mathematics of the Variation of Z_w with Distance for a Short Wire and Magnetic Dipole Antennas **279**
Appendix D Pulse or Resonant Charging of Capacitors **281**

Appendix E Some Properties of Common Transmission Lines **285**
Discussion and Some Examples 286
Attenuation Constant of a Coaxial Transmission Line 287
Impedance of Transmission Lines with Finite Conductivities or Complex Geometries 288
References 288

Index **289**

Preface

EMC as a subject has been growing for many years but has become increasingly important recently because of its wide ranging industrial implications, such as implementation of the EEC directive (89/336/EEC) on EMC in 1992. Because of this there is a need for introductory level books to help train scientists and engineers.

This book is our suggestion of how, in the first instance, to present this important but very complex subject to final year undergraduates and first year post-graduates. The particular difficulty with EMC is that very small, apparently trivial, changes to a circuit or a system can have the most far reaching effects, a point which is seldom stressed.

When we were planning the book two interrelated ideas dominated our approach. The first was that it should be an introductory level text, explaining where some commonly used equations, approximations, and techniques come from, rather than just quoting them. The second idea was to form a bridge between conventional electromagnetism texts and EMI books for working graduates starting in the field of EMC. To these ends we have concentrated on presenting basic ideas and incorporated extensive cross-referencing within the text. Also, we have not avoided repetition of important points. We stress that each topic mentioned in this book has an implied invitation to further reading and we have indicated other books, articles and journals which we believe to be of special value. We did not want either to produce another electromagnetic text in the standard mould to compete with the many excellent books already in print, or a detailed EMC handbook, of which there also are many.

We would like to thank colleagues in the Department of Electrical Engineering and Electronics and the Department of Physics at the University of Liverpool for useful discussions, in particular John Dainton, Les Wilkinson and Tony Smith. Brian Austin and David Edwards read parts of the manuscript and made useful suggestions for changes.

Special thanks are due to Freda Cheyne and Sandra Collins who drew the diagrams with such care.

Many commercial firms were very helpful. In particular we thank British Aerospace PLC, Rainford Racks Ltd., RayProof Ltd., Siemens AG, Siemens-Plessey Assessment Services Ltd, General Electric Inc, Emerson & Cuming (UK) Ltd., Pilkington Glass PLC, Schaffner EMC Ltd., Electro-Metrics Inc., Asea Brown Boveri Ltd., and muRata MFG. Co. Ltd. for either useful talks on measurements and/or permission to include copyright material.

Finally we thank Ann-Marie Franks, our Commissioning Editor at John Wiley & Sons, for all her help and encouragement.

PAC
MAH
1991
University of Liverpool

1 An Overview of Electromagnetic Compatibility

1.1 HISTORICAL PERSPECTIVE AND BASIC DEFINITIONS

Electromagnetic Compatibility (EMC) is a field of study applying basic physics to complex electrical and electronic systems with the aim of enabling them to co-exist harmoniously. If this is achieved then they perform their tasks in a satisfactory manner. It has been given a precise definition which will be met shortly, but before this it is necessary to examine a little the historical development of the subject.

The electromagnetic interference of one particular system on another part of itself or some other system has been known since work on electrical systems started about a century ago. However it is only since the Second World War that this problem has become of general interest. All the evidence suggests moreover that it will become a major environmental concern in the next twenty years or so as the use of electronic equipment spreads throughout every aspect of our lives.

In order to discuss this problem it is necessary to define some of the terms and abbreviations that will be used throughout the book. In particular the word 'interference' used indiscriminately in early literature has become replaced by the word 'disturbance' to describe the phenomena (noise, unwanted signal, etc) causing the degradation (interference) of the system in question [1]. The abbreviation EMI stands for Electromagnetic Interference and it is defined as the degradation of the performance of a device, equipment or system by an electromagnetic disturbance. RFI stands for Radio-Frequency Interference and this is defined as the degradation of the reception of a wanted signal caused by a radio frequency disturbance. These abbreviations are commonly used and involve the concept of 'disturbance' so it is important at this point to make some more precise definitions:

Electromagnetic disturbance is any electromagnetic phenomenon that may degrade the performance of a device, equipment or system, or adversely affect inert or living matter. An electromagnetic disturbance may be electromagnetic noise, an unwanted signal or a change in the propagation medium itself.

Radio-frequency disturbance is electromagnetic disturbance having components in the radio-frequency range.
Electromagnetic radiation is the propagation through space of energy in the form of electromagnetic waves, which continue to propagate even if the original source is 'switched off'. By extension the term 'electromagnetic radiation' is often used to cover reactive phenomena between closely spaced circuits due to magnetic and electric fields generated by currents and charges, usually discussed in terms of inductance and capacitance.

In free space electromagnetic waves propagate with the velocity of light c (299 792 458 $\mathrm{m\,s^{-1}}$ exactly), which is often approximated to $3\times10^8\,\mathrm{m\,s^{-1}}$. It is useful to remember that this corresponds to travelling a distance of about 1 m in 3 nanoseconds or 1 foot in 1 nanosecond. If the frequency of the wave is f Hz, then the period $T=1/f$ seconds and the wave travels a distance c/f metres in this time; this distance is called the wavelength, $\lambda=c/f$. The angular frequency $\omega=2\pi f$ appears very often in formulae, and is sometimes called 'the frequency'. In this book we shall adhere strictly to the proper naming convention. The relation between wavelength and the period is shown in Figure. 1.1. Using this formula for the wavelength, it is instructive to look at a table of frequency against wavelength.

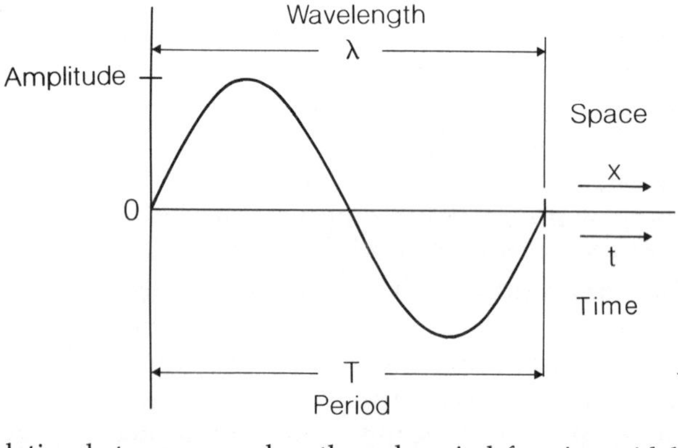

Fig. 1.1 Relation between wavelength and period for sinusoidal waveform

Table 1.1. Frequencies and wavelengths

Frequency (Hz)	Free-space λ(m)	
60	5000 000	
1000	300 000	
1000 000 (1 MHz)	300	(AM radio band)
100 000 000	3	(FM band)
1000 000 000 (1 GHz)	0.3	

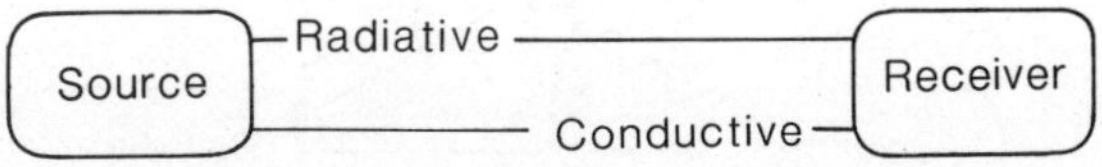

Fig. 1.2 Schematic diagram of idealised coupling modes between a source and a receiver

It is a useful skill to be able to convert quickly from frequency to wavelength. The reasons are that many electromagnetic effects vary with frequency, but the validity of the approximations which can be used to make calculations for design purposes depend on the physical size of the system compared with the wavelengths of the dominant fields present. This means that when a problem in EMC is being addressed it is vital to hold simultaneously in ones mind the ideas that there will probably be a *frequency range* over which problems are most severe and if this is the case then there will also be a corresponding *distance scale* over which different approximations are made in order to perform calculations.

Having established the idea that frequency and size play important roles, one must consider if the ways in which different systems interact can be generalised. Consider, as a first approximation, that all interference problems have a source, a number of coupling paths or modes and a receiver. One model is shown diagrammatically in Figure 1.2, where two coupling modes are illustrated. The radiative mode refers to direct field coupling between two otherwise isolated systems. The conductive mode refers to situations where the dominant interaction is caused by current flow between the source and receiver. In real situations the interaction of two systems could in principle be obtained from a full solution of Maxwell's equations with the appropriated boundary conditions. However, this is seldom possible in practice, so a number of approximations are used. These approximations lead to the coupling modes of illustrated in Figure 1.2, the basis of which will be discussed in later chapters. In particular it will become clear that in some situations a further division can be made of the radiative mode into both reactive (capacitive and inductive) and true radiative coupling.

In general the source, as well as the coupling modes, will have a complex behaviour which is dependent on frequency. The receiver also will have a complicated, and not necessarily obvious, response as a function of frequency. To tackle EMC problems between these two it is necessary to identify the frequencies present using harmonic (or Fourier) analysis of any repetitive waveforms, and consider both harmonic and intermodulation distortion of these waveforms where non-linear elements are present; these topics will be discussed in detail in Chapter 5. Suffice it to say that repetitive waveforms and non-linear circuit elements can generate harmonics (integer multiples of some fundamental frequency) and 'beats', i.e. sums and differences of mixed signals, which lead to unexpected frequency components in a noise signal. Finally one should remember that a particular problem may be caused by more than one source and one receiver. In most of what follows the simplification is made of assuming a single source and receiver.

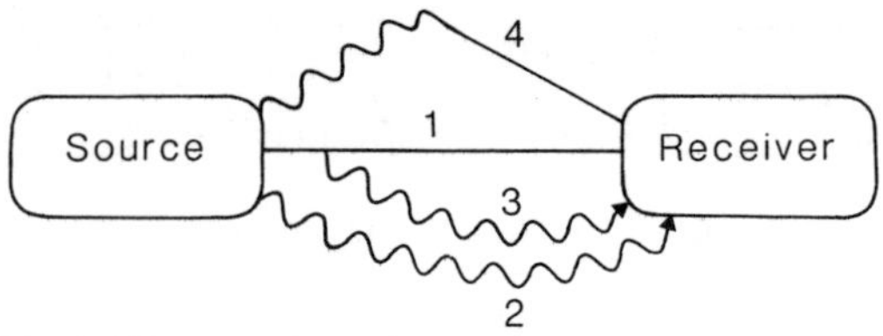

Fig. 1.3 Schematic diagram showing a selection of more complex coupling modes

It is of course clear from Figure 1.2 that many other possibilities for coupling modes are implicit in the diagram. A couple of these are shown in Figure 1.3. The conduction mode is shown linked to a single conductor but of course this is generally two or more conductors along which signals, power and noise will travel. The conduction mode can be further subdivided into common-mode and differential-mode. A common-mode disturbance refers to a common noise current or voltage in all the cables of a system relative to its earth. A differential-mode disturbance means that the different noise currents and/or voltages occur on different conductors of a cable system relative to the signal earth.

1.2 EXAMPLES OF EMC PROBLEMS

To give some idea of the sorts of EMC problem met in practice it is worth considering a few cases from the recent past. An international pay phone manufacturer had been making and successfully selling phones in one area of the world. However, on breaking into a new market (the USA) it was found occasionally that the phones were installed in locations (the tops of skyscrapers) where they were exposed to microwave radiation from nearby communication systems. In this event the operation of the phones was degraded and this was because they had been put into an electromagnetic environment for which they had not been tested.

A few years earlier a recently modified radio controlled crane in a steel works had unexpectedly tipped a ladle of molten steel over workers, killing one and injuring others [2]. Investigation showed that in some circumstances the new radio link had interfered with the original push-button controller, thus causing the accident. The change in circumstances which produced the fatal effect was traced to some temporary scaffolding which had enhanced the radio signal in some locations, thus causing the problem. This is an example of a complex system being subtly changed, altering the electromagnetic environment.

A third example lies in the field of automotive electronics. Modern display systems use high frequency logic (Fourier analysis of the repetitive waveforms involved shows that currents and voltages contain components greater than the

10th harmonic with appreciable amplitudes) capable of interfering with the car radio and communications systems. Careful specification of the allowable emitted radiation is difficult since the location of antennas, the electromagnetic design of the car interior/exterior, and the noise output of the display determine whether the complex system will cause problems. In particular it makes it difficult to fit one display to many different cars unless its emissions specification is made very low. There are many areas where difficulties arise as well be seen later in the book. To give some idea of the range of problems the following topics are of concern at the present time [3]. The order in this list does not reflect their relative importance and such a list will change constantly.

(1) *Military systems.* Problems here involve protection against EMP and electronic countermeasures including the proposed development of microwave weapons [4] and Stealth technology [5]. Considerations of the array of communications and radars in battle conditions is also important.

(2) *Electrostatic discharges.* Here a major concern is the destruction of microchips during handling. However, electrostatic discharges to aircraft and automobiles also pose problems which have to be dealt with satisfactorily.

(3) *Mains signalling.* The use of the public mains supply to transmit RF signals (in the range 3–8.5 kHz and 40–150 kHz) is increasing, as is the chance of any associated interference problem.

(4) *Disturbance of the mains supply.* These arise from many causes and will be discussed later in this chapter.

(5) *Spectrum utilisation.* The continued expansion of legal and illegal users of the limited radio-frequency spectrum leads to many RFI problems.

(6) *Ignition and detonation hazards.* Emission from radio stations can give rise to potential hazards in petrochemical and similar industrial plants (e.g. refueling of aircraft) where flammable vapours may be present. A possible hazard for the future are the uses of CB radios, especially if they exceed the maximum legally allowed transmitter power, in petrol stations and near quarries where electrical detonators are used in blasting operations.

(7) *Malfunctioning* of electrical and electronic systems as a result of various sorts of electrical interference (see the example above).

(8) *Data security.* Modern computer systems operate with fast logic generating high frequency harmonics which when radiated are capable of being detected (decoded) some distance from the installation. This emission poses a threat to supposedly secure systems. For example, it is not difficult to pick up the radiation generated by the video circuits within a remote VDU which has not be specially protected and feed this into another terminal, to see what is being displayed.

1.3 FORMAL DEFINITION OF EMC

In order to set up ways of dealing with complex systems and to assist in their development the concept of Electromagnetic Compatibility has been developed. The IEEE [1] has defined this as:

Electromagnetic Compatibility, EMC, is the ability of a device, equipment or system to function satisfactorily in its electromagnetic environment without introducing intolerable electromagnetic disturbances to anything in that environment.

EMC must exist both inter-system and intra-system and the systems may be, for example, a radio, a total electronic control system or, for that matter, a whole city. Within this system all the electrical and electronic devices must co-exist harmoniously. EMC can be considered in terms of a simplified version of Figure 1.2, as shown in Figure 1.4. This figure illustrates that any system acting as a receiver must work adequately when exposed to specified electromagnetic disturbances (specification levels in). At the same time the same system must not be the source of electromagnetic interference above certain specified levels (specification levels out). The crucial point is that many of these specified levels are enforced by legislation, and such legal requirements can be expected to grow in the future, with the possibility of severe penalties for infringement.

The two lines shown in Figure 1.4 are a reminder that, in general, there are two possible coupling paths both into and out of any given system, although in a given situation one may not need to consider both of them. Usually the designer of a sub-system does not have the freedom to specify his own EMC limits but has to design and construct equipment to externally imposed limits, some of which, as we have already noted, may be legal requirements, others purely advisory. However in some cases there may be no specifications, in which case the design work has to be done with some care to ensure the system can be used in a normal environment. Even in situations where product design is not involved, the techniques of EMC control can be used to help solve electrical noise problems in any electrical system or experiment.

To make EMC more precise the IEEE have introduced [1] for use in their EMC journal other definitions, often encountered in the literature and specifications, which can be usefully introduced at this point:

(*Electromagnetic*) *susceptibility* is the inability of a device, equipment, or system to perform without degradation in the presence of an electromagnetic disturbance. Note that susceptibility is lack of immunity.

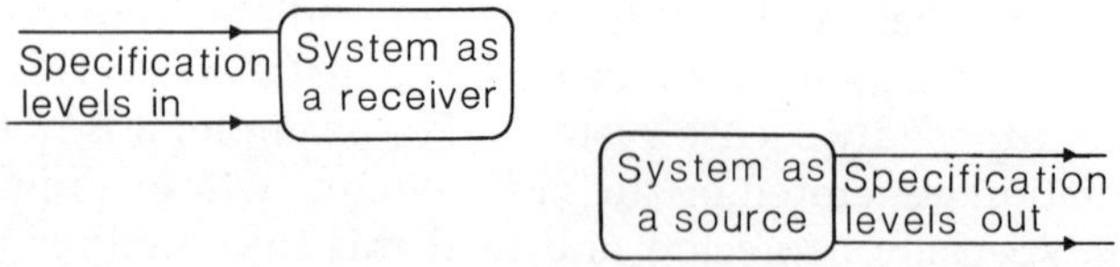

Fig. 1.4 How EMC specifications relate to different coupling modes for the same system acting as either a source or receiver

Immunity to disturbance is the ability of a device, equipment or system to perform without degradation in the presence of an electromagnetic disturbance.

(*Electromagnetic*) *compatibility level*. The specified maximum electromagnetic disturbance level expected to be impressed on a device, equipment, or system operated in particular conditions.

Immunity level is the maximum level of a given electromagnetic disturbance incident on a particular device, equipment or system for which it remains capable of operating at the required degree of performance.

Immunity limit is the specified immunity level.

Immunity margin is the difference between the immunity limit of a device, equipment or system and the electromagnetic compatibility level.

(*Electromagnetic*) *compatibility margin* is the ratio of the immunity level of a device, equipment or system to a reference disturbance level.

The relationship between these terms are shown schematically in Figure 1.5 (which has been adapted from [1]). Most EMC quantities for a real system display a probability curve; thus in this figure EL_1 represents the most probable emission level. However emission levels up to EL_2 (the emission limit) may be found with low probability. Similarly the same system will have a most probable immunity level IL_1, but with low probability may exhibit immunity

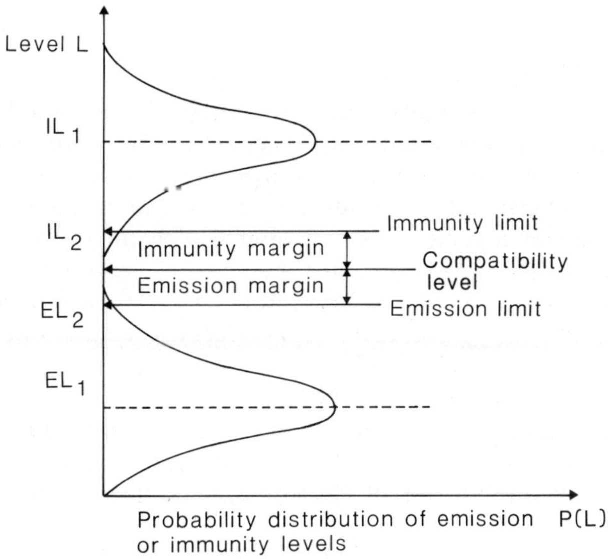

Fig. 1.5 An illustration of the concepts of immunity/emission limits and how they relate to the compatibility level, adapted from reference [1]. (Reproduced by permission of IEEE)

problems right down to IL_2, the immunity limit. As will be seen later, these various 'levels' are often specified in terms of a field strength in which the device, equipment or system must operate or which it must not produce beyond a specified distance.

It should be noted that the 'probability' mentioned here is a difficult concept to quantify and for a given system could not realistically be calculated from first principles, for example by starting from the circuit diagrams! Instead one is led to considering the statistical behaviour of samples of the system, and by measurement finding how the emission and immunity levels are distributed within these samples. From these data one might then assign a suitable probability distribution which best represents the data and calculate the appropriate probabilities. The corollaries to this approach are that the designer should have some familiarity with statistics, and quality control in the manufacture of further systems is vital to ensure that they perform in the same way as the test samples.

1.4 NOISE SOURCES

In order to orientate the reader and warn about potential problems it is worth listing, with some discussion, the most common sources of electrical noise and interference. By 'noise' is meant any unwanted signal within a system which will have characteristic features which may help to identify its source. Any system which causes electric and magnetic fields to change with time (in some cases even static fields can cause problems) has the potential to cause electrical interference and these can be divided into two types, natural and man made.

1.4.1 Natural sources

(1) *Lightning.* Electrical discharges between charged clouds and between clouds and ground can cause severe electrical interference in a number of situations. Thus designers of equipment should remember that it may be exported to countries where violent storms are more frequent than in their own country. A schematic diagram of a typical current-time waveform for a ground strike is shown in Figure 1.6. The current–time shape, relative magnitude and frequency of strikes at a particular location are all statistical in nature with big variations in all of these parameters [6, 7]. Because of this one is usually left with designing for a particular level (met say, in practice only once in 20 years). The chance remains, however, that more severe strikes could occur. Protection against structural damage to buildings leads to the introduction of lighting conductors but this rarely prevents electromagnetic problems. In general lightning results in three forms of electromagnetic disturbance:

(a) A direct strike to any conductor, such as an overhead power line, will cause a large surge throughout the system. It is very difficult to protect parts of the system close to the location of the strike since voltages in excess of

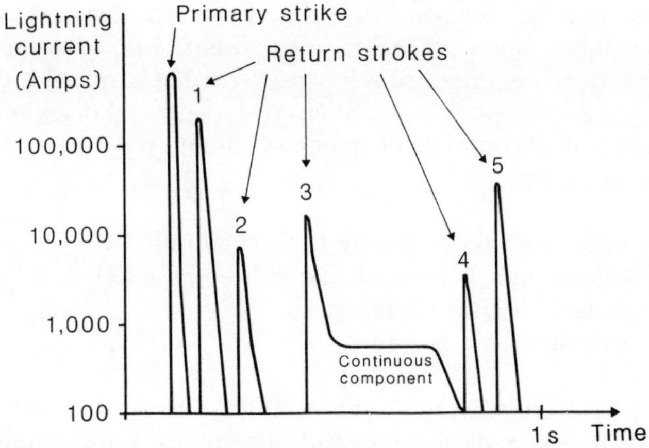

Fig. 1.6 Schematic diagram of a typical lightning current waveform. The number of secondary strikes, rise times and current levels vary from event to event. The primary strike usually has a rise time of about 1 μs (not shown to scale in the figure)

100–200 kV are generated relative to 'ground' depending on the effective ground impedance (resistive and inductive). Ground currents caused by strikes can also couple to any cable system running in that ground.

(b) Associated with any charged thunderstorm clouds there is always an electric field of order 1–10 kV m^{-1} at ground level. When the lightning strikes, discharging regions of the cloud, this field collapses causing induced transients on nearby conductors. These electric field variations, besides causing interference, are used to detect the presence of potentially dangerous storm conditions.

(c) The rapid change of current along the discharge channel emits a wide band of RF up to 50–100 MHz. This radiation is the main cause of atmospheric noise but is also used in the detection and tracking of storms.

Thus lightning can be considered to be a broadband source which is potentially destructive and has been found to be problematic in telephone and power systems, aircraft (most civil aircraft are struck at least once a year) and computer networks distributed over large sites.

(2) *Solar and cosmic sources.* Changes in the ionosphere caused by emission from the sun cause problems for radio transmission due to varying ionospheric reflection (in the 2–30 MHz bands) and for satellite communications by varying ionospheric transmission (150–500 MHz). Radiation from other cosmic sources produces an RF background which is important in the 100–1000 MHz range.

1.4.2 Man-made sources

(1) *Electrostatic discharges, ESD.* Any object moving relative to another whilst in contact (e.g. gas flowing over a solid, one solid rubbing another) has the

capacity of exchanging electrons (triboelectric effect) leaving each substance charged oppositely. Such charging can build up substantial potentials (10–25 kV) with stored energies of a few mJ. The discharging of this energy can produce fast rising current pulses which give people shocks and can damage electrical equipment. A number of major problems have been caused by ESD in recent years including

(a) explosions in supertankers during tank cleaning
(b) damage to microchips during handling by personnel
(c) explosions during aircraft refueling
(d) damage to car electronic systems.

Many EMC specifications contain ESD tests. The size of an ESD pulse is statistical in nature, so typical pulse and test currents are usually specified. A typical injected current test waveform is shown in Figure 1.7.

(2) *EMP (sometimes called NEMP)* is an electromagnetic pulse due to nuclear explosions [6]. This noise source is primarily of interest to military system designers, but as will be seen shortly it also has serious implications for civilian applications, even in non-nuclear countries.

Whenever a nuclear device is detonated a huge electromagnetic pulse is generated. The size of the pulse and the area affected is only of major concern for exoatmospheric explosions, such as the one in July 1962 when a 1.4 megaton bomb was detonated about 250 miles above Johnson Island in the Pacific, in a test called Starfish [8], which was visible 3500 miles away in Australia.

When a device is exploded beyond the atmosphere the gamma and x-ray radiation emitted travels in straight lines until it hits the top of the atmosphere, where intense currents, at two different altitudes, are set up due to ionisation [6b]. The secondary electrons form currents moving at light speed which

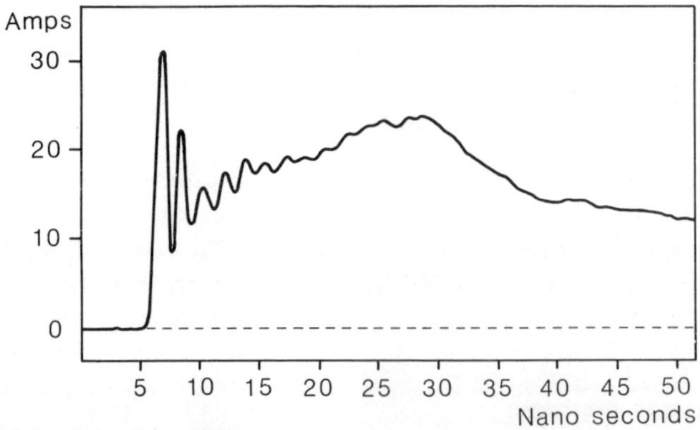

Fig. 1.7 Typical injected current pulse used to simulate ESD events. (Reproduced by permission of Schaffner EMC Ltd.)

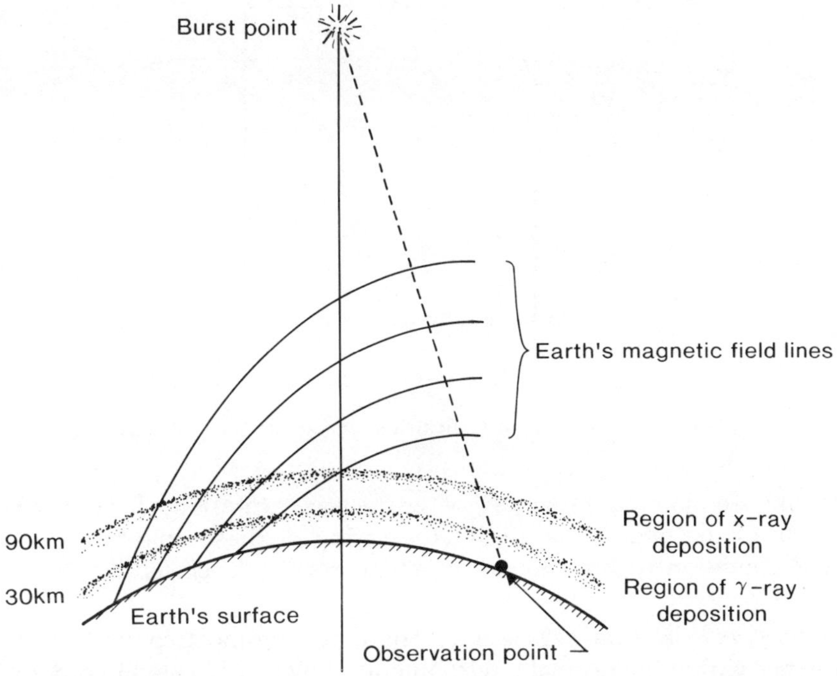

Fig. 1.8 Schematic diagram of the method by which an EMP source is produced in the upper atmosphere by a nuclear explosion in space, adapted from reference [6b]. (Reproduced by permission of AIP)

generate an extensive radio source, as shown in Figure 1.8. The magnitude of the observed electromagnetic pulse at a given point is quite sensitive to the gamma-ray and x-ray yield of the explosion, the distance to the explosion the temperature of the explosion and the angle of incidence of the radiation on the atmosphere [6b].

The radio source may be many hundreds of km in diameter and electric fields at the Earth's surface up to 50 000 $V m^{-1}$ should be expected. Such intense fields can produce interference of a severe kind, with very large induced currents in exposed conductors. Systems such as power and communication networks may be strongly threatened by such events [3, 8]. A typical test pulse used for assessing system response is shown in Figure 1.9; the rise time of an EMP pulse is approximately 10 ns. It is worth noting that such a field cannot be experienced over the *whole* area, since the required energy would be too large [8]. The extent of the radio source often means that regions many hundreds of kilometres from the explosion (whilst experiencing no direct nuclear weapon effects such a heat or blast) may nevertheless experience severe EMP. Thus both civilian and military authorities have to include EMP in their planning.

A related area, but of limited interest to civilian applications, is SGEMP (or System Generated EMP) which is experienced by rockets or satellites exposed

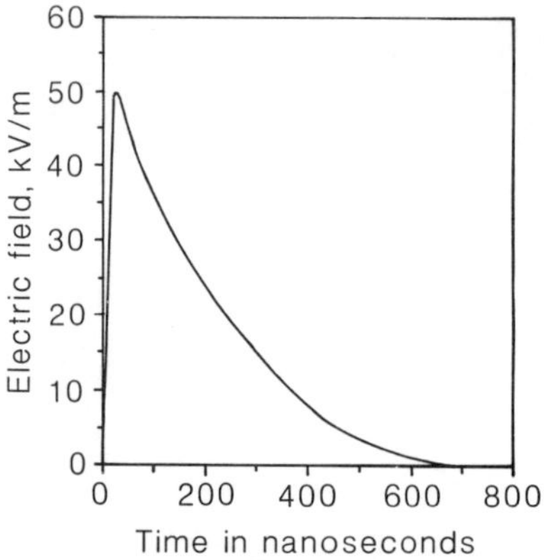

Fig. 1.9 Profile of a typical test pulse for EMP studies

to the intense gamma radiation of a nuclear explosion. In this case the interaction of the gamma rays with the casing of these objects can generate EMP-like condition within causing problems for any enclosed electronics.

(3) *Electrical and electronic sub-systems*. These are the most common sources of interference within the normal environment. They can be listed in order of their importance in polluting the general background, but however in a specific circumstance any one device may be crucial. Typical lists [9] give the following general order:

(a) automotive noise sources: ignition system, alternators, electric motors;
(b) power distribution systems: power lines, AC and DC substations, generating stations;
(c) industrial equipment: welding machines, induction heaters, circuit breakers, microwave heaters, cranes, variable speed drives using pulse width modulated inverters, local oscillators, digital equipment including computers.

Figure 1.10 (adapted from reference [9]) shows the overall trend in background radiation due to all of these causes, highlighting in particular how important automotive systems are in generating general electromagnetic pollution.

Within a piece of equipment, the actual source of the transient voltage and current can generally be traced to three sources:

(i) high frequency pulse trains in digital systems;
(ii) intended high frequency oscillator circuits;
(iii) circuit transients caused by simple switching operations.

Specific examples of these cases will be encountered in Chapters 2 and 5.

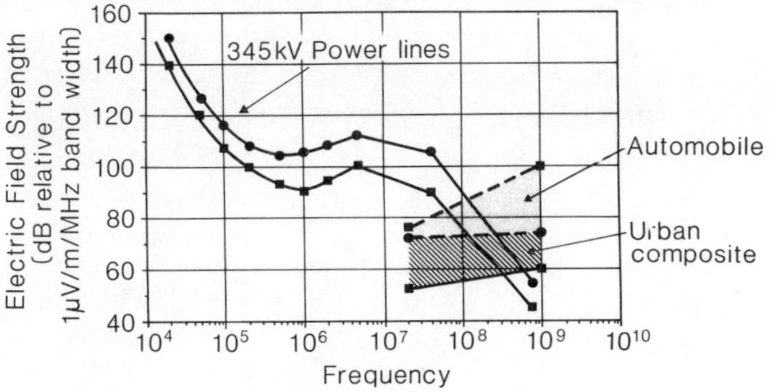

Fig. 1.10 Trends in background electromagnetic radiation. Adapted from reference [9]. See Sections 5.2.9 and 5.4.2.1 for an explanation of the units used to measure field strength. (Reproduced by permission of E. Skomal)

(4) *Variations in the mains supply voltage.* Such variations are caused by the nature of the supply, the distribution system and whatever equipment is attached, and can be sub-divided into two major groups.

(a) Low frequency variations. The voltage of a power supply will always be pulled down by loads (because of the non-zero source impedance) and although the public supply is designed to minimise this, some variation cannot be avoided (for the UK system ±6% is allowed). Inside a particular plant other voltage drops may occur and so a designer is wise to consider ±10% fluctuations are possible over a few hours. If the loads fluctuate more rapidly then, of course, the voltage fluctuations also increase in frequency. Whenever a major fault occurs a sudden voltage step appears before corrective switching can take place.

Waveform distortions from a pure sine wave also occur, usually due to non-linear loads. Typical design rules for domestic appliances allow for up to 6% harmonic distortion whereas industrial users should plan for a maximum of 10%. Varying loads on 3 phase systems may lead to unbalance so causing running problems for 3 phase motors etc.

(b) High frequency variations. Any switching process on the power system leads to fast transients. It is not possible to give an exact relationship between peak amplitude and frequency of occurrence but it has been found that spikes of double a particular amplitude are likely to be ten times less frequent than spikes of smaller amplitude [3]. Figure 1.11 (adapted from reference [10]) shows the typical spread of such values. It is clear that one might expect a voltage spike on the 220 V mains of between 1 and 6 kV once a year. Finally since mains cables can act as an antenna they may pick up RF signals as well as being the transmission medium for mains signalling.

(5) *Radio transmitters.* The electromagnetic spectrum is used extensively for communications and so the background radiation environment has to include

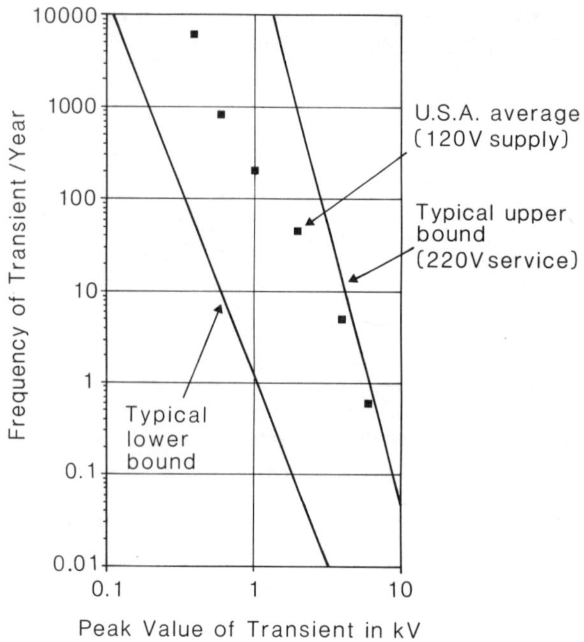

Fig. 1.11 Frequency of occurrence of transient overvoltages in 220 V systems: adapted from reference [10]. (Reproduced by permission of General Electric Co.)

all such users, both legal and illegal. A complete list of the various frequency allocations in the UK is given in a recent IEE publication [3]; however, a partial list, derived from this reference, is given in Table 1.2. There are similar, but not necessarily the same, allocations in different countries.

It is important to realise that the whole spectrum is crowded with users to that trespassing outside the allowed frequency range can have serious consequences. Designers of equipment should remember that interference from the above sources may be especially severe close to the transmitters.

Table 1.2. Some selected UK frequency allocations

9–148.5 kHz	Radio-navigation maritime
148.5–283.5 kHz	AM broadcasting
526.5–1606.5 kHz	AM broadcasting
27.6–30 MHz	CB
88–105 MHz	FM broadcasting
400–410 MHz	Satellites [MET NAV MOB]
470–582 MHz	TV broadcasting
582–600 MHz	Radar
606–854 MHz	TV broadcasting
862–934 MHz	Cellular radio base stations
935–960 MHz	Cellular radio base stations

1.5 GENERAL METHODS OF SOLVING INTERFERENCE PROBLEMS AND MEETING EMC REQUIREMENTS

The whole collection of EMC techniques which are discussed in this book can be used in two general ways. The first is to use them as a guide to problem solving in the operation or development of systems which are plagued by electrical disturbances. In this case the major initial problem is to identify correctly the source of the interference. In order to do this it is often helpful to be familiar with the characteristics of common noise sources discussed in Section 1.4. When considering a noise source it is usually useful to obtain the answers to the following questions which may be of help in deciding the guilt or otherwise of this candidate in causing the particular interference problem.

(1) How often does the noise occur; is it continuous, intermittent, regular? Does the noise source occurrence pattern coincide with that observed by the interference?

(2) When transformed from the time domain by Fourier analysis what are the relevant emitted frequencies? Is the frequency of the interference identifiable with any of these harmonics? The technique of Fourier analysis will be discussed in Chapter 5.

(3) Is the noise characterised by a narrowband or broadband relation to the system acting as a receiver? This topic will also be discussed in Section 5.2.9.

(4) Are the disturbing signal levels sufficient to cause damage to the threatened system at first exposure or after many exposures? Protective techniques to overcome overload problems will be discussed in Section 6.5.2.

It is only when the noise source has been identified and its coupling paths ascertained that the best corrective procedures can be carried out.

The second major use of EMC techniques is to enable a design to be produced which can meet certain EMC specifications. At the simplest level there are five main actions that can be undertaken to alter the electrical interference of one system by another, as follows.

(a) *Locate the sensitive circuit or receiver as far away as possible from the source.* This can work well for coupling via electromagnetic fields but not so well if the coupling path is conductive.

(b) Some field coupling is polarised so that it can sometimes be minimised by *suitable orientation of the receiver to the source.* Such a technique can only be used for fixed receivers and sources.

(c) *Introduce a barrier or shield surrounding the source and/or the receiver.* When this approach is used the correct treatment of any conductors penetrating the shield becomes of paramount importance (use of filters, proper topological connection to ground, etc.).

(d) *Make the source less noisy with careful design.*

(e) *Make the receiver less sensitive to the electrical disturbance by careful design.*

Meeting a specific EMC regulation usually requires action based on the last three techniques in this list, but obviously some of the interior improvements needed for (d) and (e) will utilise the principles contained in points (a) and (b) in some respects, although many other methods will be used as well. Some of the more complex treatments that are possible are discussed in later chapters. In particular it is worth commenting that in general actions that make a receiver less sensitive usually also make it less emissive when acting as a source but that this is not always the case.

1.6 EMC REGULATIONS AND TESTING

EMC regulations or standards exist in varying degrees of complexity and completeness in most major countries. The list of devices and systems covered by them continues to grow. The lack of a specific recommendation for a particular application does not necessarily free the designer from setting up her own standards. Economics plays a crucial role in EMC engineering and in particular for this unregulated situation. However a search for EMC regulations covering the system in question is an essential early step in any design since including the necessary EMC features from the outset is usually cheaper than having to make changes later in the product development.

A large number of bodies specify EMC regulations including the British Standards Institute (BSI) in the UK, the Federal Communications Commission (FCC) in the USA, and the Verband Deutscher Elektrotechniker (VDE) in Germany, together with a whole range of defence related organisations. The first body to propose EMC standards was the International Special Committee on Radio Interference (CISPR) which arose out of a meeting in 1933 of those concerned with radio interference problems. Nowadays 44 nations belong to CISPR and many of its publications have been adopted as National standards. The IEC (of which CISPR is a part) is the main international organisation and at the European level CENELEC, the European Committee for Electro-Technical Standards, together with the European Telecommunications Standards Institute (ETSI) are its counterpart. Because of the EEC drive towards harmonisation so as to encourage trade an increasing number of Euro-norms (EN) are being issued. For the radio interference area Table 1.3 lists the specification numbers relevant to various equipment types used in different countries. At the present time the specifications having legal force are decided by each country. From 1992 all manufactured goods (with some exceptions for a limited period) placed on the market within the EEC will have to satisfy the

Table 1.3. Some equivalent standards for different types of equipment

Equipment	CISPR publication	European Standards	USA	British Standards	Germany VDE
Emission levels					
ISM*	11	EN 55 011	47CFR Pt.18	BS4809	0871
Ignition	12	72/245/EEC	—	BS833	0879
Radio and TV	13	EN 55 013	47CFR Pt.15	BS905-1	0872
Domestic appliances†	14	EN 55 014	—	BS800	0875
Lighting (luminaries)	15	EN 55 015	—	BS5394	—
Information technology	22	EN 55 022	47CFR Pt.15	BS6527	0871
Immunity levels					
Radio and TV	20	EN 55 020	—	BS905-2	—
Instrumentation	IEC 801	HD 481	SAMA PNC 33.1 ANSI 63.12	BS6667	—

*Industrial, Scientific and Medical, e.g. RF heating equipment.
†includes power tools.

EEC Directive on EMC and this will have far-reaching implications for all companies. It is clear that knowledge of EMC will have to become far more widespread than at present. Other areas of electrical work have their own organisations which decide their EMC specifications which may be over and above any general national requirement. For example in the UK during 1988 three EMC specifications covered military aircraft namely;

(1) UK Tri-Service Defence Standard 59-41 (1986);

(2) Eurofighter Specification EFA-SP-J-000-2032 for the European Fighter Project (1988);

(3) EH Industries Specification EA-98Q010J for the Anglo–Italian helicopter programme (1986).

Thus in general the designer faces a major problem in finding out which specifications apply in each case. This is especially true if the product has to be used in a number of different countries. Information on standards may be obtained from the relevant national body (BSI for the UK) or any library holding the national standards or through various professional institutes such as the IEEE (USA) and the IEE (UK). Establishing which standards apply can be simplified by approaching an EMC consultant or seeking the help of an EMC testing laboratory. A number of relevant addresses are collected together in Appendix A which also includes the address of the NAMAS organisation in the UK. NAMAS accredits test facilities in a number of areas including EMC and can provide an up-to-date listing of all approved UK test organisations.

In order to see briefly how compliance with specifications is checked it is important to realise that any proposed tests usually fall under four general headings:

CE Conducted emission tests
CS Conducted susceptibility tests
RE Radiated emission tests
RS Radiated susceptibility tests.

Of course not all specifications will require tests under each heading; for example BS 6527:1988 only requires RE and CE tests. This British Standard defines the permitted levels of radiated and conducted emissions from Information Technology Equipment. This standard is harmonised with EN 55022 and so will be the relevant standard when the EEC Directive on EMC becomes effective. The equipment covered by BS 6527:1988 can loosely be described as all electronic equipment employing microprocessor or microcomputer devices. It is clear that this specification will have far-reaching consequences.

Radiated tests can refer to both reactive and radiation fields. They measure the fields emitted (RE) by the equipment under test (EUT) or the effect of fields (RS) on the correct operation of the EUT. Conducted emission (CE) tests measure the currents generated by the EUT on cables connected to the equipment (e.g. mains, signal, control, etc). Conducted susceptibility (CS) tests measure the effects of currents injected onto connecting cables or other parts of the EUT on the correct operation of the EUT.

As an example of the procedure followed, consider how RS and RE tests might be carried out. For example the radiated susceptibility tests require suitable sources to generate the necessary electromagnetic environment together with a measuring system. A schematic of a suitable system is shown in Figure 1.12.

In order to make a successful measurement it is necessary for a number of parameters to be defined.

(1) The measurement must be carried out in a known environment. This could be an open field site (see description later in this section), but because the

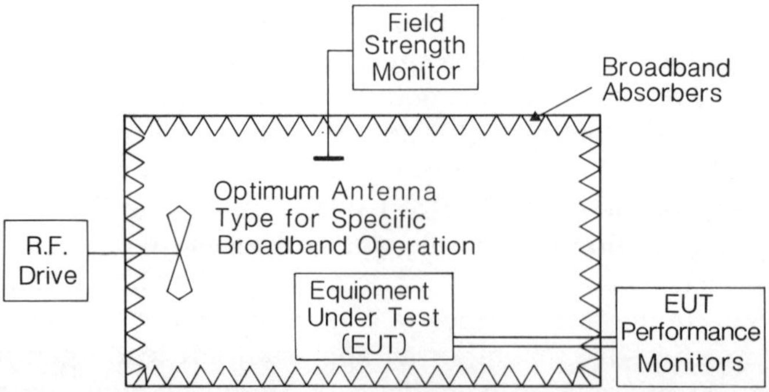

Fig. 1.12 Schematic diagram of an electromagnetically anechoic room suitable for EMC susceptibility testing

field levels used might cause interference to others, more normally an electromagnetically anechoic room or a screened enclosure is used. The resulting electromagnetic environment has to be defined and unchanging. Screened anechoic rooms would be advisable but are difficult to construct for frequencies below 200 MHz. Such enclosures use broadband pyramidal absorbers mounted on nearly all the surfaces. The working and construction of such absorbers will be discussed in Chapter 3. In any event the problem is to ensure that reflections and resonances are properly controlled.

(2) The exciting source for susceptibility testing must provide a repetitive impinging electromagnetic environment having the correct waveshape and frequency components, i.e. it must give a realistic representation of the expected disturbance. As an example of this point, reference [11] shows that it is not possible to use lightning as a suitable source of intense electromagnetic pulses for simulation of EMP effects in aircraft, due to a deficiency of high frequencies. Antennas of special sorts having a reasonable bandwidth are often used to generate the radiation and usually more than one is required to cover the frequency range of interest. Chapter 5 will discuss briefly some of the antenna systems used and some alternative methods of setting up the necessary controlled electromagnetic environment.

(3) It is necessary to know how to calculate the specification levels (susceptibility electric and magnetic field levels) from measured voltages and current, picked up by suitable receiving antennas or sensors, by using the correct antenna factors. These in general will differ from the normal free space values.

(4) The monitor used to check the behaviour of the equipment under test (EUT) must be able to tell clearly if it malfunctions in any way.

In the case of RE tests a similar screened room may be used (see MIL SPEC 59-41 (1986)) or more commonly the use of an open field test site is mandatory (see BS 6527:1988). Such a site has to be large enough to minimise the effect of reflections (except from the ground plane) and the usual arrangement together with the size requirement is shown in Figure 1.13.

In an open field test site the receiving antenna is moveable vertically to find the maximum signal (allowing for reflections from the ground plane). The size of the ellipse (distance F in Figure 1.13) is 3 m, 10 m or 30 m as specified by the test conditions. The space above the ellipse has to be free of reflecting objects. For such tests the radiation emitted from the working EUT has to be measured using a suitable antenna and receiver. In general these must have adequate sensitivity, bandwidth and be properly calibrated. In every case (for RE and RS) the effects of any background must be established and the impedance of any mains supply should be controlled. Pictures of typical test installations are shown in Figures 1.14 and 1.15.

More details of these types of test can be found in the literature [12–14]. There are also many other forms of test depending on the electromagnetic environment to be simulated. For example special tests exist for exposure to low

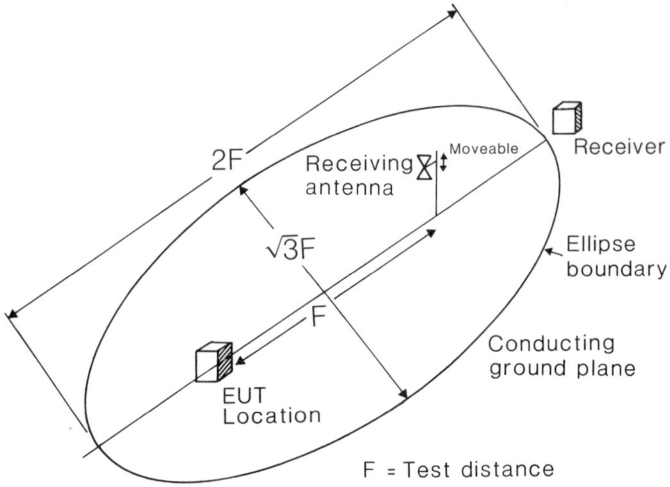

Fig. 1.13 Schematic diagram of an open field test site. The distance *F* is specified by the test conditions

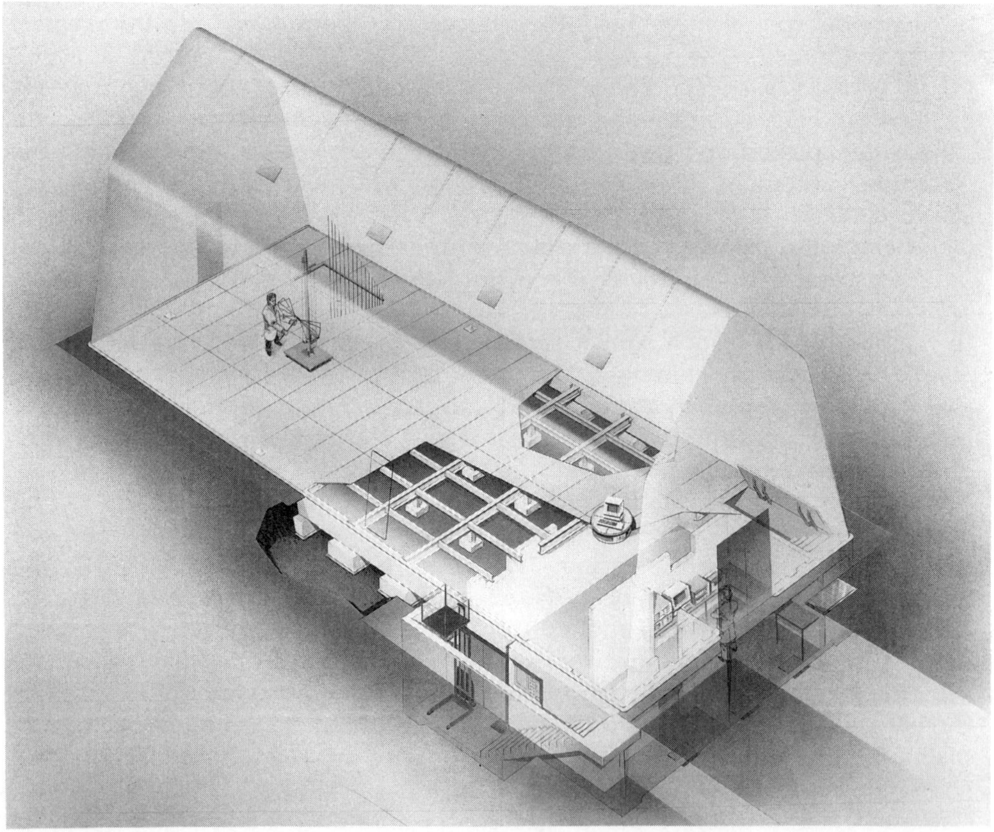

Fig. 1.14 All weather open field test site with dielectric cover. (Reproduced by permission of Siemens Plessey Assessment Services Ltd.)

Fig. 1.15 EMC tests on an aircraft in a typical anechoic room. (Reproduced by permission of Ray Proof Ltd.)

frequency magnetic fields, electrostatic discharges [15, 16], lightning and EMP [6]. Different types of conducted emission/susceptibility tests also exist requiring a variety of special equipment. One very important factor in all the tests, however, is that most of them are sensitive to some small changes in the electromagnetic boundary conditions. Such changes are often unsuspected or difficult to control so that many EMC measurements are not always reproducible at the same or at different sites. One parameter which has to be controlled, as mentioned earlier, is the impedance of the mains for conducted emissions tests. This is because the mains has a differing impedance at different frequencies and unfortunately this is affected by what is connected to the mains nearby. To ensure a standard form for the impedance frequency response a Line Impedance Stabilisation Network (LISN) is interposed between the mains and the equipment to be tested. The network is usually defined by the EMC specification. Ott [12] and Standler [17] describe two in common use.

Radiated Susceptibility tests have recently been replaced in some cases by Bulk Current Injection (BCI) tests [18–22]. This has largely been because it is costly and difficult to provide the large volumes of high fields necessary to proof test aircraft and automobiles. In the BCI method known currents are injected using transformers onto crucial bundles of cables entering a specific equipment.

It has been found that the effect of external fields is mostly to excite such currents. This might of course mean that field dependent susceptibility tests would have to be changed to current injection ones. Fortunately in most cases the link to earlier radiated susceptibility tests can be made by carrying out a low level radiated field test and measuring the currents produced on the cables to be exposed to BCI. The currents required to simulate the high level tests are then obtained by scaling. This procedure appears relatively successful below 400 MHz [22].

1.7 ORGANISATION OF THE BOOK

One of the main problems in EMC work is that theories and experiments can often be misleading. This may be because they have been chosen inappropriately. On the other hand it may be because small differences in the boundary conditions have produced big changes to the field values expected or measured in a given system. The circumstances that cause this to happen are not always easy to recognise. Because of this it is most important to have a clear understanding of the reasons for and the limits of the different theories and practices used. To this end this book presents much of the underlying field theory that underpins all of the procedures used to obtain EMC.

Chapters 2 and 3 give the necessary overview of electromagnetic theory important for applying general EMC ideas in any specific system and emphasises EMC topics not covered elsewhere. Chapter 4 is an extensive review of shielding theories highlighting the need for precision in defining the source and shield geometries if the theories are to agree with experiment. Real shielding problems are discussed in the light of this requirement. Chapter 5 introduces Fourier analysis, antennas and the various systems used to set up test fields. Finally Chapter 6 presents a simple introduction to EMC design in the light of the fundamentals introduced earlier. Good design practices can be of help in preventing some problems, and, as will be seen in many EMC guides, there are often 'rules of thumb' in particular cases but to use such techniques blindly without an appreciation of their basis is to court disaster.

REFERENCES

[1] IEEE editorial policy statement (1987) *IEEE Transactions on Electromagnetic Compatibility* **EMC-29** 202–206. See also IEC 50(161) (1990) *International Electrotechnical Vocabulary* Chapter 161 Electromagnetic Compatibility.

[2] White, D. R., Atkinson, K. and Osburn, J. D. M. (1985) Taming EMI in Microprocessor Systems, *IEEE Spectrum* **22** (December) 30–37.

[3] *Electromagnetic Interference* (1987) Report on IEE Public Affairs Board Study Group, IEE (London), September (1987).

[4] Florig, H. K. (1988) The Future Battlefield: a Blast of Gigawatts, *IEEE Spectrum* **26** 50–55.

[5] Adam, J. A. (1988) How to Design Invisible Aircraft, *IEEE Spectrum* **25** (April) 26–31.

[6] Rickets, L. W., Bridges, J. E. and Miletta, J. (1976) *EMP Reduction and Protective Techniques* Wiley, New York. See also Karzas, W. J. and Latter, R. (1965) The Detection of Electromagnetic Radiation from Nuclear Explosions in Space, *Phys. Rev.* **137B** 1369–1378.
[7] Golde, R. H. (1977) *Lightning* Vols 1 and 2 Academic Press, New York. See also Degauque, P. and Hamelin, J. (1990) *Compatibilité Électromagnétique* Dunod, Paris for an extensive discussion.
[8] Rabinowitz, M. (1987) Effect of the Fast Nuclear Electromagnetic Pulse on the Electric Power Grid Nationwide: A Different View *IEEE Trans. on Power Delivery* **PWRD-2** 1199–1208. For an overview, see this paper and the intense discussions following it, pp. 1209–1222.
[9] Skomal, E. N. (1978) *Man Made Radio Noise* Van Nostrand Reinhold, New York.
[10] *Transient Voltage Suppression Manual* (1978) 2nd edn, editor Hey, J. C. General Electric Co., Auburn, New York (Present address PO Box 591, Summervale, New Jersey). For more recent work see also Standler R. B. (1989) *Protection of Electronic Circuits from Overvoltages* Wiley, New York.
[11] Nanevicz, J. E., Vance, E. F., Radasky, W., Uman, M. A., Soper, G. K. and Pierre, J. M. (1988) EMP Susceptibility Insights from Aircraft Exposure to Lighting *IEEE Trans on EMC* **EMC-30** 436–472.
[12] Ott, H. W. (1988) *Noise Reduction Techniques in Electronic Systems* 2nd edn, Wiley, New York.
[13] Keiser, B. E. (1979) *Principles of Electromagnetic Compatibility* Artech, Dedham.
[14] Violette, J. L. N., White, D. R. J. and Violette, M. F. (1987) *Electromagnetic Compatibility Handbook* Van Nostrand Reinhold, New York.
[15] Mardiguian, M. (1984) *Interference Control in Computers and Microprocessor Based Equipment* Interference Control Technologies, Gainesville, VA.
[16] *EMC, Electromagnetic Compatibility* (1985) Schaffner Electronik AG., CH-4708 Luterbach, Switzerland.
[17] Standler, R. B. (1989) *Protection of Electronic Circuits from Overvoltages* Wiley, New York.
[18] Burbridge, R. F., Edwards, D. J., Railton, G. J. and Williams, D. J. (1990) Aspects of the Bulk Current Immunity Test *Proc 7th Int. Conf. on EMC, University of York* IEE Conference Publication 326. pp 162–168.
[19] Marshall, R. C. (1990) Convenient Current-Injection Immunity Testing *Proc 7th Int. Conf. on EMC, University of York* IEE Conference Publication 326, pp 173–176.
[20] Oberto, G., Bolla, L., Rostagno, G. and Barardo, R. (1988) Critical Examination of Bulk Current Injection Techniques: Experimental Comparison *Proc 6th Int. Conf. on EMC, University of York* IERE Conference Publication 81, pp 101–107.
[21] Audone, B., Ferrero, G., Giorcelli, L. and Vecchi, C. (1988) Critical Examination of Bulk Current Injection Techniques: Theoretical Aspects *Proc 6th Int. Conf. on EMC, University of York* IERE Conference Publication 81, pp 109–115.
[22] Carter, N. J., Redman, M. and Willis, P. E. (1988) Validation of New Aircraft Clearance Procedures *Proc 6th Int. Conf. on EMC, University of York* IERE Conference Publication 81, pp. 117–124.

2 The EMC Implications of Basic Electromagnetic Ideas

2.1 INTRODUCTION

This chapter is a review of electromagnetism, concentrating only on those aspects of the subject which are relevant in this book. It will be assumed that the reader has met vector calculus and Maxwell's equations before, as well as complex or phasor notation, but the conclusion that such a reader understood fully, or is happy with, this mathematics will not be drawn! The basic physics of how electromagnetic waves propagate through vacuum, dielectrics and conductors and how they are reflected at surfaces will be summarised. The importance of the impedance concept will be stressed, and a few EMC examples which illustrate the basic ideas will be presented. These fundamentals will then be used in Chapter 3 to examine carefully how electromagnetic waves penetrate one or more layers of material, and subsequently in Chapter 4 to address the problem of shield design.

It must be stressed that this Chapter will concentrate on expounding basic ideas and results, and must be treated as being complementary to the more traditional books on electromagnetism. Few results will be derived from first principles; instead the emphasis will be to present concepts and equations in a form which will ensure that the reader is competent to use them for realistic calculations and is aware of the significance of any approximations which may have been incorporated into the derivation. Thus this chapter will provide convenient references for important equations, written in a consistent notation, that can be used in all the other chapters.

2.2 THE BASIC FIELDS OF ELECTROMAGNETISM

There are four basic vector fields in electromagnetism which are of prime importance:

Field symbol	Units	Comment
E	N C^{-1} or V m^{-1}	Electric field
D	C m^{-2}	Electric displacement
B	tesla	Magnetic flux density
H	A m^{-1}	Magnetic field strength

The notation used throughout the book will be to use bold characters for vector quantities. The magnitude of such a vector, e.g. **E**, will be written as $|\mathbf{E}|$ or simply E where there is not danger of confusion. As well as these fields there are two other quantities which are of primary importance in electromagnetism when matter is involved in the problem. These are **P**, the electric polarisation per unit volume in a dielectric, and **M**, the magnetisation per unit volume in a magnetic medium.

The quantity **P** is related to the density of the surface charge σ_b which is bound to the dielectric by:

$$\sigma_b = \mathbf{P} \cdot \mathbf{n} \tag{2.1}$$

where **n** is an outward pointing unit vector perpendicular to the surface. Inside the dielectric **P** points in the same direction as the applied field **E** which induces the surface charge, i.e. from the negative surface charges to the positive surface charges; this is the opposite convention for the direction of **E**, which points from positive charges to negative charge.

The magnetisation **M** in magnetic materials plays a role similar to **P** in dielectrics. It is related to the magnetic properties of the atoms which make up the magnetic material by

$$\mathbf{M} = N\mathbf{m} \tag{2.2}$$

where **m** is the magnetic dipole moment of each atom, with N atoms per unit volume.

These six quantities are intimately related, and the key to understanding their significance is contained in the two following equations;

$$\mathbf{B} = \mu_0(\mathbf{H} + \mathbf{M}) \tag{2.3}$$

and

$$\mathbf{E} = (\mathbf{D} - \mathbf{P})/\varepsilon_0. \tag{2.4}$$

The crucial feature to notice are the positive and negative signs, and to appreciate that the **E** and **B** fields are the 'real' fields, in the sense that one needs to know these in order to calculate the Lorentz force **F** acting on a moving charge q, given by

$$\mathbf{F} = q(\mathbf{E} + \mathbf{v} \times \mathbf{B}). \tag{2.5}$$

The presence of matter in the form of dielectric or magnetic material changes these fields in different ways. On the one hand (in situations of interest to us) magnetic material adds to the magnetic field **H** produced inside it by external currents, whereas polarisation charges in a dielectric reduces the effect of the electric field **E** produced inside it by external free charges. If there is no material around then there is no real reason to use both **E** and **D** or **B** and **H** fields, since they are related by

$$\mathbf{D} = \varepsilon_0\mathbf{E} \tag{2.6}$$

and

$$\mathbf{B} = \mu_0 \mathbf{H} \tag{2.7}$$

but when material is involved, the formulation of electromagnetism in terms of the six quantities described gives a consistent and manageable way to understand different effects and perform calculations. In EMC work it is normal to use **H** (rather than **B**) to describe magnetic phenomena, a practice which will normally be followed throughout the book.

There is however, a significant complication to this simplistic approach. The quantities **P** and **M** are not constants but functions of **E** and **H**. This important observation reflects the fact that the magnetisation **M** and the polarisation **P** are produced by the underlaying structure of different materials which responds to the **H** and **E** fields when they are applied to the material.

As there are an almost limitless number of different possible materials, with widely differing mechanical, electrical and crystallographic properties, there arises an immediate problem of how to incorporate this vast diversity into electromagnetic theory so that it remains useful in practical situations. A partial solution to this problem is to make use of an approximation.

It is observed that in many types of common everyday materials such as wood, water, plastics, non-ferrous metals and similar linear, isotropic substances the relationship between **M**, **H** and **P**, **E** can be approximated accurately by very simple linear relationships of the form

$$\mathbf{P} = \varepsilon_0 \chi_e \mathbf{E} \tag{2.8}$$

and

$$\mathbf{M} = \chi_m \mathbf{H} \tag{2.9}$$

where χ_e is taken to be a constant, independent of **E**, called the electric susceptibility, and χ_m is called the magnetic susceptibility. Starting from these equations, the relation between **B**, **H** and **E**, **D** can be recast in the forms

$$\mathbf{B} = \mu_0(\mathbf{H} + \mathbf{M}) = \mu_0(1 + \chi_m)\mathbf{H} = \mu_0 \mu_r \mathbf{H} = \mu \mathbf{H} \tag{2.10}$$

and

$$\mathbf{D} = \varepsilon_0(1 + \chi_e)\mathbf{E} = \varepsilon_0 \varepsilon_r \mathbf{E} = \varepsilon \mathbf{E}. \tag{2.11}$$

The four new, material dependent quantities introduced here have the following names:

μ_r relative permeability	μ permeability
ε_r relative permittivity	ε permittivity.

Numerical values for these quantities have been measured in many materials and are tabulated in reference books. For example, some values of ε_r are shown in Table 2.1, adapted from [1].

For magnetic materials the situation is more complicated than for dielectrics. Many materials have χ_m with a positive value of order 10^{-5}, and are called paramagnetic materials. Others, called diamagnetic materials, have χ_m

Table 2.1. Typical values of ε_r as a function of frequency at about 25°C [1]. (Reproduced by permission of W. H. Freeman and Company).

	Frequency (Hz)		
Substance	10^2	10^6	10^{10}
Benzene	2.28	2.28	2.28
Butyl rubber	2.43	2.40	2.38
Fused Silica	3.78	3.78	3.78
Ice		4.15	3.20
Lucite	3.20	2.63	2.57
Neoprene	6.70	6.26	4.0
Polystyrene	2.56	2.56	2.54
Teflon	2.1	2.1	2.08
Water	81	78.2	34

negative with typical values between -10^{-5} and -10^{-8}. It should also be noted that many reference books tabulate the mass or molar susceptibility given by

$$\chi_{\text{mass}} = \chi_m/\rho \qquad \text{or} \qquad \chi_{\text{molar}} = A\chi_m/\rho$$

where ρ is the density and A the molecular weight. These values imply that for most materials including metals such as copper and aluminium, $\mu_r \approx 1$. The implications of this for the shielding of magnetic fields is discussed in Chapter 4.

It is crucial to note that these 'constants' are not truly constant, but almost invariably have a frequency dependence that can change their values by large factors. Thus it is vital to be aware of the appropriate frequency range involved in a particular problem in EMC. It should also be noted that these linear relationships between **B**, **H** and **E**, **D** imply that, in general, these pairs of vectors point in the same direction inside materials, with the result that the algebra encountered later when dealing with Maxwell's equations is reasonably straightforward.

2.2.1 Ferromagnetic material

There is one common but very important exception to this simple scene, and that is when ferromagnetic material is encountered. In these materials the relationships between **B**, **H** and **M** can be extremely complex, depending on such factors as direction in the material and even its past magnetic history. There is certainly no requirement that the vectors point in the same direction. Indeed inside a bar magnet, **B** and **H** point in approximately opposite directions! The functional relationship between **B** and **H**, written as **B**(**H**), can be complicated (see Figure 20.5 in [1]) and is not given by a simple linear equation like $\mathbf{B} = \mu\mathbf{H}$. The magnetisation $\mathbf{M}$ can be found from the expression

$$\mathbf{B}(\mathbf{H}) = \mu_0(\mathbf{H} + \mathbf{M}) \tag{2.12}$$

(which has general validity), yielding

$$\mathbf{M} = \mathbf{B}(\mathbf{H})/\mu_0 - \mathbf{H} \tag{2.13}$$

This equation highlights the fact that the magnetisation at a particular point inside a ferromagnetic material depends in a non-trivial way both on the applied **H** field and the **B** field generated by the rest of the bulk material. In the literature it is usual to present curves showing the dependence of the measured value of **B** on **H**. The relative permeability of the ferromagnetic material at each value of H can then be found from the formula

$$\mu_r = B/\mu_0 H. \tag{2.14}$$

For these materials the values of μ_r are often very large, of order 10^4or more, see Table 4.1, and they are suitable for magnetic field shielding, as will be discussed in Section 4.2.1. However when the applied H field exceeds a critical value, the effective value of μ_r drops to small values, of order 1–10. This effect is called saturation. For EMP studies involving large amplitude fast transient fields, the magnetisation of materials such as iron can be parameterised by a simple formula of the form [19]

$$\mu_r(H) = 1 + \frac{\mu_{ro} - 1}{1 + e^{(H - H_c)/\Delta H}}$$

where μ_{ro} is the unsaturated permeability, H_c is the critical field and ΔH is the range of H over which saturation occurs. For iron typical values are $\mu_{ro} = 10^4$, $H_c = 400\ \mathrm{A\,m^{-1}}$, $\Delta H = 50\ \mathrm{A\,m^{-1}}$. Thus $\mu_r(100) \approx 10^4$ whereas $\mu_r(1000) = 1.06$.

2.2.2 Summary

Classical electromagnetism is underpinned by a whole deeper level of physics that can be fully explained by using quantum mechanics. However, for the most part, this deeper level can be ignored when dealing with large scale, macroscopic phenomena. Using the classical fields **D**, **E**, **H** and **B**, together with the material and frequency dependent quantities μ_r and ε_r, will suffice to make calculations which are adequate for most situations discussed in this book. Provided the results are not extrapolated and applied to tiny pieces of material or to situations with extremely strong fields, or to frequencies in the infra-red or optical ranges, the underlying level of atomic and quantum physics can be safely ignored. Ferromagnetism apart, the frequency dependence of μ_r and ε_r will be the most important occasion when this lower level of physics becomes relevant, and even than it can usually be incorporated adequately by using appropriate numerical values for μ_r and ε_r at different frequencies.

2.3 MAXWELL'S EQUATIONS FROM AN EMC PERSPECTIVE

The starting point for discussing advanced electromagnetism are Maxwell's equations. These are four coupled differential equations, which define relationships between the four fields discussed in Section 2.2, and are written in the notation of vector calculus as

$$\nabla \times \mathbf{E} = -\frac{\partial \mathbf{B}}{\partial t} \tag{2.15}$$

$$\nabla \times \mathbf{H} = \frac{\partial \mathbf{D}}{\partial t} + \mathbf{J} \tag{2.16}$$

$$\nabla \cdot \mathbf{D} = \rho \tag{2.17}$$

$$\nabla \cdot \mathbf{B} = 0 \tag{2.18}$$

where ρ is the density of free electric charge and $\mathbf{J}$ the current density. The material dependent equations (i.e. not truly fundamental) given in (2.10) and (2.11) i.e. $\mathbf{B} = \mu_r \mu_0 \mathbf{H}$ and $\mathbf{D} = \varepsilon_r \varepsilon_0 \mathbf{E}$ together with a material dependent equation connecting current flow to the electric field

$$\mathbf{J} = \sigma \mathbf{E} \tag{2.19}$$

where σ is the conductivity of the medium, with SI units of $\mathrm{S\,m^{-1}}$, are the most important equations that will be needed in this book.

These famous equations form a complete set from which, in principle, all the properties of a system of electromagnetic fields, currents and charges can be derived. However in reality these equations are very difficult to solve in realistic situations of the sort considered in this book. It is instructive to look carefully at these equations to see precisely what they describe and why there are so difficult to solve.

2.3.1 Maxwell's equations

The fundamental feature to notice about Maxwell's equations is that they are true at every point in an electromagnetic system, and that they refer to properties (ρ, $\mathbf{J}$, etc.) which exist *at that point*. They describe how electromagnetic fields *change* around a particular point (i.e. they give the spatial derivatives), and show that these changes only depend on the appropriate quantities $\mathbf{J}$, ρ, $\partial \mathbf{B}/\partial t$ and $\partial \mathbf{D}/\partial t$ which exist there. This is a crucial observation, for it follows that the equations as they stand cannot be solved immediately to give the fields $\mathbf{E}$ and $\mathbf{H}$ because these fields will depend on the precise distribution of currents and charges situated in the space around the chosen point, and none of this information is described within the equations.

Take for example equation (2.15), which is the relationship between the curl of the electric field at any point, and the rate of change of the $\mathbf{B}$ field at that same point. Recall that the curl of a vector $\mathbf{E}$ at a given point can be defined as the limit of a line integral $\int \mathbf{E} \cdot d\mathbf{l}$ per unit area around a closed path which surrounds the point, as the path is shrunk to an infinitesimal size. In the limit, this line integral *may have* a finite value, and this is defined to be the magnitude of the curl of **E**, written as $|\nabla \times \mathbf{E}|$. The curl vector itself is constructed so as to be the normal to the surface defined by the path along which the line integral was evaluated. This little surface is orientated so that the line integral is the maximum possible around the point. The feature to notice is that the path integral depends on how the **E** field varies around the point in question, and because an integration has been performed, information about the precise value of **E** in this region has been lost. All that is known from this particular equation (2.15) is that the vector $\partial\mathbf{B}/\partial t$ defines, in both magnitude and direction, the vector $\nabla \times \mathbf{E}$. However, (2.15) does show that, whatever the magnitude of the **E** field, it lays in the plane perpendicular to the vector $\partial\mathbf{B}/\partial t$ at the point in question.

Similar observations apply to Equation (2.16). Curl **H** is given by the vector sum of **J** and $\partial\mathbf{D}/\partial t$, so that at the point in question the **H** vector must lay in the plane perpendicular to this resultant vector. Appendix B shows a detailed calculation which illustrates all of these points.

In a similar way equations (2.17) and (2.18) describe how the divergence of the fields behave at given point. For example the divergence $\nabla \cdot \mathbf{D}$ at a given point is the limit of the surface integral $\int \mathbf{D} \cdot d\mathbf{S}$ per unit area, as the volume enclosed by the surface shrinks to an infinitesimal size. Once again, the integration over this little surface which encloses the point depends on how the field **D** varies around the point, and that the information about the precise value of **D** at the point cannot be calculated in a direct way from equation (2.17).

The mathematical difficulty associated with Maxwell's equations is that they cannot be used directly to find the actual **E** and **H** fields at a given point in a particular situation. Given this restriction how can they be used? One possible answer is that if one can to find out or write down the value of the fields at any points, anywhere (by understanding boundary conditions), then in principle one could start from these points and work inwards step by step towards, using Maxwell's equations to calculate the new values of the fields at a nearby point from the previous adjacent values. It is not suggested that this procedure is, in general, a realistic way to proceed. This line of thinking illustrates the uniqueness theorem for the solution of Maxwell's equations, which shows that field distributions obtained from a particular set of boundary conditions are unique.

A second way of using Maxwell's equations is to integrate them over an extended region of space to produce equations which are much easier to use in some situations, as will be discussed in Section 2.7.

A further point about all equations, including Maxwell's, is obvious but can be overlooked. Readers with a traditional training in mathematics may have a subjective tendency to read them from right to left, in the sense that the left-hand side of an equation is perceived to have been obtained from the right-hand side. However all equations must be read as acting in both directions, so that if the quantities on one side of the equation change, there must be a corresponding change on the other side to compensate. Thus if at any point the curl of the electric field is non-zero, then Equation (2.15) demands that there has to be a changing **B**, $\partial \mathbf{B}/\partial t$ at that point. Similarly if at any given point $\nabla \times \mathbf{H} \neq 0$, then current **J** and/or $\partial \mathbf{D}/\partial t$ have to exist there in order to balance Equation (2.16). In studying EMC problems we shall be considering situations where some part of **E** and **H** fields at a given location are caused by systems of currents and charges remote from the local distributions. In order to balance Maxwell's equations these extra contributions to the fields cause further currents to flow or charges to appear which modify the local distributions, i.e. they cause unwanted interference. This, in essence, is the entire problem of EMC.

A final point about Maxwell's equations is that they contain time derivatives $\partial \mathbf{D}/\partial t$ and $\partial \mathbf{B}/\partial t$, which are related to curl **H** and curl **E**. Now it has been seen that the curl of a vector field is one way of describing the *structure* of the field in the vicinity of a point, since 'curl' is defined in terms of a line integral which is sensitive to the spatial details of the fields. Equations (2.15) and (2.16) show that the field structure depends on these time dependent vectors in a very direct way. If the **B** and **D** fields vary slowly (low frequency) then these terms might not be as important in determining the spatial structure of the **E** and **H** fields compared to the current density **J**. On the other hand, when **B** and **D** vary very rapidly (high frequency) these time dependent terms may have a significant, if not major, role in determining the overall field geometry. In this situation the distribution of the current density **J**, which in practical situations is normally confined within conductors, might not dominate the field distributions, and the whole problem becomes much more complicated. Thus it is clear that very important frequency dependent effects should be expected which will complicate matters, and that in either the low or high frequency case it might be possible to make different sorts of approximations to help in solving particular problems.

Although Maxwell's equations describe the spatial derivatives of fields, the actual values of the fields at any given point, which are the crucial quantities for making EMC calculations and measurements, depend on the precise way in which currents and charges are distributed in the space around it. Thus at any given point the total field may be constructed from a part which is due to remote current and charges and another part which is due to the local time derivates of the **B** and **D** fields, as in Equations (2.15) and (2.16). However it is obvious that it is possible to consider points which are far away from any currents or charges, where only the local time derivatives are important. Such a point is called the radiation field, or 'far field'. The other possibility, where the point considered is close to currents and charges, is called the 'near field'. It is vital to understand the distinction between the two, and have a precise way of knowing which approximation is appropriate in any situation. These concepts will be discussed in detail in Section 2.4.

2.3.2 Poisson's and Laplace's equations

Cases where there are time independent (static) electric and magnetic fields are of limited interest in this book. It is known from very basic *electrostatics* that there cannot be static electric fields, generated by external charges, inside a conductor or a closed conducting shell. However in the case of *magnetostatics* there is no such requirement, and static magnetic fields can penetrate inside metallic enclosures that are not specifically designed to exclude them, with the subsequent risk that these intruding fields could cause EMC problems.

In the limit when the time derivatives $\partial\mathbf{H}/\partial t$ and $\partial\mathbf{D}/\partial t$ are zero and there are no currents, the electric and magnetic fields can be defined in terms of the gradient of scalar potentials V and ϕ_m, i.e.

$$\mathbf{E} = -\nabla \cdot V$$

and

$$\mathbf{H} = -\nabla \cdot \phi_m$$

where V satisfies Poisson's equation

$$\nabla^2 \cdot V = \rho/\varepsilon$$

when the charge density is non-zero, or Laplace's equation

$$\nabla^2 \cdot V = 0$$

when the charge density is zero. The magnetic scalar potential V_m satisfies Laplace's equation

$$\nabla^2 \cdot V_m = 0$$

when there are no currents.

Solutions to Poisson's and Laplace's equations are unique for any given set of boundary conditions and numerous ways of finding solutions by either analytic or numerical methods are extensively covered in the literature. Laplace's equation for V_m will be used in Chapter 4 when the problem of shielding against static magnetic fields is discussed.

2.4 NEAR AND FAR FIELD APPROXIMATIONS AND ENERGY FLOW

The central problem of EMC is that an electromagnetic device or system can interact unintentionally with other, remote, systems. This interaction is referred to in a general way as one system 'coupling' to another. Such coupling can be via reactive mechanisms involving capacitance and/or inductance, as well as the radiative transmission of energy via electromagnetic waves between the two systems (these are not the only modes, see Figure 1.3). However there remains the problem to know which effect is appropriate in any given situation. Maxwell's equations are not of much direct help in answering this question, and it will be necessary to examine some general solutions in idealised situations to begin to understand this problem, but even it may not be easy to give a definitive answer.

To address the problem it is convenient to consider the two simplest models of interference sources or antennas, namely the short wire and the small loop, which will be assumed to be carrying alternating currents $I_0 e^{j\omega t}$. Such antennas can help to identify some key concepts and are the basic entities from which realistic sources and practical antennas can be built up.

2.4.1 Short wire antenna

The pattern of electromagnetic fields with wavelength λ around a short wire antenna of length $L \ll \lambda$ (see Figure 2.1) carrying an (unrealistic) uniform current $I_0 e^{j\omega t}$ is described by functions of distance, frequency and angle. The form quoted here are developed from those in [2] and are difficult to simplify further:

$$E_r = \frac{I_0 L e^{j(\omega t - \beta r)} \cos(\theta)}{j\omega\varepsilon_0 4\pi r}\left[\frac{2j\beta}{r} + \frac{2}{r^2}\right] \tag{2.20a}$$

$$E_\theta = -\frac{I_0 L e^{j(\omega t - \beta r)} \sin(\theta)}{j\omega\varepsilon_0 4\pi r}\left[\beta^2 - \frac{j\beta}{r} - \frac{1}{r^2}\right] \tag{2.20b}$$

$$H_\phi = \frac{I_0 L e^{j(\omega t - \beta r)} \sin(\theta)}{4\pi r}\left[j\beta + \frac{1}{r}\right] \tag{2.20c}$$

where $\beta = 2\pi/\lambda$, θ is the zenith angle to radial distance r, and the length of the short wire $L \ll \lambda$. By using relationships such as $\omega/c = \beta$ these equations can be written in various equivalent ways [3, 10].

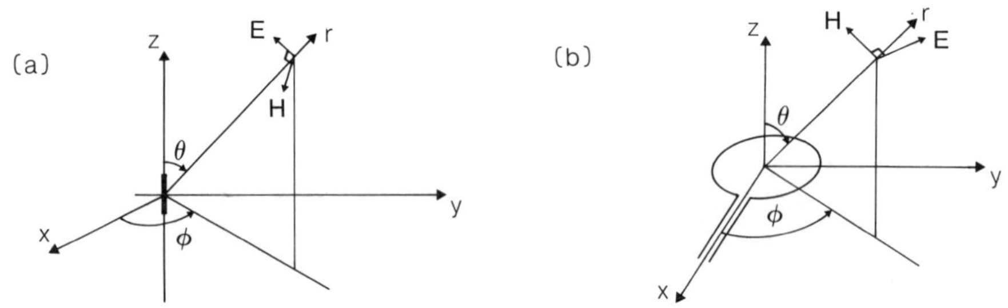

Fig. 2.1 Coordinate systems and field components for (*a*) short wire and (*b*) small loop antennas

2.4.2 Small loop antenna

A small loop of radius a carrying a uniform current $I_0 e^{j\omega t}$ radiates fields with components given by [2]

$$H_r = \frac{\pi a^2 I_0 e^{j(\omega t - \beta r)} \cos(\theta)}{4\pi r}\left[\frac{2j\beta}{r} + \frac{2}{r^2}\right] \tag{2.21a}$$

$$H_\theta = -\frac{\pi a^2 I_0 e^{j(\omega t - \beta r)} \sin(\theta)}{4\pi r}\left[\beta^2 - \frac{j\beta}{r} - \frac{1}{r^2}\right] \tag{2.21b}$$

$$E_\phi = -\frac{j\omega\mu_0 \pi a^2 I_0 e^{j(\omega t - \beta r)} \sin(\theta)}{4\pi r}\left[j\beta + \frac{1}{r}\right]. \tag{2.21c}$$

Note the similarity of form between Equations (2.20) and (2.21) but that the magnetic and electric components are exchanged.

2.4.3 The near and far field

In order to made use of Equations (2.20) given in Section 2.4.1 it is useful to be able to make some approximations which enable some of the terms to be ignored. One common approximation is to consider these fields in the limits of large and small distances, compared with the wavelengths of the radiation, from the source.

At large distances (many wavelengths), called the 'far field', the field is treated as if it was a plane electromagnetic wave propagating outwards from the source. Thus in the far field where any terms in $1/r^2$ and $1/r^3$ can be neglected compared to terms in $1/r$, the radiated field components become

$$E_\theta = -\frac{LI_0}{j\omega\varepsilon_o}\frac{e^{j(\omega t-\beta r)}\sin(\theta)\beta^2}{4\pi r}$$

$$= \frac{j\beta LI_0 e^{j(\omega t-\beta r)}\sin(\theta)}{4\pi\varepsilon_o rc} \qquad (2.22a)$$

and similarly

$$H_\phi = \frac{j\beta LI_0 e^{j(\omega t-\beta r)}\sin(\theta)}{4\pi r}. \qquad (2.22b)$$

The E_θ and H_ϕ components in the far field given by (2.22) are in phase (both have coefficients +j) while the E_r component is negligible.

The space close to the wire, where the terms in $1/r^2$ and $1/r^3$ dominate over the $1/r$ terms, is known as the 'near field' region. The near field region has a more complex structure than the far field because all three field components (E_θ, H_ϕ, and E_r) have to be considered and the phase relationships between them are different from the far field situation. However as will soon become apparent, in many EMC situations the near field approximation is valid. Thus the crucial point is to be able to identify the transition region where the near field 'turns into' the far field.

Some insight into identifying the transition region can be gained by plotting the relative contribution made by each term in the expression for E_θ, equation (2.20*b*). To do this it is convenient to write the distance $r = a(\lambda/2\pi) = a/\beta$, then since $\omega/c = \beta$ a little manipulation enables Equation (2.20*b*) to be rewritten as

$$E_\theta = jLI_0\sqrt{(\mu_0/\varepsilon_0)}\sin\theta\, e^{j(\omega t-\beta x)}\pi/\lambda^2(1/a - j/a^2 - 1/a^3)$$

i.e.

$$|E_\theta| \approx 120\pi^2 I_0 L(\sin\theta)/\lambda^2\,|(1/a - j/a^2 - 1/a^3)| \qquad (2.23)$$

where $\sqrt{(\mu_0/\varepsilon_0)} \approx 120\pi$ (see Section 2.5.1.2) and $c = 1/\sqrt{(\mu_0/\varepsilon_0)}$ is the velocity of light.

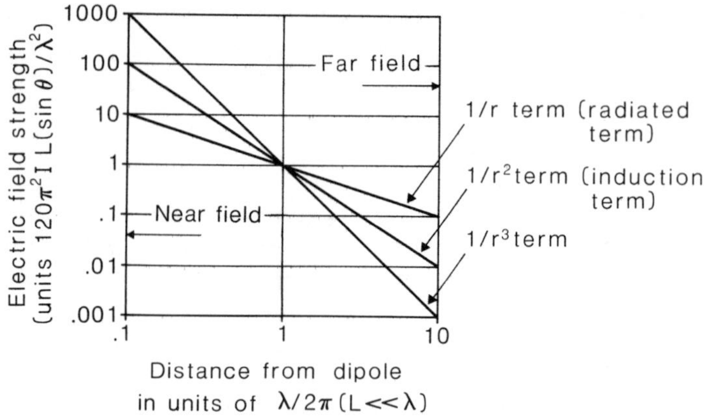

Fig. 2.2 Relative contribution to the electric field strength as a function of distance from a short wire antenna

The modulus of each of the three terms in Equation (2.23) are plotted in Figure 2.2, and it is clear that where $r \approx \lambda/2\pi$ (i.e. $a \approx 1$) all the contributions to the E_θ component of the electric field are of equal magnitude, so that neither the far field or near field approximation would be valid. This region is clearly a transition region, so that the near and far field regions are some distance away this. For a working definition it can be said that the near and far field regions should be an order of magnitude smaller or bigger than this transition distance, i.e.

$$r_{\text{near}} \leqslant 0.1(\lambda/2\pi)$$
$$r_{\text{far}} \geqslant 10(\lambda/2\pi).$$

Since this definition involves the wavelength of the radiation it is frequency dependent. EMC handbooks however often use a crude dividing line based on the value $\lambda/2\pi$, namely

$$r \geqslant \lambda/2\pi \quad \text{far field}$$
$$r \leqslant \lambda/2\pi \quad \text{near field.}$$

Some values of these various distances are given in Table 2.2. These dividing lines are useful when making approximations for shielding calculations and in deciding how close to complex systems meaningful radiation measurements can be made. In particular the dimension F, discussed in Section 1.4 and illustrated in Figure 1.13, is usually set by the lowest frequency to be measured and the need to be dominantly in the far field region.

As will be seen shortly the properties of the near and far fields are so different that it is important to realise that measurements made in one do not easily give much reliable information about the other, so it is of prime importance to understand clearly which regime is appropriate in any particular case.

Table 2.2. Near and far field estimates as a function of frequency

Frequency	$\frac{0.1\lambda}{2\pi}$	$\frac{\lambda}{2\pi}$	$\frac{10\lambda}{2\pi}$
30 MHz	15.9 cm	159.00 cm	15.90 m
300 MHz	1.59 cm	15.90 cm	1.590 m
3 GHz	0.159 cm	1.59 cm	0.159 m

2.4.4 Energy flow around a short wire antenna

The energy flow in an electromagnetic field is found by making use of the Poynting vector, **S**, defined as:

$$\mathbf{S} = \mathbf{E} \times \mathbf{H} \tag{2.24}$$

where **E** and **H** are the field vectors. Derivations found in most textbooks (e.g. Section 28.6 in reference [1]) show that the energy flow, W, outwards from within a volume enclosed by a surface A is given by the surface integral of this vector, yielding the Poynting theorem:

$$W = \int_A \mathbf{S} \cdot \mathrm{d}\mathbf{A}. \tag{2.25}$$

Based on this equation it is usual to associate an instantaneous energy flow density of $\mathbf{E} \times \mathbf{H}$ J m^{-2} with each point in the field, with **S** defining the direction of flow, even though the derivation of the Poynting theorem computes the total power flowing through the whole surface. As Jeans has pointed out [11] '... it has not been proved, and we are not entitled to assume, that there is an actual flow of energy at every point equal to the Poynting Flux'.

In simple cases such as the far field region, where the **E** and **H** fields are polarised and in phase, this yields satisfactory results, but in the near field region it is not easy to see clearly what is going on because the fields have both radial and angular dependence together with complex phase relations. For example the Poynting vector formed by the cross product of the radial and azimuthal fields $\mathbf{E}_r \times \mathbf{H}_\phi$ of the short wire antenna given by Equations (2.20) does not point radially away from the dipole. Thus one might conclude that a system situated in the near field of a transmitter may have energy incident upon it from other than the radial direction; Figure 2.3 shows schematically the energy flow around a short wire. Normally it is the time averaged Poynting vector $\mathbf{S}_{av}$ which is usually of practical interest, and this is given by (see Section 38.2.3 in reference [1])

$$\mathbf{S}_{av} = 1/2\ \mathrm{Re}(\mathbf{E} \times \mathbf{H}^*) = 1/2\ \mathrm{Re}\{(E_r\boldsymbol{r} + E_\theta\boldsymbol{\theta}) \times H_\phi^*\boldsymbol{\phi}\} \tag{2.26}$$

where **E** and **H** are the expressions for the electric and magnetic fields written as complex numbers. The term Re stands for the real part of the complex number and the * means take the complex conjugate (replace j by −j). Now if

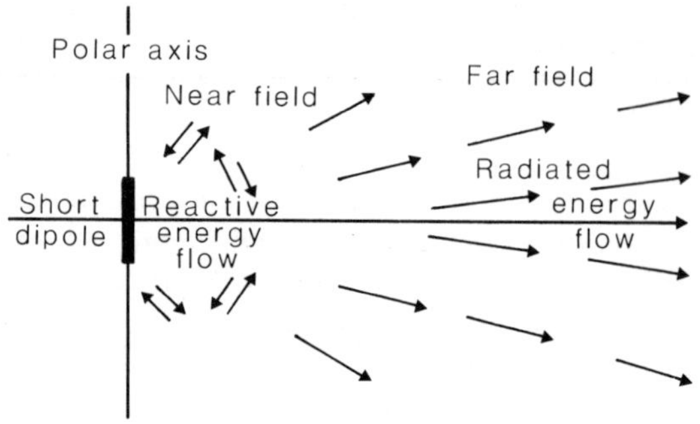

Fig. 2.3 Schematic diagram of the energy flow in the electromagnetic field around a short wire antenna

the Poynting vector around the short wire is integrated over a surface surrounding the wire, a time averaged radial outflow of energy at all distances r is found, not just in the far field. Averaged over time, energy flows out from the short wire, through the near field and escapes as radiation in the far field. The contribution to the energy flow of the E_r term actually vanishes, because it involves finding the real part of the product $E_r H_\phi^*$ which from equation (2.20) has a contribution of the form

$$\mathrm{Re}\left[\{1/cr^2 + 1/\mathrm{j}\omega r^3\}\{-\mathrm{j}\omega/cr + 1/r^2\}\right]. \tag{2.27}$$

On expanding this the real part is found to be zero, so that the radial electric field does not contribute to the time averaged power flow. Thus using E_θ and H_ϕ at all distances leads to a knowledge of the power flow.

2.4.5 High and low impedance fields

This section will introduce concept of, and stress the difference between, high and low impedance fields. This topic is important because the type of field present has a great effect on the screening behaviour of various materials. Briefly, at low frequencies it will be shown in Chapters 3 and 4 that magnetic (low impedance) fields are very difficult to screen using conducting materials such as copper, whereas electric (high impedance) fields are much easier to screen.

2.4.5.1 Fields around a short wire antenna

The previous discussion about energy flow and the mathematical equations used to describe the near and far fields do not make clear two facts. The first is that the far fields due to the short wire and small loop are very similar, with the lines of **E** and **H** for the loop being the same as the lines of **H** and **E** for the short wire (see Figure 2.1). The energy flow found from the Poynting vector is in the radial direction away from the antennas, so that the far fields of both sources can be treated identically. Any energy which couples into a system situated in the far field can be thought of as been carried by radiation from the source to the system. The oscillating **E** and **H** fields in the far field form a self-consistent, self-contained, system, which is independent of what happens back at the antenna which initially gave rise to them. The current in the antenna could be stopped without influencing further their behaviour. In the near field however, the situation is more complicated and needs some careful discussion.

Consider the mechanisms which are generating the fields in the space around a short wire antenna. An oscillating current is flowing, which will cause charges to accumulate at both ends of the wire, giving rise to a changing electric field roughly parallel to the wire. This changing electric field will, via Equation (2.15) give rise to a changing magnetic field perpendicular to it, i.e. wrapped around the wire.

As well as this, the current flowing along the wire will also give rise, as will be discussed in Section 2.6.1, to another (changing) magnetic field around the wire. This in turn will give rise to another changing electric field, which can be added to the electric field produced by the spatially separated charges at the ends of the wire. Thus in the near field of the short wire the **E** and **H** fields are produced by two different mechanisms. The dominant E_θ and H_ϕ fields are 90° out of phase because at the instant when the dipole moment is a maximum, producing the largest electric field between the ends of the wire, as in Figure 2.4, the current is zero, so there will be no magnetic field around the wire.

It might be guessed that it will be difficult to cause a large current to flow in a circuit consisting of a single length of short wire, so it is reasonable to assume that the magnetic field will be quite weak. On the other hand the maximum value of the electric field produced by the dipole charges at the ends of the wire will only depend on how much charge can be deposited there. Thus in the near field of a short wire antenna the most important field will be the electric field produced by the dipole moment. The impedance of the short wire seen by the power source driving the current is V/I, where V is the voltage and I the current supplied. As the current is low, this impedance will be high.

At this point it is possible to pre-empt an important concept to which will be discussed later, and note that the ratio of the magnitudes of the **E** and **H** fields, $Z = E/H$, is a quantity with dimensions of ohms, i.e. the same units used to measure the impedance of the wire V/I seen by the power supply driving the current in the short wire. In the case of the short wire it has been argued that

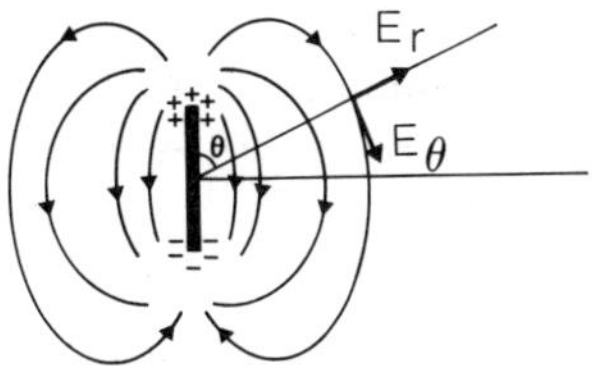

Fig. 2.4 Schematic diagram of the electric field around a short wire antenna at the instant when the current is zero

the **E** field dominates in the near field, so that the ratio E/H is large. Both the circuit impedance and the 'impedance' $Z(=E/H)$ of the electromagnetic field at any point in the immediate vicinity of the wire will have large numerical values. This sort of near field is known as a high impedance, or electric, field.

2.4.5.2 *Fields around a loop antenna*

For a magnetic dipole antenna, which consists of a loop of wire carrying an alternating current it is easy to get a large current to flow around the closed loop made from a good conductor (making the impedance V/I small), but difficult to get positive and negative electric charges to accumulate in different locations so as to provide an electric field between them. Thus in the near field region of a small current carrying loop the magnetic field dominates over the electric field. The near field of a loop antenna is called a low impedance, or magnetic, field. Both the impedance of the circuit and the ratio $Z=E/H$ are small.

It must be stressed that if the loop has a high impedance then there will be electric fields due to the potential difference driving the current around the circuit (as well the presence of small surface charges on the wires guiding the currents around the loop [4]), and these fields are shown schematically in Figure 2.5. Loops with high impedance will be surrounded by electromagnetic fields which can be thought of as being a mixture of low impedance (magnetic) and high impedance (electric) fields. This point will be discussed in more detail in Section 5.4.3.

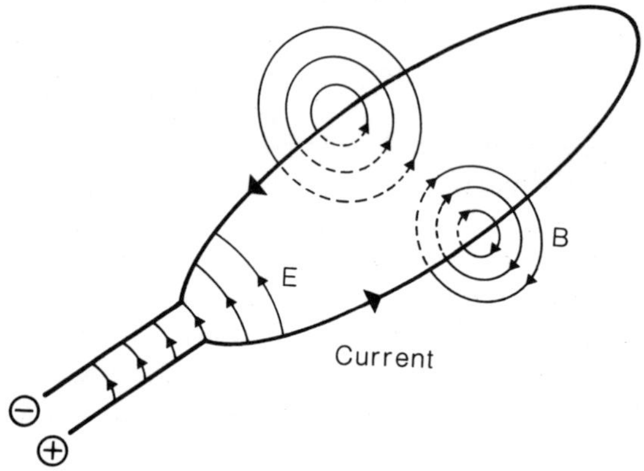

Fig. 2.5 Schematic diagram of the magnetic field present at the instant when the current is a maximum around a small loop antenna. If the loop has finite impedance then there will be electric field lines, a few of which are indicated, which will give dipole (high impedance) contribution to the fields, as will be discussed in Section 5.4.3

2.4.6 Reactive fields

In the previous section it was stated that the radiated energy flow associated with the time averaged Poynting vector was found by calculating the integral over a surface of the real part of the vector $\mathbf{E} \times \mathbf{H}^*$. This poses the question as to the interpretation which should be attached to the imaginary part of this vector. The answer is that this term corresponds to the electromagnetic field energy which is not radiated but is interchanged dynamically with the oscillating current flowing in the antenna which give rise to the fields. This state of affairs is similar to that encountered in circuit theory using discrete components such as capacitors and inductors. A capacitor in a circuit stores energy in its internal electric field during the charging process and returns it to the circuit when discharging. Similarly in the inductor the energy is stored and retrieved from the magnetic field around the inductor. In both cases there is a reactance associated with both of these fields which act as an impedance to the current flow. For the capacitance this is $1/j\omega C$ and for the inductor it is $j\omega L$. Work has to be done to establish each field. In the capacitor this work is needed to separate the charges, and in the inductor to drive the current against the back emf. This work is not lost but is converted into the energy stored in the field which is recovered when the reverse processes happen. The behaviour of the reactive fields around an antenna, even a simple short wire, is identical to this, and there is no difference between the two cases.

2.5 WAVES IN DIFFERENT MEDIA

This section will study how electromagnetic radiation behaves when it enters different media. The most general conditions are when $\varepsilon \neq \varepsilon_0$, $\mu \neq \mu_0$, $\sigma \neq 0$ and $\rho \neq 0$. In this case the full set of Maxwell's equations given by equations (2.15)–(2.18) are needed, together with $\mathbf{D} = \varepsilon\mathbf{E}$, $\mathbf{B} = \mu\mathbf{H}$ and $\mathbf{J} = \sigma\mathbf{E}$. These equations can be manipulated to yield the following pair of differential equations (see Section 28.2 in reference [1]):

$$\nabla^2\mathbf{E} = \varepsilon\mu \frac{\partial^2\mathbf{E}}{\partial t^2} + \sigma\mu \frac{\partial\mathbf{E}}{\partial t} + \nabla(\rho/\varepsilon) \tag{2.28a}$$

and

$$\nabla^2\mathbf{H} = \varepsilon\mu \frac{\partial^2\mathbf{H}}{\partial t^2} + \sigma\mu \frac{\partial\mathbf{H}}{\partial t}. \tag{2.28b}$$

The general solutions of these differential equations represent attenuated waves. If the period of the waves concerned is large compared with the relaxation time of the charge carriers in a conductor, then a very good approximation is to put $\rho = 0$, because the electrons move quickly enough to prevent charge building up in particular locations.

2.5.1 Waves in a dielectric medium

In pure dielectrics where $\sigma = \rho = 0$, Equations (2.28) simplify immediately to give:

$$\nabla^2\mathbf{E} = \varepsilon\mu \frac{\partial^2\mathbf{E}}{\partial t^2} \tag{2.29a}$$

and

$$\nabla^2\mathbf{H} = \varepsilon\mu \frac{\partial^2\mathbf{H}}{\partial t^2}. \tag{2.29b}$$

Equations similar in form to (2.29) are encountered in many branches of physics [7] and represent unattenuated waves travelling with a velocity $1/\sqrt{(\varepsilon\mu)}$. These solutions show that electromagnetic waves exist and were Maxwell's most famous prediction.

In free space the velocity of the waves is often given the special symbol $c = 1/\sqrt{\varepsilon_0\mu_0}$. To simplify the discussion without sacrificing any of the important ideas, it is easier to start by examining the behaviour of uniform, linearly polarised plane waves. These are waves where only E_y and H_z are non-zero. A wave of this type is called a principal zero-order TEM (transverse electric and magnetic) wave. The polarisation condition means that the components E_z, E_x, H_y, H_x are all zero and the uniformity condition means that the intensity of the wave is the same at all points perpendicular to the direction of motion. Mathematically this means that $\partial H_z/\partial y = 0$, as are all the derivatives $\partial/\partial y$ and $\partial/\partial z$; see Figure 2.6. The vector $\mathbf{E}\times\mathbf{H}$ points in the direction of motion of the waves. Thus the only non-zero terms remaining in (2.29) are:

$$\frac{\partial^2 E_y}{\partial x^2} = \varepsilon\mu \frac{\partial^2 E_y}{\partial t^2} \tag{2.30a}$$

and

$$\frac{\partial^2 H_z}{\partial x^2} = \varepsilon\mu \frac{\partial^2 H_z}{\partial t^2}. \tag{2.30b}$$

These are equations of the type:

$$\frac{\partial^2 \phi}{\partial x^2} = \frac{1}{V^2}\frac{\partial^2 \phi}{\partial t^2}$$

which is the general form of a wave equation [7]. As can be proved by differentiation, the wave equation is satisfied by solutions of the form $\phi(t, x) = F(t \pm x/V)$ where F is an arbitrary function. This solution represents a wave travelling in the $+x$ direction (the 'forward' direction) when the negative sign is taken, and in the $-x$ direction when the positive sign is used. For purely

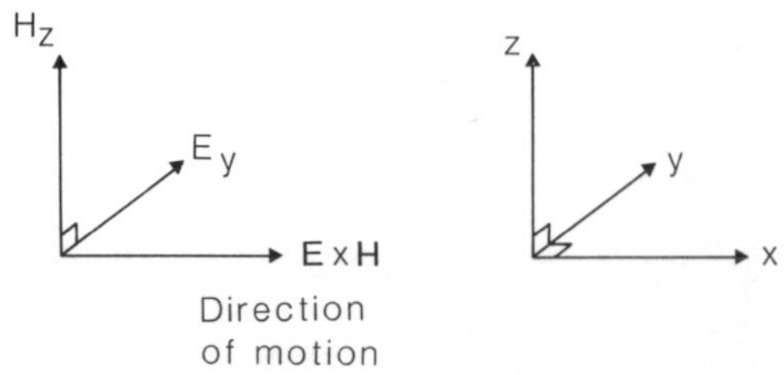

Fig. 2.6 Right-handed coordinate system used to describe a TEM wave

mathematical convenience in the remainder of this section $\phi(t, x)$ will considered to be a forward travelling wave described by a simple trigonometric function of the form $\phi(t, x) = \sin(2\pi f(t - x/V))$. This is because it is a very simple function to manipulate. As the arbitrary function F could be expanded as a Fourier series (see Chapter 5) with sinusoidal terms, making this choice will not invalidate any conclusions. Consequently a valid equation for the electric component E_y in this simple form is:

$$E_y = E_0 \sin(2\pi f(t - x/V)) \tag{2.31}$$

provided f is interpreted as the frequency of the wave, and $V = 1/\sqrt{\mu\varepsilon}$ is the phase velocity. This can be written:

$$E_y = E_0 \sin(2\pi ft - \beta x) \tag{2.32}$$

where $\beta = 2\pi/\lambda$, and λ is the wavelength. There is a similar formula for the magnetic component

$$H_z = H_0 \sin(2\pi ft - \beta x). \tag{2.33}$$

These solutions for E_y and H_z are in phase at any point or at any time.

2.5.1.1 *Refractive index*

The index of refraction is the ratio of the free space velocity to the velocity in the medium, hence

$$n = V_0/V = \sqrt{(\mu\varepsilon/\mu_0\varepsilon_0)} = \sqrt{(\mu_r\varepsilon_r)}. \tag{2.34}$$

As an example, for distilled water (which can be treated as a dielectric because it is a very poor conductor; at 18 °C, $\sigma = 4 \times 10^{-8}$ S cm^{-1}) $\mu_r = 1$ and $\varepsilon_r \approx 81$. Equation (2.34) predicts $n = 9$. When visible light is used (with frequency of order 10^{14} Hz) it is found that $n = 1.33$. This difference between theory and experiment is because the value of ε_r used here was measured with static fields, whereas in reality, as has already been noted, ε_r is a function of frequency. For water ε_r drops rapidly above 1 or 2 GHz, see Table 2.1

2.5.1.2 *Characteristic impedance of a dielectric*

Equations (2.32) and (2.33) for a principal zero-order TEM wave can be written as

$$E_y = E_0 \sin(\omega t - \beta x) \tag{2.35}$$

$$H_z = H_0 \sin(\omega t - \beta x) \tag{2.36}$$

by using $\omega = 2\pi f$. In phasor notation these become

$$E_y = E_0 e^{j(\omega t - \beta x)} \tag{2.35a}$$

$$H_z = H_0 e^{j(\omega t - \beta x)}. \tag{2.36a}$$

For the TEM wave Maxwell's Equation (2.15) becomes

$$\frac{\partial E_y}{\partial x} = -\mu \frac{\partial H_z}{\partial t}. \tag{2.37}$$

Thus substituting (2.35)

$$-\beta E_0 \cos(\omega t - \beta x) = -\mu \, \partial H_z / \partial t.$$

Integrating this with respect to t

$$H_z = \frac{\beta E_0}{\mu\omega} \sin(\omega t - \beta x) = \frac{\beta E_y}{\mu\omega}. \tag{2.38}$$

Since E_y and H_z are in phase it is possible to define:

$$Z = E_y/H_z = \mu\omega/\beta = (E_0/H_0) = \mu/\surd(\varepsilon\mu)$$

i.e.

$$Z = \surd(\mu/\varepsilon). \tag{2.39}$$

The quantity Z given by Equation (2.39) is a constant for each dielectric and is known as the characteristic impedance of the medium. This is because it has units of ohms and depends on the values of μ and ε which are material dependent. In Section 2.4.5 the idea was introduced that the ratio of the magnitudes E/H for a system of electromagnetic fields, called the wave impedance, is a useful quantity. Here it is seen that an impedance Z can also be defined in terms of the physical properties of the medium in which the fields reside. As vacuum is treated as a 'medium' with its own properties, in free space this characteristic impedance (defined for plane waves or any waves far from the source) is given the special symbol

$$Z_0 = \surd\mu_0/\varepsilon_0 \tag{2.40}$$

and has a numerical value of 376.7303 Ω; this is often approximated by $Z_0 \approx 120\pi$ Ω. It is a real quantity, which reflects the fact that the **E** and **H** fields in TEM electromagnetic waves in free space are in phase. For dielectrics the two Equations (2.39) and (2.40) can be combined to yield

$$Z = Z_0\surd(\mu_r/\varepsilon_r). \tag{2.41}$$

Since for dielectrics $\mu_r \approx 1$ and $\varepsilon_r > 1$ (see Table 2.1), their characteristic impedances are real and somewhat smaller than Z_0. Shortly, a quite different expression for the impedance of conductors (Equation (2.79)) will be derived and it will be shown that their characteristic impedances are complex numbers with magnitudes considerably smaller than Z_0.

2.5.2 The impedance of the near field

Referring to the far field Equations (2.22*a*), (2.22*b*) of the short wire antenna it will be easily seen that the ratio $|\mathbf{E}|/|\mathbf{H}| = E_\theta/H_\phi = 1/(c\varepsilon_0) = Z_0$. However it is clear from the general equations for the field around the wire (2.20*a*)–(2.20*c*) that this cannot be true in general. In the near field where the $1/r^2$ and $1/r^3$ terms dominate the ratio $|\mathbf{E}|/|\mathbf{H}|$ has a much more complicated behaviour and is not $\approx Z_0$, but it is still of interest. In general, the ratio $|\mathbf{E}|/|\mathbf{H}|$ is called the wave impedance, Z_W, normally written in complex number notation, $Z_W = |Z_W| e^{j\phi}$. The fact that Z_W is complex simply means that **E** and **H** will not be in phase. In many cases this phase difference will not be of interest and important results can be obtained by working with $|Z_W|$.

As an example consider the result obtained if Equations (2.20*a*)–(2.20*c*) for the field around a short wire antenna are used to calculate $|Z_W|$ at any point in the space around the short wire. The mathematical details can be seen in Appendix C but for the short wire antenna it is found that in the far field where $r \gg \lambda/2\pi$,

$$|Z_W| \approx Z_0$$

whereas in the near field where $r \ll \lambda/2\pi$ then

$$|Z_W| \approx Z_0(\lambda/2\pi r)$$

which is bigger than Z_0, see Figure 2.7.

The idea that the near field of a short wire antenna has a high value of impedance, hence is dominantly an electric field, is contained within the equations describing the fields, which where themselves were derived using Maxwell's equations.

Similarly the impedance of the near field of a small loop antenna (see Section 2.4.2 and Appendix C) is given by

$$|Z_W| \approx Z_0(2\pi r/\lambda)$$

which is smaller than Z_0, as shown in Figure 2.7.

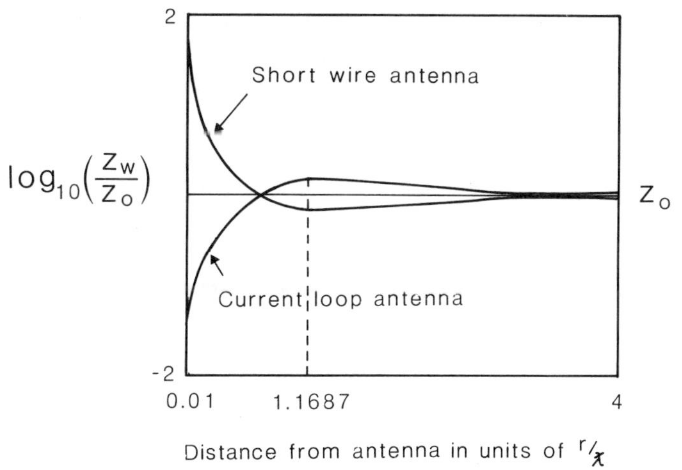

Fig. 2.7 Relative wave impedance as a function of r/λ (where $\lambda = \lambda/2\pi$) for a high impedance field produced by a short wire antenna and a low impedance field produced by a small current loop

2.5.3 The importance of the concept of impedance

It should be appreciated at this stage that assigning an impedance to electromagnetic fields (as in the previous section) or to materials (as for dielectrics in Section 2.5.1.2) is not an irrelevant operation but leads to a calculation technique of great power and usefulness. The point is that propagating electromagnetic energy, either in free space or confined within a transmission line etc, can be partially reflected whenever it is incident upon a change of impedance, and there are very useful relations between the reflected and transmitted parts of the wave which are written in terms of the impedances.

To see this consider a semi-infinite boundary between two regions with impedances Z_1 and Z_2 onto which a TEM wave travelling in the $+x$ direction falls at normal incidence, as shown in Figure 2.8. In general some of the wave will be reflected and the remainder transmitted. The directions of the **E** and **H** vectors shown in Figure 2.8 have been drawn to ensure that the vector $\mathbf{E} \times \mathbf{H}$ always points in the direction of motion, but are otherwise arbitrary, so that the reflected magnetic vector H_r is drawn pointing in the opposite direction to the incident magnetic vector H_i. At the interface there are boundary conditions which must be satisfied [1] which are

(*a*) The tangential component of E must be continuous (no surface charge on the boundary). In this instance this implies $E_i + E_r = E_t$.
(*b*) The tangential component of H must be continuous (provided there is no current at the boundary). Taking into account the directions of H_r and H_i this yields $H_i - H_r = H_t$.

Since $E_i/H_i = Z_1$, $E_t/H_t = Z_2$ then

$$E_i/Z_1 - E_r/Z_2 = (E_i + E_r)/Z_2. \tag{2.42}$$

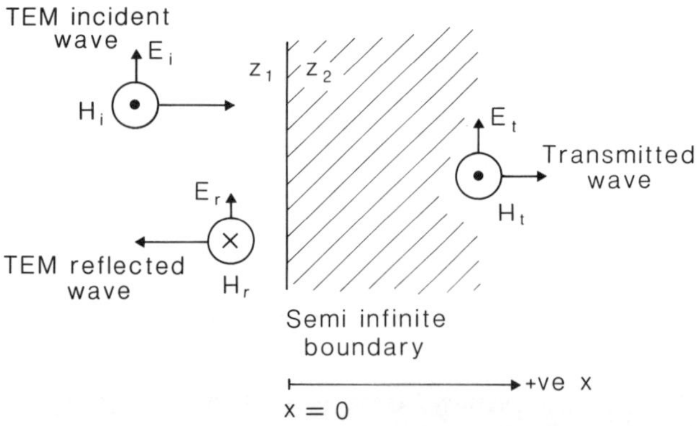

Fig. 2.8 Incident, reflected and transmitted components of a TEM wave incident onto a semi-infinite boundary, where Z_1 and Z_2 are the characteristic impedances of the two media. The circle with dot in it indicates a vector pointing out of the page, that with a cross a vector pointing into the page

The reflection coefficient is defined to be the ratio of the reflected amplitude of the electric component to the incident amplitude, i.e.

$$\rho = E_r/E_i \tag{2.43}$$

so that

$$\rho = \frac{Z_2 - Z_1}{Z_2 + Z_1}. \tag{2.44}$$

The transmission coefficient is defined as $\tau = E_t/E_i$, hence

$$\tau = \frac{2Z_2}{Z_2 + Z_1}. \tag{2.45}$$

With these definitions of τ and ρ

$$\tau = 1 + \rho. \tag{2.46}$$

The relationships (2.44) and (2.45) can be used when the impedances are real or complex and will form the core of the approach to EMC shielding in Chapter 4. For the moment it should be noticed that when the impedance Z_2 is much smaller than Z_1, the transmission coefficient is small and the reflection coefficient $\rho \approx -1$. In the limit when $Z_2 = 0$ there is perfect reflection. This is an indication that the characteristic impedances of good reflectors of electromagnetic waves, such as metals, are very small.

Finally it should be recalled that the derivation of (2.44) and (2.45) only considered plane waves at normal incidence. The general equations which apply to plane waves at all angles of incidence are known as the Fresnel equations, which are similar to these quoted here, but include terms in $\cos\theta_I$ and $\cos\theta_T$, where θ_I and θ_T are the angles of incidence and transmission; see Section 30 in [1] for full details.

For non-plane waves, E/H will be different at each point in space. However in a number of situations E/H will only be a function of the distance from a specific small source (loop, dipole, pair of parallel wires etc.) so that a fixed distance r (for a cylindrical or spherical boundary at the medium change) these same rules of reflection and transmission can be applied. For situations where the boundary of the medium change has no such symmetry, calculation of the reflection and transmission coefficients becomes more difficult.

2.5.4 Impedance in front of a boundary

As an example of the usefulness of the concept of impedance consider the case of reflection from a change of impedance of a forward travelling TEM wave, written in phasor notation (2.35*a*) as $E_0 e^{j(\omega t - \beta x)}$, incident upon a perfect reflector (a conductor with $\sigma = \infty \text{S m}^{-1}$), as in Figure 2.8. None of the incident field penetrates the conductor, so that at the surface, which is defined to be the origin $x = 0$, the transmission coefficient

$$\tau = 2Z_2/(Z_2 - Z_1) = 0$$

which means that effectively $Z_2 = 0$. The reflection coefficient

$$\rho = (Z_2 - Z_1)/(Z_2 + Z_1) = -1.$$

Thus the reflected wave, travelling in the $-x$ direction, is given by

$$\rho E_0 e^{j(\omega t + \beta x)} = -E_0 e^{j(\omega t + \beta x)}. \tag{2.47}$$

The total electric field E_y in front of the reflector is made up from the sum of the incident and reflected amplitudes. Hence, at any point in front of the reflector:

$$E_y = E_0 e^{j(\omega t - \beta x)} - E_0 e^{j(\omega t + \beta x)}$$
$$E_y = E_0(e^{-j\beta x} - e^{j\beta x})e^{j\omega t}$$

i.e.

$$E_y = -2E_0 j \sin(\beta x)e^{j\omega t} \tag{2.48a}$$

(note the negative sign), which can be also written

$$E_y = -2E_0 \sinh(\beta x)e^{j\omega t}. \tag{2.48b}$$

At $x = -L$

$$E_y = 2E_0 j \sin(\beta L)e^{j\omega t} \tag{2.49}$$

Similarly for the magnetic component the total field at a point x is

$$H_z = H_0 e^{j(\omega t - \beta x)} - H_0 e^{j(\omega t + \beta x)}. \tag{2.50}$$

For the medium in front of the reflector having impedance Z_1, $H_0 = E_0/Z_1$ for the incident wave and $H_0 = -E_0/Z_1$ for the reflected wave. Thus

$$H_z = \frac{E_0}{Z_1}(e^{-j\beta x} + e^{j\beta x})e^{j\omega t}$$

i.e. at $x = -L$

$$H_z = 2E_0 \cos(\beta L)e^{j\omega t}/Z_1. \tag{2.51}$$

Equations (2.49) and (2.51) correspond to a standing wave pattern set up in front of a perfect reflector, as both the E and H components of the field have spatial and time dependence. At $x = -L$ in front of a perfect reflector the wave impedance $Z_w = E/H$ is given by:

$$E_y/H_z = Z_w = Z_1 \frac{2E_0 j \sin(\beta L)e^{j\omega t}}{2E_0 \cos(\beta L)e^{j\omega t}} \tag{2.52}$$

i.e.

$$Z_w = jZ_1 \tan(\beta L). \tag{2.53}$$

For a non-perfect reflector, the total electric and magnetic components E_y and H_z at a point at $x = -L$ in front of the reflector is given by:

$$E_y = E_0(e^{j\beta L} + \rho e^{-j\beta L})e^{j\omega t}$$

and

$$H_z = E_0(e^{j\beta L} - \rho e^{-j\beta L})e^{j\omega t}/Z_1.$$

Thus using (2.44) for the reflection coefficient ρ yields the result

$$Z_w = Z_1 \frac{(Z_2 + jZ_1 \tan(\beta L))}{(Z_1 + jZ_2 \tan(\beta L))} \tag{2.54}$$

for the wave impedance a distance L metres in front of a non-perfect reflector with characteristic impedance $Z_2\ \Omega$. Note that the most general form of (2.54), which allows for absorption, will be discussed in Chapter 3. In Equation (2.54) $\beta = 2\pi/\lambda$ where λ is the wavelength of the incident TEM wave in the medium with impedance Z_1.

2.5.4.1 *Half-wave dielectric windows*

By way of illustration, it is convenient to use Equation (2.54) at this stage in two cases of interest. The first case concerns a piece of dielectric material with thickness L, made from material which has an impedance Z_1, as shown in Figure 2.9. If matters can be so arranged that the wave impedance at the surface of the material at $x = -L$, call it Z_L, is equal to the impedance of the incoming wave Z_2, then the reflection coefficient will be zero and there should be perfect transmission of the incident wave through the surface at $x = -L$.

Consider reflection from the surface at $x = 0$, so that at the point $x = -L$

$$Z_2 = Z_L = Z_1 \frac{(Z_2 + jZ_1 \tan(\beta L))}{(Z_1 + jZ_2 \tan(\beta L))} \tag{2.55}$$

i.e.

$$jZ_1Z_2 \tan(\beta L) = jZ_1^2 \tan(\beta L). \tag{2.56}$$

As $Z_1 \neq Z_2$ this can only happen if $\tan(\beta L) = 0$, i.e.

$$L = n\pi/\beta \tag{2.57}$$

where n is an integer. Since $\beta = 2\pi/\lambda$, the smallest L to satisfy (2.57) is equal to $\lambda/2$, where λ is the wavelength in the dielectric. Dielectric material of just this

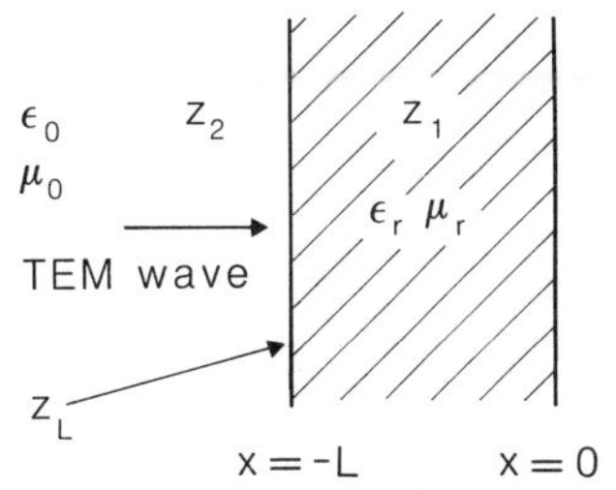

Fig. 2.9 Parameters which need to be considered when a TEM wave passes through a piece of material of thickness of L metres

thickness will allow perfect transmission of the incident radiation through the surface at $x = -L$ even though there is a change of impedance between the two different media, a result of practical use in building radomes. For a given thickness of material perfect transmission through the surface is obtained at one wavelength (frequency) only. This example will be considered again in more detail in Chapter 3.

2.5.4.2 *Quarter-wave coatings*

An extension of the technique discussed in the previous section is to coat a surface of material of characteristic impedance Z_2 with a coating having a different characteristic impedance Z_1 and a thickness $\lambda/4$. This is illustrated in Figure 2.10. In this situation no reflections are possible if the wave impedance at the surface of the coating at $x = -L$ is equal to Z_0, the impedance of free space, i.e.

$$Z_0 = Z_L = Z_1 \frac{(Z_2 + jZ_1 \tan(\beta L))}{(Z_1 + jZ_2 \tan(\beta L))} \tag{2.58}$$

which can be rewritten as

$$Z_0 = Z_L = Z_1 \frac{(Z_2/\tan(\beta L) + jZ_1)}{(Z_1/\tan(\beta L) + jZ_2)}.$$

When $\beta L = \pi/2$ (i.e. $L = \lambda/4$), then $\tan(\beta L) = \infty$, so that the impedance of the coating that produces this condition is given by:

$$Z_0 = Z_1^2/Z_2$$

or

$$Z_1 = \sqrt{(Z_2 Z_0)}. \tag{2.59}$$

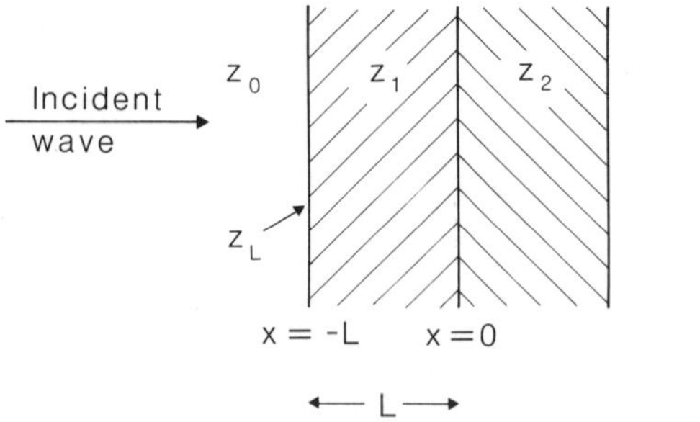

Fig. 2.10 A dielectric coating of thickness $L(=\lambda/4)$ and characteristic impedance Z_1 ohms deposited onto a material with impedance Z_2 ohms. If $Z_1 = \sqrt{(Z_2 Z_0)}$there will be perfect transmission of the incident radiation through the coating. Note in this case λ is the wavelength of the radiation in the coating

2.5.5 Summary of the impedance concept

The examples discussed in the previous sections show that, if an impedance can be assigned to, and calculated from, the properties of the media through which electromagnetic fields propagate, it becomes possible to manufacture materials with specific and very useful electromagnetic properties. Also the impedance concept provides a very useful and general way to discuss how electromagnetic waves are transmitted or reflected at boundaries where there are changes of impedance, one of the major concerns in EMC for designing shielding. Bearing this in mind, the next section will consider the case of waves travelling in an arbitrary medium, with the intention of deriving specific expressions for the impedance of both conductors and dielectrics in terms of their physical properties. But before doing this a final comment is appropriate.

Equation (2.54) for the wave impedance in front of a non-perfect reflector involves not only the characteristic impedances Z_1 and Z_2 but a term in $\tan(\beta L)$ which arises from the interference effects of the incident and reflected components of the electromagnetic fields. Nevertheless by taking this expression at face value a result of great practical importance was derived. This example holds a valuable lesson. The concept of wave impedance is a fundamental way of describing the total electromagnetic environment at a given point. From now on this philosophy will be whole-heartedly adopted and electromagnetic fields will be discussed in terms of their wave impedances. In particular it has been seen that the near field can be described as either a low or high impedance field, and waves with such different properties will be seen to have very different reflection effects when they are incident on a material. This whole approach will be used in Chapters 3 and 4 to tackle shielding problems in the near field.

A material which allows perfect transmission can also be considered to be a perfect absorber. Absorbers are of great importance in EMC as they are commonly found on the walls of electromagnetically anechoic chambers, and ideally they should absorb perfectly all frequencies which impinge upon them as any reflected energy may rise to spurious signals or troublesome resonances. The two examples discussed above are of limited use as they are perfect absorbers at one frequency only, but they suggest how improved absorbers could be made. For absorbing over a range of frequencies a large number of different thickness coating can be arranged in series, or tiles made with variable density carbon from front to back, which gives a smoothly varying impedance. For highly efficient absorbers, cones or pyramids can be made from similar materials. Such objects are seen lining the walls of a typical electromagnetically anechoic room. Absorbers will be discussed in more detail in Chapter 3.

2.5.6 Plane waves in an arbitrary medium

2.5.6.1 The propagation constant γ

It is necessary to clarify what is meant by a conductor and a dielectric. In an arbitrary material the transmitted wave must obey the Maxwell Equation (2.16)

$$\nabla \times \mathbf{H} = \mathbf{J} + \partial \mathbf{D}/\partial t$$

with $\mathbf{J} = \sigma \mathbf{E}$, i.e., neither the conduction current or the displacement current can be neglected at this stage. As before, consider the case of a TEM wave ($E_x = 0$, $\partial E_y/\partial z = 0$ etc.) so that Equation (2.16) becomes

$$-\frac{\partial H_z}{\partial x} = \sigma E_y + \varepsilon \frac{\partial E_y}{\partial t}.$$

Assuming a sinusoidal variation for **E** the wave can be written as $E_y = E_0 \mathrm{e}^{\mathrm{j}\omega t}$; hence

$$-\frac{\partial H_z}{\partial x} = (\sigma + \mathrm{j}\omega\varepsilon) E_y. \tag{2.60}$$

In Equation (2.60) σ arises from the effect of the conduction current and $\mathrm{j}\omega\varepsilon$ from the displacement current. When $\sigma \gg \omega\varepsilon$ the conduction current dominates and when $\sigma \ll \omega\varepsilon$ the displacement current dominates. Under this conditions when

$$\sigma/\omega\varepsilon \leqslant 1/100$$

such materials are called dielectrics and when

$$\sigma/\omega\varepsilon > 100$$

they are called conductors. In any material, for TEM waves Equations (2.16) and (2.15) are written

$$-\frac{\partial H_z}{\partial x} = \sigma E_y + \varepsilon \frac{\partial E_y}{\partial t} \tag{2.61}$$

and

$$\frac{\partial E_y}{\partial x} = -\mu \frac{\partial H_z}{\partial t}. \tag{2.62}$$

These equations can be solved most conveniently for E_y; differentiating (2.61) by $\partial/\partial t$ and (2.62) by $\partial/\partial x$ and adding the two subsequent equations to remove the term in $\partial^2 H_z/\partial x\, \partial t$ (which equals $\partial^2 H_z/\partial t\, \partial x$ because H_z is a function of x and t only) yields

$$\frac{1}{\mu}\frac{\partial^2 E_y}{\partial x^2} - \varepsilon \frac{\partial^2 E_y}{\partial t^2} - \sigma \frac{\partial E_y}{\partial t} = 0. \tag{2.63}$$

With harmonic variation $E_y = E_0 \mathrm{e}^{\mathrm{j}\omega t}$, Equation (2.63) becomes

$$\frac{\partial^2 E_y}{\partial x^2} - \gamma^2 E_y = 0 \tag{2.64}$$

where

$$\gamma^2 = (\mathrm{j}\omega\mu\sigma - \omega^2 \mu\varepsilon). \tag{2.65}$$

A solution to (2.64) for a wave travelling forward in the $+x$ direction is

$$E_y = E_0 e^{-\gamma x} e^{j\omega t}. \qquad (2.66)$$

Equation (2.66) is the general formula to describe the electric component of a uniform plane wave travelling through any medium. The complex number γ is called the propagation constant and can be written $\gamma = \alpha + j\beta$ where the real part α is called the attenuation constant and the imaginary part β is the phase constant. If the real part is non-zero then attenuation of the wave occurs when it travels through the medium. The field E_0 is the electric field at the point $x = 0$ *in the medium*. If $x = 0$ is the surface of the medium then E_0 must not be confused with the amplitude E_i of any wave incident onto the surface; these two are related via the transmission coefficient $\tau = E_0/E_i$.

The H_z component in an arbitrary medium can be found by using the impedance Z_w, since $H_z = E_y/Z_w$. In general Z_w is complex, so can be written as

$$Z_w = |Z_w| e^{j\phi}. \qquad (2.67)$$

Note that H_z will have the same spatial and time dependence as E_y, but in general there will be a phase lag ϕ between the E and H components,

$$H_z = (E_0/|Z_w|) e^{-\gamma x} e^{j(\omega t - \phi)}. \qquad (2.68)$$

As will be seen shortly, in a good conductor $\phi = 45^\circ$.

2.5.6.2 *Skin depth*

For a conducting medium, where $\sigma \gg \omega\varepsilon$ Equation (2.65) yields the approximation

$$\gamma^2 = j\omega\sigma\mu$$

so that

$$\gamma = \sqrt{j}\sqrt{\omega\sigma\mu}$$

hence

$$\gamma = \pm(1 + j)\sqrt{(\omega\sigma\mu/2)}. \qquad (2.69)$$

This derivation uses the relation $(1 + j)^2/2 = j$, so that $\sqrt{j} = (1 + j)/\sqrt{2}$.

Using Equations (2.66) and (2.69), the electric field at a point x in a conducting medium is in general given by

$$E_y = E_0 e^{-(1+j)\sqrt{(\omega\sigma\mu/2)}x} e^{j\omega t}. \qquad (2.70)$$

Now define a quantity ∂ (with dimensions of distance) as

$$\partial = 1/\sqrt{(\omega\sigma\mu/2)}. \qquad (2.71)$$

Then Equation (2.70) can be written as

$$\begin{aligned} E_y &= E_0 e^{-(x/\partial)(1+j)} e^{j\omega t} \\ &= E_0 e^{-(x/\partial)} e^{j(\omega t - x/\partial)} \end{aligned} \qquad (2.72)$$

At a distance $x = \partial$ inside the conductor the field falls to e^{-1} of its surface value and ∂ is called the skin depth. The term in $e^{j(\omega t - x/\partial)}$ can be compared with the expression for a wave travelling in free space, $e^{j(\omega t - \beta x)}$, where $\beta = 2\pi/\lambda$, so that the wavelength in the conductor is

$$\lambda = 2\pi\partial \tag{2.73}$$

which is considerably smaller than the free space wavelength. To illustrate this, consider the real part of (2.72)

$$E_y = E_0 e^{-(x/\partial)} \cos(\omega t - x/\partial). \tag{2.72a}$$

In Figure 2.11 the ratio E_y/E_0 given by (2.72*a*) is plotted against x (in units of ∂) and compared with two wavelengths ($= 4\pi\partial$) of the undamped cosine term. It is clear that the amplitude of the wave decreases very rapidly with distance, and after only one wavelength has reduced by a factor $e^{-2\pi} \approx 0.0019$.

As a specific example, consider the skin depth in copper, for which $\mu_r = 1$, $\mu = 4\pi \times 10^{-7}\,\mathrm{H\,m^{-1}}$ and $\sigma = 5.8 \times 10^7\,\mathrm{S\,m^{-1}}$. Using these values Table 2.3 can be computed. Because the skin depth is so small thin copper sheets can be effective screens against high frequency radiation.

As an example of propagation of steady waves through a conductor consider the problem of communicating with submarines when they are under water. For sea water at low frequencies $\varepsilon_r = 81$, $\mu \sim 4\pi \times 10^{-7}$, σ is of order $4\,\mathrm{S\,m^{-1}}$, so at 3 kHz the skin depth ∂ given by equation (2.71) is $\partial \approx 5$ m. This reduces to ≈ 2.8 m at 10 kHz; hence low frequency is favoured for deeper penetration. Also a receiving antenna should be at least 1/6 wavelength to be effective. In air at 3 kHz, such an antenna would need to be about 10 000 m long, much too

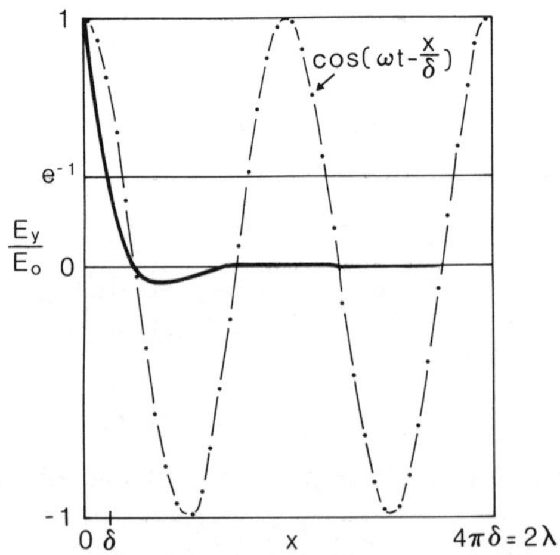

Fig. 2.11 Relative amplitude of the electric component (solid line) of a TEM wave as a function of distance inside a conductor. For comparison the dotted lines show the unattentuated wave. The skin depth ∂ metres is marked, and the wavelength of the TEM wave in the conductor is $2\pi\partial$ metres

Table 2.3. Skin depth in copper at different frequencies

Frequency	Skin depth (m)
60 Hz	8.5×10^{-3}
1 MHz	6.6×10^{-5}
30 000 MHz	3.8×10^{-7}

big for surface ship, but underwater at this frequency the wavelength $2\pi\partial$ is about 30 m, so it is easy to have an effective antenna for low frequency waves on a submarine.

2.5.6.3 *Current distributions in conductors*

Since $\mathbf{J} = \sigma\mathbf{E}$ the solutions for the $\mathbf{E}$ fields in the conductors found previously also give the current distributions. Thus using (2.72) gives

$$J_y = \sigma E_0 e^{-x/\partial} e^{j(\omega t - x/\partial)}. \qquad (2.74)$$

This current density is concentrated near the surface of the conductor, (where the field producing the current is largest) since the term $e^{-x/\partial}$ falls off as distance x into the metal increases. The skin depth is frequency dependent, becoming smaller as the frequency increases (see Table 2.3). This can have very important consequences and if the frequency is high enough one can get situations where different currents can flow on opposite sides of a thin metal sheet separating electromagnetic fields of different frequencies and phases.

In situations where the induced current density in a metal is not uniform a simple argument will show that this must lead to a rise in the apparent values of the resistance of the conductor. Consider a wire carrying a current i distributed with a uniform current density, and think of the wire as being split into two parts, with resistance R_1 and R_2, but with each part carrying half the current. The power dissipated is

$$P = i^2(R_1 + R_2)/4.$$

If the current distribution is non-uniform (but the total current remains the same), then one part, R_1 say, must carry a smaller current $(i/2 - \zeta)$, and the other part a larger current $(i/2 + \zeta)$. In this case the power dissipated is

$$P = (i/2 - \zeta)^2 R_1 + (i/2 + \zeta)^2 R_2$$
$$P = i^2(R_1 + R_2)/4 + \zeta^2(R_1 + R_2) + i\zeta(R_1 - R_2).$$

Choosing $R_1 = R_2$ leads to

$$P = i^2(R_1 + R_2)/4 + \zeta^2(R_1 + R_2). \qquad (2.75)$$

In equation (2.75) there is an extra term compared with the equation which corresponds to the uniform current density case, as if the total resistance of the wire has increased.

Note that from Table 2.3 at 60 Hz the skin depth in copper of 8.5 mm means that in thick conductors carrying large currents, such as bus-bars, the bulk of the current is carried within the first few centimetres of the surface. For cost effectiveness these should have a rectangular (or hollow) cross section rather than a square one, and that the effective resistance of the conductor will be higher than the DC case.

Similarly the resistance of the conductors in a transmission line or waveguide will increase with frequency which will change their characteristics, see Section 3.5.3.

2.5.7 Wave propagation in a good conductor

As the wavelength in the conductor $= 2\pi\delta$, the velocity of propagation is correspondingly reduced to $f2\pi\delta$ compared with that in free space. This is the 'phase velocity' and represents the velocity at which a point of constant phase travels through the conductor. Energy travels at the group velocity, v_g, which in a conductor is twice the phase velocity (see Section 29.1 in [1]). In good conductors these velocities are very low. For example, at 1 MHz the wavelength in copper is about 0.414 mm, with a corresponding phase velocity of 414 $\mathrm{m\,s^{-1}}$. This is an order of magnitude slower than the velocity of sound in copper! The supposed time for a steady wave to penetrate a thickness x would be $x/v_g = x/2\omega\delta$ seconds, which is frequency dependent and linearly dependent on the thickness, and it might be assumed that this was true in all situations. However the above results have been derived assuming a *steady state* TEM wave and in Section 2.5.9 it will be shown that the appropriate transit time for the penetration of a *pulse* of electromagnetic energy through a good conductor is independent of frequency and proportional to the *square* of the thickness.

2.5.8 Intrinsic impedance of conductors

For TEM waves in any medium the electric field component is described by equation (2.66) which is

$$E_y = E_0 e^{-\gamma x} e^{j\omega t}.$$

For the magnetic component assume the same general form but with a phase lag of ϕ such that

$$H_z = H_0 e^{j(\omega t - \phi)} e^{-\gamma x}. \tag{2.76}$$

The wave impedance $Z_w = E_y/H_z$ can be found in terms of the physical properties of the materials by making use of Equation (2.62) and substituting for E_y and H_z from (2.66) and (2.76). Making these substitutions gives

$$\gamma E_y = j\omega\mu H_z \tag{2.77}$$

hence the expression for the impedance of the medium is given by the complex quantity

$$\mathbf{Z}_w = j\omega\mu/\gamma \tag{2.78}$$

where the propagation constant γ, given by Equation (2.65), is also complex. Some references [1] write (2.78) as:

$$Z_W = \omega\mu/k$$

where

$$k^2 = (\omega^2\mu\varepsilon - j\omega\mu\sigma).$$

Equation (2.78) enables an impedance to be assigned to any material, which is a function of the properties of the material (μ, σ, ε) and the frequency of the wave. As this impedance is a complex number, in general the **E** and **H** fields are not in phase inside materials. For good conductors $\sigma/\omega\varepsilon \gg 100$, thus the approximation for γ (Equation (2.69)) can be used to yield the expression

$$\gamma \approx (1+j)\sqrt{(\omega\sigma\mu/2)} = (1+j)/\partial$$

so that in good conductors

$$Z_W \approx j\omega\mu\partial/(1+j) \approx (1+j)/\sigma\partial. \tag{2.79}$$

The magnetic component is related to the electric component by $H_z = E_y/Z_W$, so in a good conductor

$$\begin{aligned} H_z &= \sigma\delta E_y/(1+j) \\ &= \sigma\delta E_y(1-j)/2 \\ &= (\sigma\delta E_y e^{-j\pi/4})/\sqrt{2} \\ &= (\sigma\delta E_0 e^{j(\omega t - \pi/4)} e^{-\gamma x})/\sqrt{2}. \end{aligned} \tag{2.80}$$

The term $e^{j(\omega t - \pi/4)}$ in (2.80) means H_z lags E_y by $\pi/4$. It is instructive to calculate Z_W for copper as a function of frequency. Using $\sigma = 5.8 \times 10^7$ S m^{-1} and the skin depths from Table 2.3, together with Equation (2.79), Table 2.4 can be computed. The impedance of a good conductor is less than unity to very high frequencies and considerably smaller than the impedance of free space, $Z_0 \approx 120\pi$ Ω. As a consequence of this the magnitude of the magnetic intensity **H** in conductors is much bigger than the magnitude of **E**, and the energy carried by the wave in a conductor is dominantly magnetic (see reference [1], Section 29.1). This can be compared with the free space conditions, where the magnitude of the magnetic field is $H = E\varepsilon_0 c = E/Z_0 \approx E/120\pi$, smaller than E. In

Table 2.4. Impedance of copper at different frequencies

Frequency	Impedance, Ω	|Ω|
60 Hz	2.0×10^{-6} (1 + j)	2.858×10^{-6}
1 MHz	2.6×10^{-4} (1 + j)	3.69×10^{-4}
30 000 MHz	0.0454 (1 + j)	0.064

free space the magnitude of the magnetic flux density

$$B = \mu_0 H = E/c$$

whereas in a conductor, using (2.80), the maximum is

$$B = \mu H = \mu_r \mu_0 H = (\mu_r \mu_0 \sigma\, \delta E e^{-j\pi/4})/\surd 2.$$

Taking as an example sea water at 3 kHz, where $\sigma \sim 4\ \mathrm{S\,m^{-1}}$ and $\partial \approx 5$ m

$$\mu_r \mu_0 \sigma \delta/\surd 2 = (1 \times 4\pi \times 10^{-7} \times 4 \times 5)/\surd 2$$

so that

$$B \approx 1800 \times 10^{-8} E e^{-j\pi/4}.$$

This is very different from the free space case because the **B** field in the sea water is calculated to be some 5000 times larger than the **B** field in the incident wave.

2.5.9 Diffusion

Inside a metal, the relation between the current density and the electric field is $\mathbf{J} = \sigma \mathbf{E}$. Using this and Maxwell's equations, together with the vector identity

$$\nabla \times (\nabla \times \mathbf{B}) = \nabla(\nabla \cdot \mathbf{B}) - \nabla^2 \mathbf{B}$$

the following differential equation can easily be derived [14]:

$$\partial \mathbf{B}/\partial t = (1/\sigma\mu)\nabla^2 \mathbf{B}. \qquad (2.81)$$

This equation has the same form as that which describes diffusion processes, which in one dimension is:

$$\partial u/\partial t = D \partial^2 u/\partial x^2 \qquad (2.81a)$$

where the constant $D = (1/\sigma\mu)$ is called the diffusion coefficient. This differential equation has been extensively studied [15], and the quantity $(1/\sigma\mu)$ is called the magnetic diffusity. There are different solutions to this equation depending on the boundary conditions, and two limiting cases can be noted here.

The first solution is when all the diffusing substance is deposited at $x = 0$ when $t = 0$, and allowed to diffuse in the $+x$ direction (and is a conserved quantity). In this case the concentration $C(x, t)$ is given by

$$C(x, t) = \frac{C_0}{\surd(\pi D t)} \exp(-x^2/4Dt) \qquad (2.82)$$

where C_0 is the amount deposited at $x = 0$.

Another solution is obtained if the amount of diffusing substance is initially extended such that $C = C_0$ for all $x < 0$, and is

$$C(x, t) = C_0(1 - \mathrm{erf}(x/\surd(4Dt))/2 \qquad (2.83)$$

where $\mathrm{erf}(y)$ is called the 'error function', defined by the definite integral:

$$\mathrm{erf}(y) = 2/\surd\pi \int_0^y \exp(-n^2)\ \mathrm{d}n$$

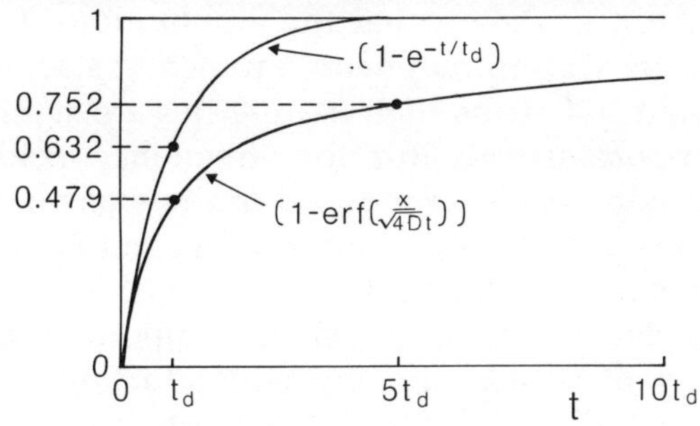

Fig. 2.12 The function $1-\mathrm{erf}(x/\surd(4Dt))$ vs t encountered in the solution of the diffusion equation. This diagram has been drawn putting $D=1$, $x=1$. For comparison the function $1-\exp(-t/t_d)$, where t_d is the diffusion time, is also shown

n being a dummy variable. The function $1-\mathrm{erf}\ (x/\surd(4Dt))$, which is the important part of solution (2.83), may not be very familiar, so in Figure 2.12 it is drawn as a function of t, using x and D both set to unity. It builds up in an exponential-like way, but at a slower rate; for comparison the function $(1-\exp(-t/t_d))$ is also plotted.

The solution (2.83) involves the term $x/\surd(4Dt)$. Thus it is possible to introduce a time t_d given by

$$t_d = x^2/D = \sigma\mu x^2 \tag{2.84}$$

constant for fixed x, called the diffusion time. It is a characteristic rise time for the solution (2.83), since after t_d seconds the function has risen from zero to $(1-\mathrm{erf}(1/2))=0.479$. However, even after a 'long time' such as $5t_d$, the function has only reached $\approx 75\%$ of its final value, which is very different from exponential behaviour.

These solutions can be applied to the way in which a fast transient magnetic field, such as are encountered in EMP, penetrates a conductor. As will be seen in Chapter 5, Fourier analysis shows that an isolated pulse contains a continuous distribution of frequency components, each of which would propagate through a material with a different group velocity and have a different skin depth. This complexity is best handled by treating the magnetic field as a diffusing 'substance' obeying the diffusion equation. When an electromagnetic pulse is incident upon a conductor, the energy of the pulse which penetrates into it and is not reflected is carried inside dominantly by the magnetic component, and this diffuses through the metal. The diffusion time $t_d = \sigma\mu x^2$ can be taken as a working estimate of the time it takes the magnetic field produced by EMP to penetrate through a conducting shield of thickness x metres. For typical conductors t_d can be substantial, e.g. a copper shield of 1 mm thickness has $t_d \approx 70\ \mu$s, whereas for 1 cm thickness $t_d \approx 0.007$ s, so that the energy arrives inside the shield long after the pulse has passed by. The details of exactly what happens depends on the pulse amplitude, width, the thickness x of the material and its magnetic properties. This is discussed in reference [19], where it is shown that the amplitude of the diffused-through transmitted field through a ferromagnetic conducting slab, of conductivity σ, is proportional to $(\omega\mu_r\sigma^2x^3)^{-1}$. Thus high frequencies are significantly attenuated, and diffusion is important only for low frequency fields. However other parameters, such as eddy currents induced in the walls of the cavity, and the shape of the shield, can have effects 5 orders of magnitude bigger; see reference [16] and the discussion in Section 4.3.2.

Two further examples of where diffusion is important are in the design of large bandwidth current shunts and in many situations occurring in plasma physics.

2.6 INTEGRAL FORMS OF MAXWELL'S EQUATIONS

2.6.1 Faraday's and Ampère's laws

The differential form of Maxwell's Equations (2.15)–(2.18) which have been used up to now have been satisfactory to describe electromagnetic waves, their interactions with the media through which they propagate, and to introduce the concept of wave impedance. However to make further progress it is necessary to carry out the operation which was hinted at in the general discussion of Section 2.3 and derive from Maxwell's equations another set of equations which describe the relationships between fields that are spread out over an extended region of space. These new equations are often called the integral forms of Maxwell's equations and are obtained by applying Stokes' theorem. This theorem in vector algebra relates the surface integral of the curl of a vector field **V** over a surface **S** to the line integral of **V** along the edge C (contour) of the surface (see Appendix B for details), i.e.

$$\int_C \mathbf{V}\cdot d\mathbf{l} = \int_S (\nabla\times\mathbf{V})\cdot d\mathbf{S}$$

where $d\mathbf{S}$ is a vector area. The theorem can be used to relate field properties over a surface to those along a given line. Applying it to (2.15) and (2.16) yields:

$$\int_C \mathbf{E}\cdot d\mathbf{l} = \int_S (\nabla\times\mathbf{E})\cdot d\mathbf{S} = -\int_S (\partial\mathbf{B}/\partial t)\cdot d\mathbf{S} \tag{2.85}$$

$$\int_C \mathbf{H}\cdot d\mathbf{l} = \int_S (\nabla\times\mathbf{H})\cdot d\mathbf{S} = \int_S (\mathbf{J}+\partial\mathbf{D}/\partial t)\cdot d\mathbf{S}. \tag{2.86}$$

Consider each of these forms in turn. Since

$$-\int_S (\partial\mathbf{B}/\partial t)\cdot d\mathbf{S} = -\partial/\partial t\int_S \mathbf{B}\cdot d\mathbf{S} = -\partial\Phi/\partial t \tag{2.87}$$

where Φ is the flux, we obtain from (2.85) Faraday's and Lenz's law

$$\int_C \mathbf{E}\cdot d\mathbf{l} = -\partial\Phi/\partial t \tag{2.88}$$

which relates the line integral of the electric field around a closed path to the rate of change of the magnetic flux through the surface defined by that line. This line integral of an electric field is a voltage, called the electromotive force (emf) which could drive a current if the closed loop was made from a conducting wire. The direction of the conventional current which would flow would be such as to set up its own magnetic field to oppose the changing flux. For example, in Figure 2.13, if the applied magnetic field **B** is increasing, the induced current i would be in the direction shown so as to produce a field in the opposite direction to the applied field; this is the result of Lenz's law enshrined in the negative sign in (2.88). The electrons in a real conductor will of course move in the opposite direction to the conventional current flow.

Turning to Equation (2.86) it should be noted that it is normally seen in the form known as Ampere's circuit law;

$$\int_C \mathbf{B} \cdot d\mathbf{l} = \mu \int_S \mathbf{J} \cdot d\mathbf{S} = \mu I \tag{2.89}$$

where I is the total current flowing through the surface defined by the closed contour C. This integral form of Maxwell's equations gives a useful relationship between a current I and an extended, even global, property of the magnetic field, the line integral $\int \mathbf{B} \cdot d\mathbf{l}$, which that current produces. However this equation does not easily give the value of the magnetic field at a given point, unless the contour is chosen to be a simple curve which enables the line integral to be evaluated analytically; see Figure 2.14.

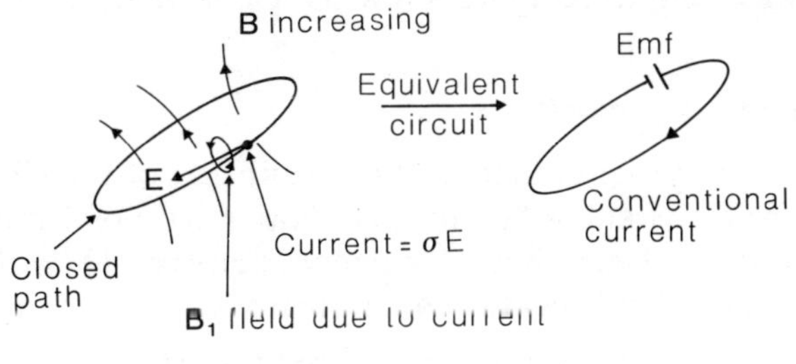

Fig. 2.13 Faraday's Law and Lenz's Law. The induced current produces a magnetic field $\mathbf{B}_1$ which opposes the changing magnetic field, **B**. Also shown is the equivalent circuit which relates the emf to the current direction

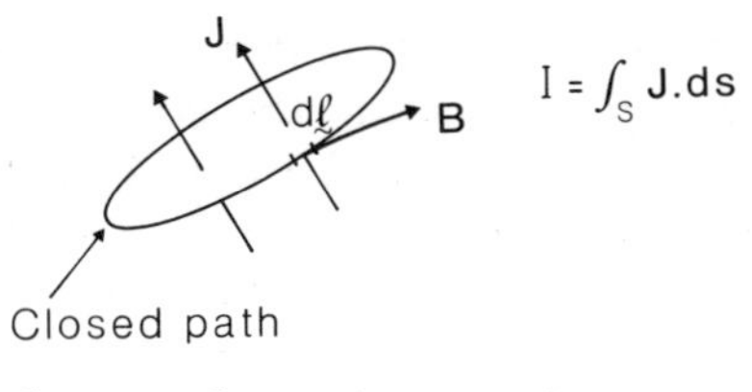

Fig. 2.14 A closed path around a region carrying a current density **J** A m^{-2}

Clearly (2.89), which is derived from (2.86), is true for static fields when $\partial \mathbf{D}/\partial t = 0$, provided the contour C passes only through media where the relationship $\mathbf{B} = \mu \mathbf{H}$ is a valid approximation. However one also sees this equation applied when high frequency currents are involved, which means that the term $\partial \mathbf{D}/\partial t$ in (2.83) has been neglected. It is instructive to see why this approximation is made, and whether this is a valid one to make under all circumstances.

In a typical conductor the term $\mathbf{J} + \partial \mathbf{D}/\partial t$ can be written

$$\mathbf{J} + \partial \mathbf{D}/\partial t = \sigma \mathbf{E} + \varepsilon\, \partial \mathbf{E}/\partial t \tag{2.90}$$

where $\mathbf{E}$ is the field which is causing the current to flow. If this is alternating it can be written as $\mathbf{E} = \mathbf{E}_0 e^{j\omega t}$ hence

$$J + \partial \mathbf{D}/\partial t = (\sigma + j\omega \varepsilon_r \varepsilon_0)\mathbf{E}_0 e^{j\omega t}. \tag{2.91}$$

This is the same term that appears in Equation (2.60), derived when considering the motion of electromagnetic waves through material. The j implies a 90° phase lead of the displacement current over the conduction current. In good conductors $\sigma \approx 10^7\ \mathrm{S\,m^{-1}}$; thus:

$$\mathbf{J} + \partial \mathbf{D}/\partial t = (10^7 + j8.854\,18 \times 10^{-12} \omega \varepsilon_r)\mathbf{E}.$$

This implies that displacement current is negligible unless the product $\omega \varepsilon_r \approx 10^{18}$! Even at frequencies of 100 GHz, where $\omega = 2\pi \times 10^{11}$, we would require ε_r to be of order 10^7 for this condition to be satisfied, a most unlikely circumstance. Thus in many, but not all, situations effects due to the current density term $\mathbf{J}$ probably swamp those due to $\partial \mathbf{D}/\partial t$, which can be dropped from the equation, so leading to Ampère's law as it is usually written.

2.6.2 Electric fields in conductors

This is an appropriate place to look at the magnitude of the electric field inside a conductor where a current is flowing. A copper wire of 1 mm^2 cross section carrying a current of 1 mA has a current density, $|\mathbf{J}| = 10^3\ \mathrm{A\,m^{-2}}$. With $\sigma = 5.7 \times 10^7\ \mathrm{S\,m^{-1}}$, the field inside the conductor is

$$|\mathbf{E}| = |\mathbf{J}|/\sigma = 1.8 \times 10^{-5}\ \mathrm{V\,m^{-1}}.$$

This is very weak; in fact it is comparable to the field 1 cm away from a single electron [4]! Even so, a field as small as this can cause an appreciable current to flow in copper and this is a clue as to why Faraday's law is so important. The line integral taken around a closed contour, the electromotive force (emf), is given by (2.88) and this can easily be made large, even if the $\mathbf{B}$ field is of modest magnitude or slowly varying, by making the contour large enough. Also the emf can be amplified by a large factor, N, by making the wire into a coil with N turns, so that the flux linkage becomes $N\Phi$. The understanding of Faraday's law has lead to widespread applications and in some sense is the basis of our whole modern civilisation as it is the principle by which our electrical supply is generated.

2.7 SOME ILLUSTRATIVE EMC EXAMPLES

2.7.1 Interference in a small loop

As an example of the use of Faraday's and Ampère's laws in an EMC problem consider the case of a small loop situated near wires carrying an alternating current. This current will produce a magnetic field **B** and a changing flux through the loop, which in turn will generate unwanted noise in the loop. The intention is to find ways to minimise the noise.

Consider the small loop to be made from a single turn of wire, see Figure 2.15, of area A m^2 and situated a distance r metres away from the centre of a pair of wires carrying a current i amperes. Because the dimensions of the loop are small and it is a long way from the pair of wires it is reasonable to make the assumption that $|\mathbf{B}|$ does not vary significantly across the loop, so use the magnitude at its centre, B_{centre} to estimate the total flux. Using Ampère's law (2.89) the formula for $|\mathbf{B}|$ at a distance r from a straight wire is found by choosing a circular path a distance r from the wire. By symmetry, the field around this path is constant and parallel to **dl** all around the loop, so

$$\int \mathbf{B} \cdot \mathbf{dl} = \mu_0 i$$

becomes

$$B \int \mathrm{d}l = \mu_0 i$$

so that

$$B = \mu_0 i / 2\pi r. \tag{2.92}$$

In the case of the two wires, with currents in opposite directions, the field at the centre of the small loop is the difference between the two individual fields:

$$B_{\text{centre}} = \frac{\mu_0 i}{2\pi}\left(\frac{1}{(r-a)} - \frac{1}{(r+a)}\right)$$
$$\approx \frac{\mu_0 i 2a}{2\pi(r^2 - a^2)} \approx \frac{\mu_0 i 2a}{2\pi r^2} \tag{2.93}$$

when $r \gg a$. Hence from Faraday's law the induced voltage,

$$V_{\text{noise}} = -\partial\phi/\partial t = -A(\partial B/\partial t).$$

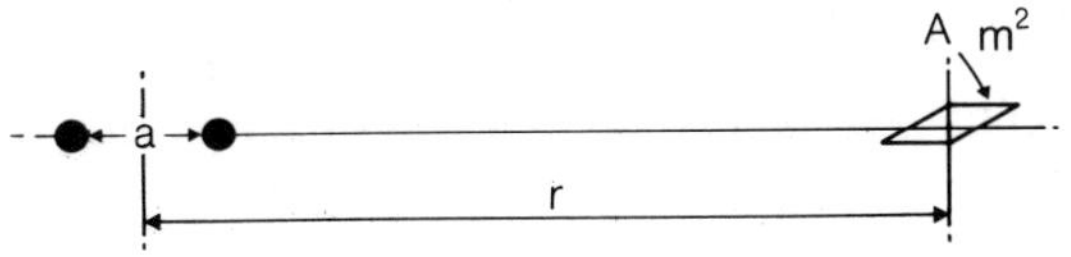

Fig. 2.15 Small square loop of area A m^2 in the plane of a pair of long thin parallel wires with separation a. Each wire carries an equal and opposite current

If $i = i_0 \sin(\omega t)$ then

$$V_{\text{noise}} = \frac{\mu_0 a A \omega i_0 \cos(\omega t)}{\pi r^2}. \quad (2.94)$$

To keep V_{noise} small requires ω, A, i_0 and the spacing between the wires, a, to be as small as possible, and r large. Notice that the noise is 90° out of phase with the current i_0 and is a maximum when i_0 is zero. Also, if the loop has N turns, the noise voltage will be N times bigger than that for the single turn calculated here.

In a realistic situation one may only have control over the wire spacing, a, as the other parameters may be fixed by other considerations. One way to minimise the noise in this situation is to make the effective area of the loop as small as possible by turning it on its side so making the **B** field parallel to the plane of the loop and reducing the flux. Such a solution to an EMC problem is an example of using method (b) listed in Section 1.5 to reduce interference. The general lesson to be learnt from this example is that pairs of wires carrying the same alternating current should be kept as close together as possible, and if possible twisted around each other, to reduce inductive pick-up in nearby loops. Another important observation is that the noise voltage generated in the loop increases with frequency. This has the effect of magnifying the importance of any small high frequency current components in the system. This will be discussed again in Section 5.2.7.

2.7.2 Interpreting measurements at different distances

In the previous EMC problem the coupling between the loop and the circuit carrying the current was inductive with the total magnetic field generated by the wires providing the necessary flux linkage. This example can be conveniently used to illustrate another lesson which should be remembered when one is attempting to track down interferences sources, and that is to warn about the difficulty of interpreting measurements made at different distances from a whole system.

Far from the pair of wires separated by distance a the magnetic intensity is given by Equation (2.93) $B = \mu_0 i 2a/2\pi r^2$ whereas at a distance r from a single wire Equation (2.92) gives $B = \mu_0 i/2\pi r$. Consider a situation where there are two pairs of wires, with different separations and carrying different currents, adjacent to each other, as in Figure 2.16. Such arrangements can be found in power lines on PCB boards.

For simplicity consider the magnitude of the magnetic field measured by a probe as it is moved from some point in Figure 2.16 not far from the closely spaced wires to some remote point about $2b$ from the centre. To keep the calculation as simple as possible consider the probe to move in the plane of the wire pairs. The curves in Figure 2.16 show how the total field outside the wires varies with distance r, and it is clear that there are two contributions, one with short range from the closely spaced pair and one with longer range from the widely separated pair. To draw this field distribution it has been assumed that

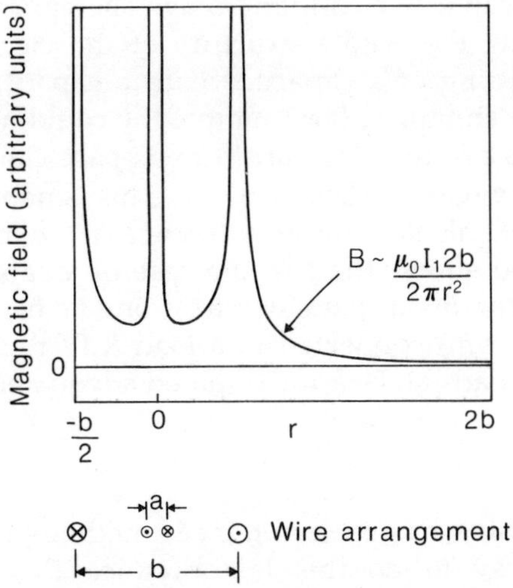

Fig. 2.16 Magnitude of the magnetic field as a function of distance in the plane of two sets of long parallel pairs of wires with different separations and carrying different currents. The innermost pair of wires with separation *a* metres is assumed to carry a current $10I_1$ compared to I_1 in the outermost pair with separation *b* metres. Each wire in a particular pair carries an equal and opposite current

the current I_1 in the pair of wires with the larger separation is 10 times smaller than the current in the wires with the smallest separation, i.e. $I_1 = I_2/10$ but and also that the relative wire spacings are such that $b = 20a$.

A long way off the biggest signal detected with a suitable probe will be due to the pair of wires with the largest separation but smallest current. However, if the same probe is moved near to the central pair of wires, with the smallest separation but carrying the larger current, a large signal will also be detected. In such a situation where *only* these two readings are taken one might conclude that the two signals detected by the probe have a common origin.

This example illustrates two points of general importance. Firstly it is not necessarily true that the biggest 'close-to' signal has the same origin as the largest signal detected at much larger distances, and secondly it is not easy to deduce the properties of the far field from measurements made in the near field. For example, the far field behaviour of complex antenna systems can be predicted from near field measurements, but very large numbers (thousands) of such measurements are needed together with complex mathematical techniques [12] in order to infer far field behaviour. These difficulties apply both to reactive and radiative coupling.

2.7.3 Capacitive and inductive coupling

If in the near field the wave impedance is high then electric field dominates and the coupling between the source and the receiver can be approximated adequately by using capacitive coupling only. The appropriate coupling can be modelled by replacing the whole structure of the electric field by a lumped circuit element consisting of a capacitor with a capacitance calculated by the normal electrostatic techniques. For example the coupling between two parallel wires of diameter d a distance D apart in free space can be discussed in terms of a capacitance $\pi\varepsilon_0/\cosh^{-1}(D/d)\,\mathrm{F\,m^{-1}}$. This enables traditional circuit analysis to be used to calculate the interference in a circuit. Ott [17] discusses such calculations, and shows that, in the case of a capacitive coupling of C_{12} farads between two circuits of parallel wires, one being driven by a voltage V_1 and the other (the victim) connected to a load $R\ \Omega$, then the noise voltage V_n generated across the load can be approximated adequately in most situations by

$$V_n = \mathrm{j}\omega R C_{12} V_1. \tag{2.95}$$

Once such an expression has been obtained, methods to reduce the noise can be planned. One way to do this is to reduce C_{12}; further discussion of capacitive coupling can be found in Section 4.2.3, and the effect of shielding on such coupling is discussed in Chapter 6.

In a similar way if, in a low impedance near field, the magnetic component dominates, the coupling between the source and the receiver can be approximated by a lumped circuit approximation mutual inductance. The very fact that circuits have to be closed loops can make them vulnerable to inductive coupling, with the induced emf, hence any noise, depending directly on the area of the circuit. However the calculation of the natural inductance between practical circuit geometries is difficult in general; some examples and useful references are given by Ott [17].

2.7.4 Switching transients

Troublesome transient voltages are encountered every time a device is switched on or off. There are two types, which occur on different time scales. The first type occurs after the closing of the switch (which connects the power supply to the device) causes a voltage pulse to propagate along the wires, as a wave along a transmission line. Reflections from changes of impedance combine very quickly to produce complex waveforms on very short times scales, typically of order $2d/v$ seconds, where d is the length of the circuit and v the speed of wave, which is comparable with the speed of light. Thus for a circuit 10 m long, the time scales are of the order of several tens of nanoseconds. The evolution of these waveforms are analysed by applying transmission line theory to the circuits, which will be discussed in Chapter 3.

The second type of transients typically occur a few microseconds after the switch contacts have closed or been opened, and arise from the interchange of stored energy between inductive and capacitive properties of the circuits. Such late transients can be studied by normal circuit analysis, and it is convenient to do this first. A detailed review of this material can be found in references [5, 18].

2.7.4.1 *Switching a transformer on*

In Figure 2.17 there is stray capacitance C_s between the primary and secondary windings of a transformer, and the power line has its own capacitance C_1. If the switch is closed *when the line voltage is not zero*, C_s will be 'pulse charged' from C_1, which will be at some initial voltage V_0. The simplified equivalent circuit shown in Figure 2.18 can be used to analyse the situation. In this figure, L is the effective inductance of the wires in the circuit and *not* the inductance of the transformer coil, which is by-passed by the stray capacitance. Similarly, R is the resistance of the wires in the circuit and not that of a discrete component. It follows that in typical real circuits both R and L will be small, whereas the total capacitance of the circuit, C, given by

$$\frac{1}{C}=\frac{1}{C_s}+\frac{1}{C_1}$$

could be quite large, so that the approximations $R^2 \ll 4L/C$ and $R/\omega L \ll 1$ (a high Q circuit) are often valid. In such a situation it can be shown (see Appendix D for details) that the maximum voltage V_c that can be generated across the stray capacitance C_s is given by

$$V_c=\frac{2V_0}{\gamma+1} \tag{2.96}$$

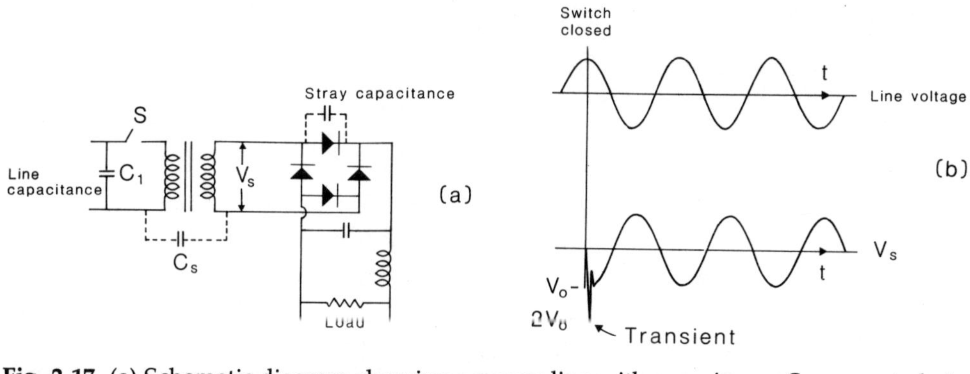

Fig. 2.17 (a) Schematic diagram showing a power line with capacitance C_1 connected via a switch S to a transformer with stray capacitance C_s, and to a load via a bridge network. Adapted from reference [18]. (*b*) Maximum amplitude voltage transient on transformer output caused by closing the switch S. (Reproduced by permission of General Electric)

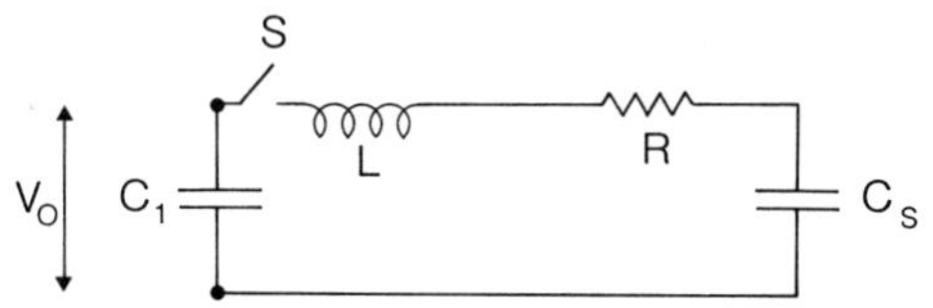

Fig. 2.18 Equivalent circuit for Figure 2.17(*a*) when closing switch S

where $\gamma = C_s/C_1$. In the limit when $\gamma \ll 1$ (i.e. the stray capacitance is very small compared to the line capacitance), $V_c \approx 2V_0$. This means that when the transformer is switched on, a transient voltage pulse of up to twice the line voltage, *but only twice the line voltage,* can be generated. Further oscillations of the *LRC* circuit could generate further smaller transient voltage spikes, until the stored energy $1/2C_1V_0^2$ is dissipated in the resistance R. These noise spikes can be transmitted onwards to the load via the stray capacitance across the rectifying diodes.

2.7.4.2 *Switching a transformer off*

The transients shown in Figure 2.19 are caused when the same circuit as in the previous example (Figure 2.17) is switched off. If the switch is opened when the line voltage is going through zero, the flux in the secondary of the transformer is a maximum because the current through the primary is 90° out of phase with the line voltage. However the rate of change of flux at this instant is near zero, so the output voltage from the secondary is also small (Faraday's law). When the switch is opened, interrupting the magnetising current, there is substantial energy stored in the inductance L of the secondary coil, which can only be transferred to the stray internal capacitance C_s of the transformer. A simplified equivalent circuit for this aspect of Figure 2.17 is shown in Figure 2.20. By conservation of energy,

$$Li^2/2 = C_2V_s^2/2$$

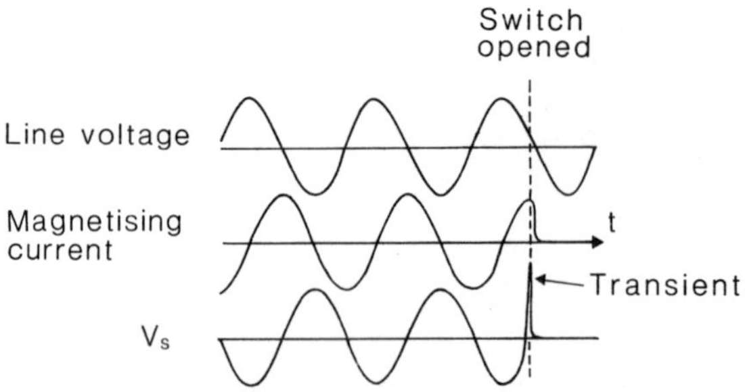

Fig. 2.19 Transient produced by opening switch S in Figure 2.17(*a*), adapted from reference [18]. (Reproduced by permission of General Electric Co.)

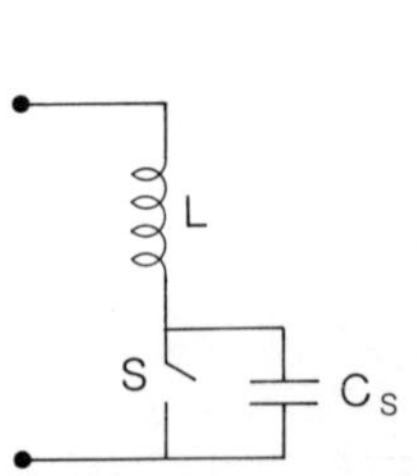

Fig. 2.20 Equivalent circuit for Figure 2.17(*a*) when opening switch S

i.e.

$$V_s = i\sqrt{(L/C_s)} \tag{2.97}$$

where i is the current which was flowing in the secondary circuit when the switch was opened. Clearly if L is large (as it can be with an iron cored transformer) and C_s is small, then the amplitude of the pulse V_s can be substantial, *and can exceed the line voltage by very large factors*, which is different from the first example considered in Section 2.7.4.1 where only twice the line voltage could be produced. As in that example, these large and potentially damaging voltage spikes can be transmitted to the load via the stray capacitance across the diodes.

Similar examples can be found in all cases where a sudden interruption to a current through an inductance (such as a fuse blowing) leaves the inductance with energy stored in the magnetic field which, unlike energy stored in the electric field of a capacitor, cannot continue to be stored without a current being present. Once initiated, the high voltage can cause arcing and sudden breakdown across switch contacts, phenomena which themselves generate further voltage spikes. See [17] for further discussion.

A final point is that, because of the high voltages that can be generated by interrupting, rather than starting, a current through an inductance, the transients produced by *closing* a switch are often very different from those produced *opening* it.

2.7.4.3 *Very early time transients*

When a power line switch is closed, a voltage step is propagated in both directions along the wires, which act as a transmission line, towards the device and the power supply providing the line voltage. To analyse this situation requires the use of transmission line theory and not lumped circuit theory. Furthermore, when this propagating pulse meets any change of impedance, reflection of part of this wave will occur, which will lead eventually to complex waveforms oscillating up and down the wires within time scales much less than 1 μs.

These waveforms can have high frequency components and often their amplitudes will be of the same order, or smaller than, the line voltage. To understand why this is so it is useful to examine the case of a voltage source V_0 with internal resistance R_s switched into a circuit consisting of a transmission line of length d and characteristic impedance $Z_0 \ll R_s$ terminated with a pure resistance $R \gg Z_0$, as shown in Figure 2.21(a). This switching operation will send a charging wave along the line towards R. Eventually the voltage across R will rise exponentially towards [5]

$$V_R = V_0R/(R_s + R)$$

with a time constant $\tau = R_{eff}C$, where C is the capacitance of the line and R_{eff} is

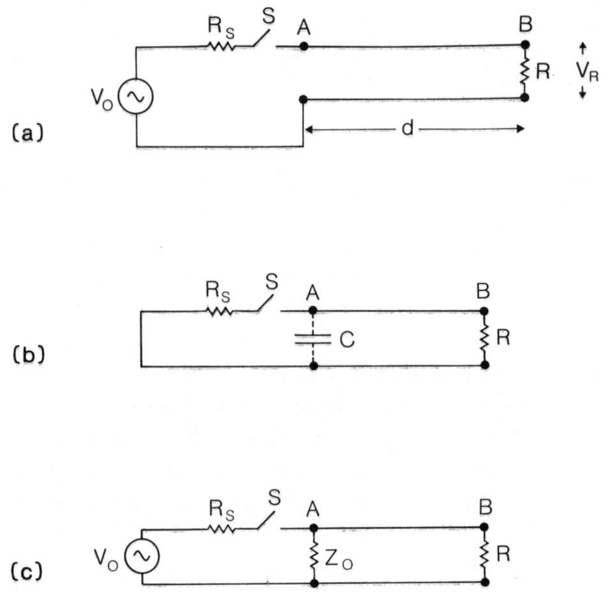

Fig. 2.21 (*a*) Equivalent circuit for switching a voltage source V_0 with internal resistance R_s via a switch S into a transmission line of length *d*, terminated with a resistance *R*. (*b*) Equivalent circuit for the transmission line with capacitance *C* terminated with a resistance *R*. (*c*) Equivalent circuit for the transmission line with characteristic impedance Z_0 terminated with resistance *R*

the effective resistance of *R* and R_s in parallel; see the equivalent circuit in Figure 2.21(*b*).

However, this rise in voltage across *R* will not be a smooth exponential but will consist of a series of steps with a time interval of $2d/v$ seconds between them, due to the charging wave reflecting up and down the line. Figure 2.21(*c*) shows the equivalent circuit seen by current flowing into the transmission line. The steps in voltage at the point *A* each have amplitude

$$V_A = V_0 Z_0 / (R_s + Z_0)$$

and at *B* by

$$V_B = V_A R / (R + Z_0)$$

so that the voltage steps across *R* will each be

$$V_R = V_0 \frac{R Z_0}{(R_s + Z_0)(R + Z_0)}. \tag{2.98}$$

If the terminating load is inductive then the analysis is more complicated and the waveforms become very complex, see reference [5]. Nevertheless, the maximum amplitude of the waveforms will be of the order of the divided down line voltage, $V_0 Z_0 / (Z_0 + R_s)$.

2.8 SUMMARY

In this chapter a number of ideas and concepts, including the interrelationships between the electromagnetic fields **E** and **H**, wave impedance, near and far fields and the role played by the material dependent quantities **B**, **D** and **J** have been introduced. The point of doing this is to try to present a core of vital ideas which could act as a conceptual framework on which to hang the detailed mathematical formalism that is needed eventually to make EMC engineering and design decisions. It may be helpful at this stage to recall explicitly those ideas which are central to this whole field.

The first is that *the frequency range* encountered in EMC problems is important because it concerns both the appropriate numerical value to be used for 'constants' such as μ_r and ε_r, and the properties of any radiation field impinging upon or generated by any given system; is the field a 'near field' or a 'far field'?

Secondly at low frequencies the 'near field' can be either dominantly an electric field or a magnetic field, and that different effects are produced within systems immersed in such fields. Thus it will clearly be of the greatest importance to identify correctly the type of field present in any given situation, as both the setting up and the interpretation of EMC tests will depend on understanding this point.

Thirdly there is a relationship between the electrical impedance of a circuit generating an electromagnetic field and the numerical ratio E/H of that field. Both quantities have the same units, ohms, and the ratio E/H for an electromagnetic field is called the wave impedance. It will be shown in Chapter 3 that this concept is very fruitful and gives rise to a powerful method of making calculations for EMC shielding, as well as a general way of discussing and calculating the reflection and flow of electromagnetic energy at points where there are any change to the local electromagnetic environment.

REFERENCES

[1] Lorrain, P., Corson, D. P. and Lorrain, F. (1987) *Electromagnetic Fields and Waves* 3rd edition, W. H. Freeman and Co., New York.

[2] Wolff, E. A. (1988) *Antenna Analysis* 2nd edition, Artech House Inc., Norwood, (originally published by Wiley, New York (1966)).

[3] Ramo, S., Whinnery, J. R. and Van Duzer, T. (1965) *Fields and Waves in Communication Electronics* Wiley, New York.

[4] Rosser, W. G. (1970) Magnitude of surface charge distributions associated with electric current flow, *Am. J. Physics* **38** 265.

[5] Vance, E. F., Graf, W. and Nanevicz, J. E. (1981) *Unification of electromagnetic specifications and standards* SRI International, Defense Nuclear Agency, Washington D.C. DNA 5433F-1.

[6] Johnk, C. T. A. (1988) *Engineering Electromagnetic Fields and Waves* 2nd edition, Wiley, New York.

[7] Main, I. G. (1984) *Vibrations and Waves in Physics* 2nd edition, Cambridge University Press, Cambridge.

[8] Hayt, W. H. Jr. (1967) *Engineering Electromagnetics* 2nd edition, McGraw-Hill, New York.

[9] Ricketts, L. W., Bridges, J. E. and Miletta, J. (1976) *EMP Radiation and Protective Techniques* Wiley, New York.

[10] Lo, Y. T. and Lee, S. W. (ed.) (1988) *Antenna Handbook: Theory, Applications and Design* Van Nostrand Reinhold, New York.

[11] Sir James Jeans, (1946) *The Mathematical Theory of Electricity and Magnetism* 5th edition, Cambridge University Press, Cambridge.

[12] Special Issue on Near-Field Scanning Techniques, (1988) *IEEE Trans. Antennas and Propagation* **AP-36** 725–902.

[14] Gartenhaus, S. (1964) *Elements of Plasma Physics* Holt Rinehart and Wilson Inc. New York.

[15] Crank, J. (1975) *The Mathematics of Diffusion* 2nd edition, Clarendon Press, Oxford.

[16] Lee, K. S. H. and Bedrosian, G. (1979) Diffusive Electromagnetic Penetration into Metallic Enclosures *IEEE Trans. Antennas and Propagation* **AP-27** 194–198.

[17] Ott, H. W. (1988) *Noise Reduction Techniques in Electronic Systems* 2nd edition, Wiley, New York.

[18] *Transient Voltage Suppression Manual* (1978) 2nd edition, ed. J. C. Hey, General Electric Co., Auburn, New York (1978) (Present address PO Box 591, Summervale, New Jersey).

[19] Karzas, W. J. and Tse Chin Mo (1978) Linear and Nonlinear EMP Diffusion through a Ferromagnetic Conducting Slab *IEEE Trans. Antennas and Propagation* **AP-26** 118–129.

3 Waves in Guiding Structures and Layered Materials

3.1 IMPEDANCE IN LOSSY MATERIALS

In Chapter 2 it has been shown that a general way to describe the electric component of a TEM wave travelling in the x direction through a medium is

$$E_y = E_0 e^{-\gamma x} e^{j\omega t} \tag{3.1}$$

where γ is the complex propagation constant given by

$$\gamma = \alpha + j\beta = \sqrt{(j\omega\mu\sigma - \omega^2\mu\varepsilon)}. \tag{3.2}$$

The real and imaginary parts of γ, called the attenuation constant α and the phase constant β, can be written [12]

$$\alpha = \omega\sqrt{(\mu\varepsilon/2)}\sqrt{(\sqrt{(1 + (\sigma/\omega\varepsilon)^2)} - 1)} \tag{3.3a}$$

$$\beta = \omega\sqrt{(\mu\varepsilon/2)}\sqrt{(\sqrt{(1 + (\sigma/\omega\varepsilon)^2)} + 1)}. \tag{3.3b}$$

Both of these have dimensions m^{-1} but are commonly given the units of neper m^{-1} and radians m^{-1}. Equation (3.1) can be written

$$E_y = E_0 e^{-\alpha x} e^{j(\omega t - \beta x)}$$

so that α neper m^{-1} describes the way the wave is damped and βx radians is the phase change which takes place when the wave propagates x metres. Note that in free space where $\sigma = 0$, $\mu = \mu_0$ and $\varepsilon = \varepsilon_0$ then

$$\gamma = \alpha + j\beta = j\omega\sqrt{(\mu_0\varepsilon_0\varepsilon)} = j2\pi f/c = j2\pi/\lambda.$$

Also, in a pure dielectric where $\sigma = 0$ but $\varepsilon_r > 0$ then

$$\gamma = \alpha + j\beta = j\omega\sqrt{(\mu_0\varepsilon_0\varepsilon_r)} = j2\pi/\lambda'$$

where λ' is the wavelength of the TEM wave in the dielectric. In materials where $\sigma \approx 0$, little damping takes place since $\alpha \approx 0$. When $\alpha > 0$, the skin depth $\partial = 1/\alpha$. The intrinsic impedance of a material is $Z = j\mu\omega/\gamma$, which for a given material with fixed values of σ, μ and ε_r is only a function of ω. As a reminder of this, the impedance will be written:

$$Z(\omega) = j\mu\omega/\gamma(\omega). \tag{3.4}$$

However, as has been shown, this is not the whole story. If a TEM wave is incident on the boundary between two media (see Figure 2.8) with non-zero intrinsic impedances $Z_1(\omega)$ and $Z_2(\omega)$ then at $x = -L$ metres in front of the boundary (i.e. $-L$ is actually within the medium with intrinsic impedance $Z_1(\omega)$ where the wavelength is λ m) the wave impedance is written (see Equation (2.54)) as

$$Z_w(\omega) = Z_1(\omega)\,\frac{\{Z_2(\omega) + jZ_1(\omega)\tan(2\pi L/\lambda)\}}{\{Z_1(\omega) + jZ_2(\omega)\tan(2\pi L/\lambda)\}}. \tag{3.5}$$

Thus Equation (3.4) for the impedance inside a material becomes modified near a boundary and must not be applied naively in all situations. However (3.5) itself needs a little modification, as it was derived using $E_0 e^{j(\omega t - \beta x)}$ to describe the TEM wave, and it was only later that (3.1) was derived as the most general form. Using (3.1) Equation (3.5) can be written more generally as

$$Z_w(\omega) = Z_1(\omega)\,\frac{\{Z_2(\omega) + jZ_1(\omega)\tan(\gamma L)\}}{\{Z_1(\omega) + jZ_2(\omega)\tan(\gamma L)\}}. \tag{3.6}$$

Equation (3.6) can be manipulated and recast in several forms. Using the relations j tan $\theta = \tanh\theta = \sinh\theta/\cosh\theta$ then

$$Z_w(\omega) = Z_1(\omega)\,\frac{\{Z_2(\omega) + Z_1(\omega)\tanh(\gamma L)\}}{\{Z_1(\omega) + Z_2(\omega)\tanh(\gamma L)\}} \tag{3.6a}$$

or

$$Z_w(\omega) = Z_1(\omega)\,\frac{\{Z_2(\omega)\cosh(\gamma L) + Z_1(\omega)\sinh(\gamma L)\}}{\{Z_1(\omega)\cosh(\gamma L) + Z_2(\omega)\sinh(\gamma L)\}}. \tag{3.6b}$$

As noted above when γ is purely complex (i.e. no absorption) there is no difference between the values of $\tanh(\gamma L)$ and $\tan(2\pi L/\lambda)$ in free space, or between $\tanh(\gamma L)$ and $\tan(2\pi L/\lambda')$ in a pure dielectric but for completeness it is preferable to use (3.6*a*) or (3.6*b*) rather than (3.5).

The reflection coefficient for the *electric* component of the wave at the boundary is

$$\rho_{12} = \frac{(Z_2(\omega) - Z_1(\omega))}{(Z_2(\omega) + Z_1(\omega))} \tag{3.7}$$

and its transmission coefficient

$$\tau_{12} = \frac{2Z_2(\omega)}{(Z_2(\omega) + Z_1(\omega))} \tag{3.8}$$

where the subscripts 12 mean that the incident TEM wave is travelling in the $+x$ direction, from material 1 into material 2. Sometimes the reflection and transmission coefficients, ρ^H and τ^H of the *magnetic* component of the wave are used. Since (see Section 2.5) the reflected electric and magnetic amplitudes are related by $E_r = -Z_1(\omega)H_r$, then:

$$\rho^H_{12} = \frac{(Z_1(\omega) - Z_2(\omega))}{(Z_2(\omega) + Z_1(\omega))} \tag{3.7a}$$

and

$$\tau_{12}^{H}=\frac{2Z_1(\omega)}{(Z_2(\omega)+Z_1(\omega))} \tag{3.8a}$$

so that $\rho_{12}^{H}=-\rho_{12}$ and $\tau_{12}^{H}=1+\rho_{12}^{H}$. These sets of equations are the ones used to calculate the way in which TEM waves interact with material boundaries, and they will be developed to study multi-layered media. They have been written as functions of ω to stress that for fixed materials and structures there are very important frequency dependent effects.

The only potential mathematical difficulty is that each of these expressions is a complex function, and readers will need to be confident in their abilities to manipulate such expressions. To this end the next few sections will present some simple case studies based on the properties of real materials, to illustrate these calculations, in particular to show the dramatic differences in the electromagnetic environment around and inside dielectrics and good conductors.

3.1.1 TEM waves incident on boundaries

The initial case studies show in detail what happens when TEM waves at normal incidence impinge on dielectrics and good conductors, represented by choosing polystyrene and copper as examples. The calculations are done using the general (complex number) equations listed in Section 3.1.

There are appearing on the market a number of 'supercalculator' packages which run on personal computers and make it very easy to perform these complex number calculations. In a review of this field [1], the program MATHCAD [2] was noted as being close to the ideal. The following calculations, and many others in this book, were done using version 2.03 of this program.

Case study 3.1.1a This case study concerns a TEM wave of frequency 1 GHz in air incident onto a single interface made of polystyrene. The reflection and transmission coefficients are calculated and the wave impedance as a function of distance in front of the surface is plotted. Finally the maximum wave impedance in front of the surface is found and compared with Z_0.

Given $f=10^9$ Hz, $\lambda=0.3$ m, $\varepsilon_0=8.85\times10^{-12}\,\mathrm{F\,m^{-1}}$, then for polystyrene at this frequency $\sigma=0$, $\mu_r=1$, $\varepsilon_r=2.56$ and taking $Z_0=377\ \Omega$. Thus using (3.2)

$$\gamma(\omega)=\surd(-\omega^2\mu\varepsilon)=33.5\mathrm{j}$$

and so using (3.4)

$$Z(\omega)=236\ \Omega.$$

The reflection and transmission coefficients for the electric component are found using (3.7) and (3.8):

$$\rho_{12}(\omega)=\frac{236-377}{236+377}=-0.23$$

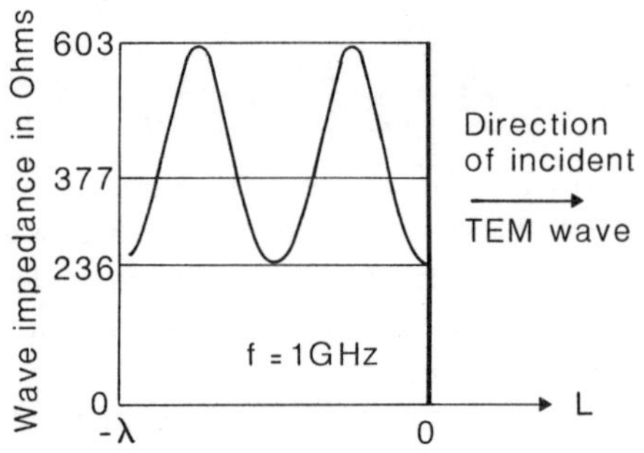

Fig. 3.1 Wave impedance as a function of position in front of a semi-infinite sheet of polystyrene with characteristic impedance 236 Ω. The surface of the material is at $L = 0$ and the material extends to the right.

(note the − sign) and

$$\tau_{12}(\omega) = \frac{2 \times 236}{236 + 377} = +0.77.$$

Notice that the impedances used here are the intrinsic impedances of the material. The wave impedance a distance L in front of the boundary (in free space) is found by plotting the modulus of the complex function $Z(\omega, L)$ against L. As there is no absorption ($\sigma = 0$) equation (3.5) can be used, with L as well as ω being treated as a variable, such that

$$Z(\omega, L) = Z_0 \frac{\{Z(\omega) + jZ_0 \tan(2\pi L/\lambda)\}}{\{Z_0 + jZ(\omega)\tan(2\pi L/\lambda)\}}.$$

This is shown in Figure 3.1 where the curve shows the wave impedance between $x = -\lambda$ and the boundary at $x = 0$, and for comparison Z_0 and the intrinsic impedance of the material (236 Ω) are drawn as straight lines. Notice that at $x = 0$ the impedance is continuous across the boundary. The maximum wave impedance Z_{max} in front of the boundary is when $L = (2n + 1)\lambda/4$, with $n = 0, 1, 2, \ldots$, i.e.

$$Z_{max} = Z_0^2/Z(\omega) = 603\ \Omega.$$

Case study 3.1.1b For comparison purposes Case study 3.1.1a is repeated but with the TEM wave travelling *from within the polystyrene* out into free space. Using the same numerical values as before, $Z(\omega) = 236\ \Omega$, but now the wavelength in the polystyrene is $\lambda' = \lambda/\sqrt{\varepsilon_r} = 0.188$ m and the reflection coefficient is

$$\rho_{21}(\omega) = \frac{Z_0 - Z(\omega)}{Z_0 + Z(\omega)} = +0.23$$

and

$$\tau_{21}(\omega) = \frac{2Z_0}{Z_0 + Z(\omega)} = +1.23.$$

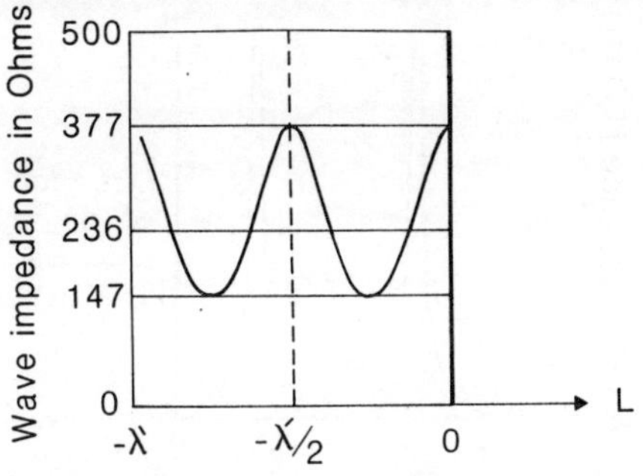

Fig. 3.2 Wave impedance as a function of position inside a semi-infinite sheet of polystyrene with characteristic impedance 236 Ω. The surface of the material is at $L = 0$ and the material extends to the left

Also the impedance in front of the boundary, i.e. within the polystyrene, is

$$Z(\omega, L) = Z(\omega)\,\frac{\{Z_0 + Z(\omega)\tanh(\gamma(\omega)L)\}}{\{Z(\omega) + Z_0\tanh(\gamma(\omega)L)\}}.$$

The modulus of the wave impedance $|Z(\omega, L)|$ is shown as the curve in Figure 3.2 between $x = -\lambda'$ and $x = 0$, and again for reference purposes Z_0 and the intrinsic impedance of the material (236 Ω) have been drawn as straight lines. Notice that if the thickness is made equal to $\lambda'/2$ (shown as a dotted line) then the impedance at each boundary would be Z_0, making the reflection coefficients zero; there would be perfect transmission. The maximum wave impedance, Z_{max}, is when $L = n\ \lambda'/2$ for $n = 0, 1, 2, \ldots$ and it equals Z_0, 377 Ω. There is a minimum wave impedance at $L = (2n+1)\ \lambda'/4$ which is given by $Z_{min} = Z(\omega)^2/Z_0 = 147.7\ \Omega$ which is smaller than the intrinsic impedance of the polystyrene.

Case study 3.1.1c A 1 MHz TEM wave is incident on copper. The skin depth, the intrinsic impedance of copper at this frequency and the reflection and transmission coefficients are calculated. The wave impedance as a function of distance in front of the surface is plotted, and the maximum impedance in front of the surface is compared with Z_0.

For copper $\sigma = 5.8 \times 10^7\ \Omega\,\mathrm{m}^{-1}$, with $\varepsilon_r = 1$ and $Z_0 = 377\ \Omega$ then

$$\gamma(\omega) = 1.51 \times 10^4 + 1.51 \times 10^4 \mathrm{j}$$

so that the intrinsic impedance of copper at this frequency is

$$Z(\omega) = 2.61 \times 10^{-4} + 2.61 \times 10^{-4}\mathrm{j}\ \Omega.$$

The skin depth $\partial = 1/\alpha = 1/1.51 \times 10^4 = 6.6 \times 10^{-5}$ m. Note the same result for the impedance is obtained by using the approximate formula given by Equation (2.79), i.e. $Z(\omega) = (1+\mathrm{j})/\sigma\partial$. The reflection and transmission coefficients are

$$\rho_{12} = -1.0 + 1.384 \times 10^{-6}\mathrm{j}$$

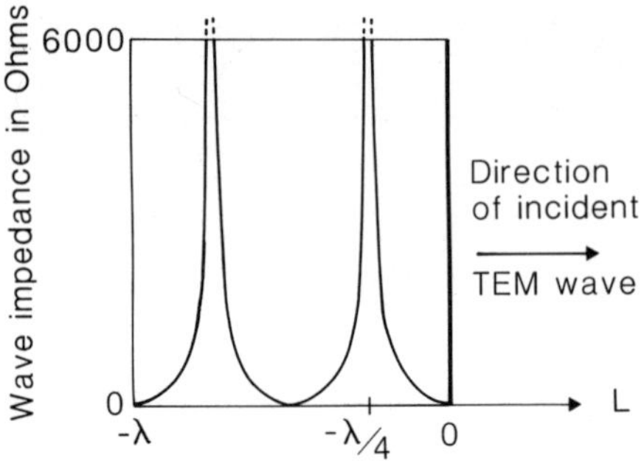

Fig. 3.3 Wave impedance as a function of position in front of a sheet of copper with very small characteristic impedance. The surface of the copper is at $L=0$ and extends to the right

(note the – sign), and similarly

$$\tau_{12} = 1.384 \times 10^{-6} + 1.384 \times 10^{-6}\mathrm{j}.$$

To four figures, $|\rho_{12}| = 1.000$ with a phase change of essentially π and $|\tau_{12}| = 2 \times 10^{-6}$ with a phase change of $\pi/4$. This confirms that good conductors make excellent reflectors of electromagnetic waves. Using (3.5) the wave impedance in front of the copper sheet can be plotted as a function of position, as shown in Figure 3.3. Note that in Figure 3.3 there are wave impedance maxima in front of the copper when $L = (2n+1)\lambda/4$ for $n = 0, 1, 2, \ldots$ with a magnitude $|Z_{max}| = |Z_1^2/Z(\omega)| = 3.8 \times 10^8\ \Omega$, which is about a million times bigger than Z_0 and too big to plot in Figure 3.3.

3.2 TRANSMISSION OF A TEM WAVE

In this section the effective transmission coefficient τ_{eff} of a wave travelling through a piece of material with finite thickness is calculated in two stages.

3.2.1 A first approximation to the transmission coefficient

Consider a TEM wave to be incident on an infinite sheet of material of thickness L m. In general, part of the wave will be transmitted through the first surface (with transmission coefficient τ_{12}), travel a distance L to the next surface, changing by a factor ($e^{-\gamma L}$) and part of that will emerge at the far side (transmission coefficient τ_{23}). If E_{in} is the amplitude of the wave incident onto the sheet and E_{out} is the amplitude which emerges from the far side of the material then $E_{out} = \tau_{\mathrm{eff}} E_{in}$, where τ_{eff} is the effective transmission coefficient for the whole slab of material. To a first approximation it is clear τ_{eff} can be written:

$$\tau_{\mathrm{eff}} = \tau_{12}\tau_{23}e^{-\gamma L}. \tag{3.9}$$

Apply this to a TEM wave in free space, with $f = 1$ GHz, incident onto a slab of polystyrene. By choosing the thickness $L = \lambda'/2$ then $|e^{-\gamma L}| = 1$ (since $\gamma L = j\pi$) and using the values for the transmission coefficients calculated in the case studies in the previous section then $\tau_{eff} = (0.77) \times (1.23)$, so that $|\tau_{eff}| = 0.947$. An immediate problem is apparent. Because the thickness $L = \lambda'/2$, perfect transmission through the material with $\tau_{eff} = 1.0$ would be expected, since the wave impedance at the boundaries in the material match Z_0, so the reflection coefficients at each boundary should be zero (see Case study 3.1.1b)). There is something missing from the approximation. Effects due to multiple reflections within the slab, generally called re-reflection effects, have been ignored and these are important.

3.2.2 Re-reflection effects

Consider a three-layer region as shown in Figure 3.4, with intrinsic impedances Z_1, Z_2 and Z_3, and with a TEM wave incident. Within the material with impedance Z_2, which has thickness L m, there will be two reflection coefficients which are of immediate interest, ρ_{21} at the left hand boundary for waves which are moving in the $-x$ direction from region 2 towards region 1, and ρ_{23} at the right-hand boundary for waves travelling in the $+x$ direction. These are given by

$$\rho_{23} = \frac{Z_3 - Z_2}{Z_3 + Z_2}$$

and

$$\rho_{21} = \frac{Z_1 - Z_2}{Z_1 + Z_2}$$

where the impedances used in this formulae are given by Equation (3.4). Consider the electric vector E_1 of a TEM wave travelling in the $+x$ direction

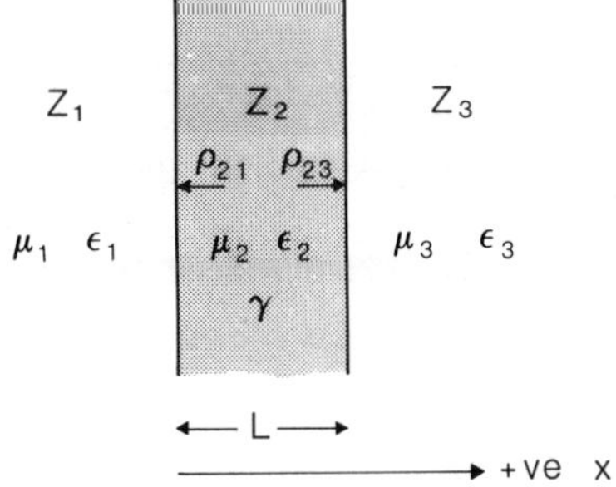

Fig. 3.4 A three layer medium with different characteristic impedances. The reflection coefficients ρ_{21} and ρ_{23} are for TEM waves incident on each boundary in the direction shown by the arrows. The propagation constant of the finite thickness central region (region 2) where re-reflection effects occur is γ

which starts out from the left-hand boundary. On reaching the right-hand boundary it has become

$$E_1 e^{-\gamma L}.$$

It is then reflected, becoming

$$\rho_{23} E_1 e^{-\gamma L}$$

and returns to the starting place, becoming

$$(\rho_{23} E_1 e^{-\gamma L}) e^{-\gamma L}$$

and it is reflected again, becoming

$$\rho_{21}(\rho_{23} E_1 e^{-\gamma L}) e^{-\gamma L}.$$

Thus after one complete set of reflections the initial wave amplitude E_1 has become a new, attenuated wave

$$E_1' = E_1 \rho_{21} \rho_{23} e^{-2\gamma L}.$$

But the same fate awaits E_1', and after one further cycle it too will become

$$\begin{aligned} E_2' &= E_1' \rho_{21} \rho_{23} e^{-2\gamma L} \\ &= E_1 (\rho_{21} \rho_{23} e^{-2\gamma L})^2 \end{aligned}$$

and so on. In the steady state the sum of all the waves travelling from left to right is

$$\begin{aligned} E_{\text{total}} &= E_1 + E_1' + E_2' + \cdots \\ &= E_1 (1 + \rho_{21} \rho_{23} e^{-2\gamma L} + (\rho_{21} \rho_{23} e^{-2\gamma L})^2 + \cdots). \end{aligned}$$

Recalling that when $|x| < 1$

$$1/(1 - x) = 1 + x + x^2 + x^3 + \cdots$$

and that since $E_1 = \tau_{12} E_{\text{in}}$, where E_{in} is the amplitude of the wave on the boundary, the series can be summed to give

$$E_{\text{total}} = \frac{\tau_{12} E_{\text{in}}}{(1 - \rho_{21} \rho_{23} e^{-2\gamma L})}.$$

This is the total amplitude of the wave *which starts out from the left-hand boundary*. This wave propagates a distance L to the second boundary where it arrives with amplitude $E_{\text{total}} e^{-\gamma L}$. This is transmitted through this boundary with transmission coefficient τ_{23}, so the wave E_{out} which, in the steady state, finally emerges into region 3 is given by

$$E_{\text{out}} = \tau_{23} E_{\text{total}} e^{-\gamma L}.$$

Hence

$$\tau_{\text{eff}} = \frac{\tau_{12} \tau_{23} e^{-\gamma L}}{(1 - \rho_{21} \rho_{23} e^{-2\gamma L})}. \tag{3.10}$$

Equation (3.10) is the general expression for the effective transmission coefficient for the *electric* component of the TEM wave through a thickness L of any material, and it contains an extra term $(1-\rho_{21}\rho_{23}e^{-2\gamma L})^{-1}$ compared with the first approximation given by Equation (3.9).

For the transmission of the *magnetic* component, (3.10) can be rewritten in the form [3]

$$T_H = p_H(1-q_H e^{-2\gamma L})^{-1} e^{-\gamma L} \tag{3.10a}$$

where p_H is the product of the magnetic field transmission coefficients and q_H is the product of the magnetic field reflection coefficients at the appropriate boundaries; see Equations (3.8*a*), (3.7*a*). If the impedances on either side of the central layer are the same (i.e. $Z_1 = Z_3$) then the electric and magnetic field transmission coefficients are equal. Returning to the example in Section 3.2.1 with the half-wave thickness slab of dielectric, where $Z_1 = Z_3 = Z_0$ so that $\rho_{21} = \rho_{23} = 0.231$ and $|e^{-2\gamma L}| = 1$.

The extra term $1/(1-\rho_{21}\rho_{23}e^{-2\gamma L}) = 1/(1-(0.231)^2) = 1.0564$ so the first approximation for the transmission coefficient (Equation (3.9)) can be modified to give

$$\tau_{\text{eff}} = 0.947 \times 1.0564 = 1.00$$

as expected. If you wish to check this calculation, remember that $\rho_{21} = -\rho_{12}$, and that ρ_{12} itself has a negative value. This result is a remarkable illustration of the non-trivial nature of electromagnetic fields, in that perfect transmission through a half-wave dielectric window (with no absorption) can be traced back to the presence of multiple internal reflections within the window! However there is no real mystery to this, as it is simply the consequence of the fact that waves can interfere, and that for a half wave dielectric window interference removes the reflected waves. This said, Equation (3.10) has general validity, and therefore can be used in cases where there is dispersion and absorption, as in conducting materials.

The reader should notice that two different approaches to doing these calculations have been met. The first is to use (3.10) with the reflection and transmission coefficients calculated using the intrinsic impedances of the materials given by (3.4) and the wave impedance of the incident wave. The alternative approach is to use (3.6*a*) to calculate the wave impedance at each boundary, together with the wave impedance of the incident wave, to calculate ρ and τ at each boundary using (3.7) and (3.8). Both approaches should yield the same answers, but in more complicated situations one method may be easier to use than the other. In Chapter 4 the detailed behaviour of the reflection, transmission, and absorption terms, calculated using (3.10*a*), will be examined in different situations. Section 3.3 is concerned with understanding how to make materials and structures which absorb radiation, and there it is more convenient to calculate the reflection coefficient using the wave impedance approach.

3.2.3 Decibels, shielding effectiveness and nepers

In EMC work the effective transmission coefficient is normally presented in terms of a shielding effectiveness, SE, measured in decibels. As the energy carried by any wave is proportional to the square of the amplitude, then:

$$\begin{aligned} \mathrm{SE} &= 10 \log_{10}(E_{\mathrm{in}}^2/E_{\mathrm{out}}^2) \text{ dB} \\ &= 20 \log_{10}(E_{\mathrm{in}}/E_{\mathrm{out}}) \text{ dB} \end{aligned} \tag{3.11}$$

where E_{in} is the field strength before a shield is installed and E_{out} is after it is installed. This can be written

$$\mathrm{SE} = 20 \log_{10}(1/|\tau_{\mathrm{eff}}|) \text{ dB}. \tag{3.12}$$

In Table 3.1 SE in dB are tabulated for different $1/|\tau_{\mathrm{eff}}|$, and one can see that an SE of 60 dB corresponds to a reduction of the amplitude of the incident field by a factor of 1000. Real shields have SE between ≈ 20 and 120 dB. Anything less than 20 dB is probably not worth the effort, whereas SE of 160 dB corresponds to the incident field being attenuated by a factor of 100 *million*, which is almost impossible to achieve with real shields, or even measure. The very best screened rooms have SE $\approx$ 120 dB. This is discussed in more detail in Chapter 4.

When a wave with amplitude E_1 travels a distance x metres through a medium with attenuation constant α the amplitude is reduced to $E_2 = E_1 e^{-\alpha x}$, thus

$$E_1/E_2 = e^{\alpha x}$$

i.e.

$$20 \log_{10}(E_1/E_2) = 20\alpha x \log_{10}(e) = 8.6859\alpha x.$$

Thus when $\alpha x = 1$ neper, SE = 8.6859 dB; this is the relationship between dB and neper.

Table 3.1. Shielding effectiveness in dB

$1/\lvert\tau_{\mathrm{eff}}\rvert = E_{\mathrm{in}}/E_{\mathrm{out}}$	SE(dB)
1	0
10	20
100	40
1000	60
10 000	80
100 000	100
1000 000	120
10 000 000	140
100 000 000	160

3.2.4 Transmission through conductors

In Case study 3.1.1c it was shown that good conductors have very high reflection coefficients and also that any wave within the conductor is attenuated dramatically after travelling a distance of only a few skin depths. These two factors make it clear that in general good conductors have very large values of SE and it is very difficult to transmit a wave through them. However there are three particular case studies which are worth looking at in some detail; transmission through thin conducting films at microwave and optical frequencies and transmission of a low impedance wave through a thick conductor. In all these cases the values $\mu_r = 1$ and $\varepsilon_r = 1$ are used.

Case study 3.2.4a Transmission through a thin conducting film at 100 MHz. Consider a thin conducting film of thickness $L = 10^{-7}$ m (0.1 μm) with a high conductivity $\sigma = 10^7$ S m^{-1} situated in free space with impedance Z_0 on either side of it. If a TEM wave of frequency 100 MHz is incident normally onto this film, then using (3.2) for γ and (3.4) for the impedance of the media, the various terms in Equation (3.10) are

$$\gamma = 6.28 \times 10^4(1 + \mathrm{j})$$
$$\tau_{12} = 3.33 \times 10^{-5}(1 + \mathrm{j})$$
$$\tau_{23} = 2 - 3.33 \times 10^{-5}\mathrm{j}$$
$$\rho_{21} = \rho_{23} = 1 - 3.33 \times 10^{-5}\mathrm{j}$$
$$\mathrm{e}^{-\gamma L} = 0.994 - 0.006\mathrm{j}$$
$$(1 - \rho_{21}\rho_{23}\mathrm{e}^{-2\gamma L})^{-1} = 40.1 - 39.58\mathrm{j}.$$

Thus

$$\tau_{\mathrm{eff}} = 0.005 - 7.02 \times 10^{-8}\mathrm{j}$$

and using (3.12)

$$\mathrm{SE} = 45.55 \text{ dB}.$$

Thus even very thin conducting films can be effective shields. This topic is discussed in more detail in Section 4.6.2. See also Figure 4.33.

For very thin conducting films where the $L \ll \delta$ (skin depth), and the intrinsic impedance $Z \ll Z_0$, a considerable and instructive simplification [5] can be made to Equation (3.10). From Equation (2.69)

$$\gamma \approx (1 + \mathrm{j})/\delta$$

and the term $|\gamma L| \ll 1$, such that (3.10) can be approximated to

$$\tau_{\mathrm{eff}} \approx \frac{\tau_{12}\tau_{23}}{1 - \rho_{21}\rho_{23}(1 - 2\gamma L)}$$

which, by using (2.79) for the impedance of the film

$$Z \approx (1 + \mathrm{j})/\sigma\delta$$

gives, after some manipulation

$$\tau_{\text{eff}} \approx \frac{1}{1 + Z_0/(2R_s)}$$

where $R_s = 1/\sigma L$ is the surface resistance of the sheet. Thus

$$\text{SE} \approx 20 \log_{10}(1 + Z_0/(2R_s)) \text{ dB}. \tag{3.13}$$

In this case study $R_s = 1\,\Omega\,\text{m}^{-2}$, hence

$$\begin{aligned}\text{SE} &\approx 20 \log_{10}(1 + 60\pi) \text{ dB} \\ &\approx 45.55 \text{ dB}.\end{aligned}$$

This shows that for thin films SE for radio and microwaves is independent of frequency and can be improved by having high conductivity films. For (3.13) to be valid R_S must satisfy [5]

$$\mu\omega L/2 \ll R_s \ll Z_0^2/(\mu\omega L). \tag{3.14}$$

Case study 3.2.4b Transmission through metals at optical frequencies. Consider the same film as in the previous case study with $\sigma = 10^7\,\text{S}\,\text{m}^{-1}$ and thickness $L = 10^{-7}$ m. Recomputing SE at a frequency of 6×10^{14} Hz using (3.10) yields SE = 146 dB. Equation (3.13) cannot be used at these frequencies because the conditions (3.14) are not satisfied since $\mu\omega L/2 \gg R_s$. Taken at face value this result implies very effective shielding so that it would seem that such thin films are opaque, but *this result is seriously in error*. At these ultra-high frequencies, which in this case correspond to a visible wavelength of 500 nm, very thin conducting films are transparent to light.

The mistake which has been made is to assume that at optical frequencies the conductivity is a real constant. At optical frequencies the period of the wave is comparable to the mean time between collisions, τ_{drift}, of the conduction electrons moving through the material. It is inappropriate to discuss this in detail, but an excellent introduction to this subject can be found in Chapter 32 of reference [4]. Using a simple model of conduction, the conductivity can be written

$$\sigma' = \sigma/(1 + j\omega t_{\text{drift}})$$

where

$$t_{\text{drift}} = m\sigma/Nq^2$$

and m = mass of an electron, N = number of electrons/m^3, and q is the charge on an electron. Thus the conductivity is complex and frequency dependent. For copper, $t_{\text{drift}} \approx 2.4 \times 10^{-14}$ s so that at radio and microwave frequencies the imaginary part of the conductivity usually can be neglected. However at the optical frequency $f = 6 \times 10^{14}$ Hz, using $\sigma = 10^7\,\text{S}\,\text{m}^{-1}$, and taking

$t_{drift} = 2 \times 10^{-14}$ s, then

$$\sigma' = 1.76 \times 10^3 - 1.33 \times 10^5 \text{ j S m}^{-1}.$$

Using this in Equation (3.10) to calculate γ, etc, gives SE $\approx$ 14 dB, i.e. the thin film is effectively transparent to light.

Notice that at this frequency the real part of the conductivity has been reduced by a factor of order 10^4 compared to its 'low frequency' value, and there is a large negative imaginary component. As the period of the wave and t_{drift} are comparable in this frequency region, rapid variations in the value of σ', hence SE, should be expected at adjacent optical and infra-red wavelengths. This fact is used to make glass coated with thin conducting films which transmit visible light but reflect longer wavelengths, including infra-red. There is also a similar effect with radio transmission through the ionosphere; long waves are reflected, short waves pass through. As well as these effects it should be realised that the 'thin' film used in these case studies is in fact rather thick compared to realistic, optically transparent, films where typical thicknesses are less than 0.01 μm (100 Å). In such very thin films the thickness is less than the mean free path of the conduction electrons ($\approx$0.042 μm for copper at room temperature), which has the effect of reducing the film conductivity compared to the conductivity of the bulk material; this changes the SE at microwave frequencies. A discussion of the design and performance of realistic transparent thin film shields is given in references [5], with more discussion in Section 4.4.

Case study 3.2.4c Transmission of a low impedance field through a conductor. Consider a copper shield of thickness 1 mm situated $r = 10$ cm from an electromagnet carrying a 50 Hz current. This case might be treated as being that of a shield situated in the low impedance near field of a magnetic dipole transmitter. The wave impedance in the near field (see Appendix C) is $Z_w \approx Z_0 (2\pi r/\lambda) \approx 3.9 \times 10^{-5}$ Ω, which can be compared with the intrinsic impedance of copper at this frequency

$$Z = 1.8 \times 10^{-6}(1 + \text{j}) \ \Omega.$$

Using (3.10) gives SE = 6.6 dB; increasing the thickness to 1 cm only makes SE = 22 dB.

These results imply that even thick copper is ineffective as a magnetic shield. On the other hand it must be remembered that Equation (3.10) is derived assuming TEM waves at normal incidence, which certainly is not true in the near field. Also eddy currents, which will be induced in the copper, will produce their own magnetic fields which may have some effect. As will be seen in Chapter 4 when magnetic shielding is discussed more carefully, the general point that non-magnetic conducting shields are unsatisfactory for screening low frequency magnetic fields is, in fact, nearly always valid. The fundamental reason for this is that the very low intrinsic impedance of good conductors is nearly matched by the low wave impedance of the incident electromagnetic field, so the reflection coefficient is small.

3.3 THE REFLECTION COEFFICIENT FROM MULTI-LAYERED MEDIA

This section will develop the formula for the reflection coefficient ρ from multi-layered media in order to use it in case studies about absorbers, such as those used in electromagnetically anechoic rooms.

Consider a planar, multi-layered, structure consisting of layers numbered 1 to k, each with thicknesses d_i, different conductivities σ_i, permittivities ε_i, and permeabilities μ_i as shown in Figure 3.5. Each layer has different intrinsic impedances $Z_i(\omega)$ given by Equation (3.4) and propagation constants $\gamma_i(\omega)$. Layer k is taken to be semi-infinite. A TEM wave with wave impedance Z_0 travelling in the $+x$ direction is incident on the layer 1. The reflection coefficient is:

$$\rho = \frac{Z_{w1} - Z_0}{Z_{w1} + Z_0} \tag{3.15}$$

where Z_{w1} is the wave impedance at the first boundary, which from (3.6a) is just

$$Z_{w1} = Z_1 \frac{\{Z_{w2} + Z_1 \tanh(\gamma_1(\omega)d_1)\}}{\{Z_1 + Z_{w2} \tanh(\gamma_1(\omega)d_1)\}}$$

where Z_{w2} is the wave impedance at the second boundary,

$$Z_{w2} = Z_2 \frac{\{Z_{w3} + Z_2 \tanh(\gamma_2(\omega)d_2)\}}{\{Z_2 + Z_{w3} \tanh(\gamma_2(\omega)d_2)\}}$$

where Z_{w3} is the wave impedance at the third boundary,

$$Z_{w3} = Z_3 \frac{\{Z_{w4} + Z_3 \tanh(\gamma_3(\omega)d_3)\}}{\{Z_3 + Z_{w4} \tanh(\gamma_3(\omega)d_3)\}}$$

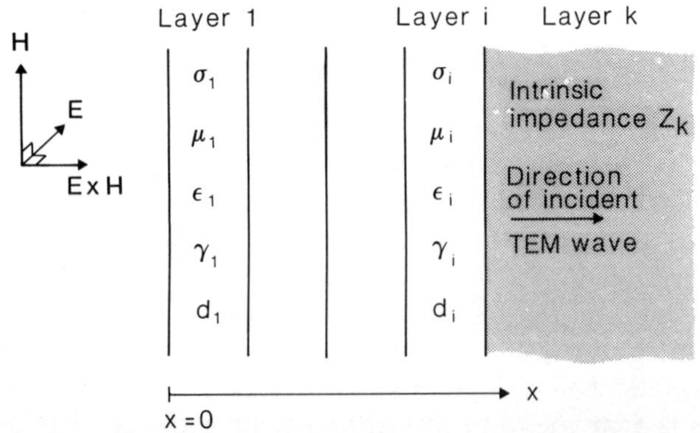

Fig. 3.5 A multi-layer medium where the electrical properties and thickness of each layer i is different. The final layer, layer k, is assumed to be semi-infinite. A free-space TEM wave is incident onto the surface of layer 1 at $x = 0$

and so on up to the last but one layer, layer $k-1$,

$$Z_{wk-1} = Z_{k-1} \frac{\{Z_k + Z_{k-1} \tanh(\gamma_{k-1}(\omega) d_{k-1})\}}{\{Z_{k-1} + Z_k \tanh(\gamma_{k-1}(\omega) d_{k-1})\}} \tag{3.16}$$

where in this final term Z_k is the intrinsic impedance of the last layer.

These formulae can be evaluated in the same way as the case studies in Section 3.2. To calculate ρ one proceeds as follows.

(*a*) Assign values of σ, ε, μ and thickness d to each layer.
(*b*) Choose a value for ω and the wave impedance Z_0 of the incident wave.
(*c*) Calculate the propagation constant γ_i and intrinsic impedance Z_i for each layer at this frequency.
(*d*) Start working backwards, first finding the wave impedance of the last but one layer $k-1$, Z_{wk-1}, using equation (3.16), then use this to find Z_{wk-2} and so on until Z_{w1} is found.
(*e*) Calculate the final value of ρ using Equation (3.15).

The behaviour of this model is discussed in detail in the next section.

3.4 DESIGN OF ABSORBERS

3.4.1 Introduction

In order to make materials which absorb electromagnetic radiation and not just reflect it back into the environment, they must be lossy so that the energy in the wave can be dissipated within the material. Energy can be absorbed by joule heating when currents flow in conductors, but these have low intrinsic impedances and normally act as good reflectors. One way out of this impass is to compromise and make layered materials which have variable properties, such that their surface impedance is as close as possible to the incident wave impedance, but then change their intrinsic impedance inside by gradually increasing their conductivity, keeping the reflection coefficient at the boundary of each layer as low as possible but allowing Joule heating to dissipate progressively the energy in the wave. This section will examine an idealised model of a layered absorber, with a view to understanding how to do the necessary calculations, and get a feeling for such parameters as their thickness and effectiveness.

Before this, it is useful list some of the factors which could be important and make a guess about a possible good design. Designers of absorbers have some control over the choice conductivity and ε_r, how they can be distributed within the absorber, the thickness of each layer, the total thickness of the absorber, the number of layers etc, and hence should try to put these factors in some order of importance.

3.4.2 Some factors in absorber design

In any practical absorber a large number of factors are important (including cost and safety of materials) but in order to examine the general properties it is worth bearing the following points in mind.

(1) It is a general principle in physics that waves are unaffected by structures considerably smaller than a wavelength. This means that any successful absorber is likely to have a thickness of order of one wavelength. The scattering from a surface depends not only on its absorbing properties but also on its shape, but this factor is not discussed in this section.
(2) The skin depth plays an important role in describing the distance waves penetrate into conductors. After 1 skin depth the wave amplitude which has *survived* is $E_0 e^{-1}$.
(3) Dielectric materials have intrinsic impedances $= Z_0/\sqrt{\varepsilon_r}$. As most dielectrics have $\varepsilon_r \geqslant 2$ (see Table 1.1) this makes it difficult to have an impedance big enough to be very near to Z_0. The effective dielectric constant might be lowered towards $\varepsilon_r \approx 1$ by having a foam of material with low average density, but low mechanical strength.
(4) Absorbers are often mounted on the walls of screened rooms which have metal walls. If this is the case the conductivity of the layers adjacent to the walls should be high to obtain better impedance matching at that boundary and not contribute unnecessarily to any reflection.

3.4.2.1 *A hypothetical absorber*

Consider the following model of a hypothetical, planar, multi-layered absorber made from 50 layers of non magnetic material ($\mu_r = 1$) with the following properties

(*a*) ε_r varies from 1.5 in layer 1 to 2.56 in the last one.
(*b*) The conductivity of the layers varies from $\sigma \approx 0$ in layer 1 to $\sigma \approx 5.8 \times 10^5\ \mathrm{S\,m^{-1}}$ in the last one (which is 100 times smaller than copper; graphite has $\sigma \approx 10^5\ \mathrm{S\,m^{-1}}$).
(*c*) The thickness d_k of each layer is made equal to 2 'skin depths', where the skin depth is calculated at a frequency of 1 GHz, by using Equation (2.71) for the skin depth of a good conductor. This gives a 'wrong' value for the required thickness in the first two or three layers (it underestimates it), in the sense that in these layers where the conductivity is very low $1/\alpha$ is not equal to the thickness, but this parametrisation is good enough for the present purposes.

Given this recipe, it is of interest to find the reflection coefficient and the total thickness of the absorber. This question can best be answered by building a computer model of the absorber. In such a computer model the conductivity and dielectric constant of layer k could be parametrised by the formulae:

$$\sigma_k = 5.8 \times 10^5 (k/k_{\max})^5$$

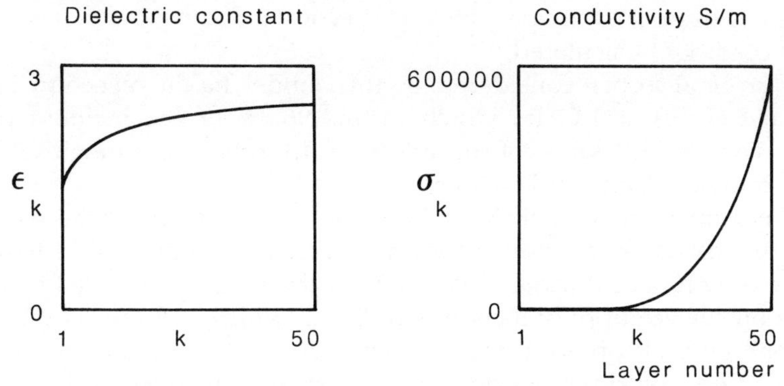

Fig. 3.6 The assumed parametrisation of the electrical properties in each layer of the hypothetical multi-layered absorber considered in Section 3.4.2.1

and

$$\varepsilon_k = 1.5 + 1.06\ \log_{10}(k)/\log_{10}(k_{\max})$$

so in the last layer where $k = k_{\max}$, $\sigma = 5.8 \times 10^5\ \text{S m}^{-1}$ and $\varepsilon_r = 2.56$. Graphs of these as a function of layer number are shown in Figure 3.6. By changing these parametrisations different arrangements of electrical properties could be tried.

Once the electrical properties are chosen, the propagation constant for each layer is found using Equation (3.2), and the wave impedance Z_k at each boundary calculated using the chain of equations (3.15) to (3.16), in that order. In Table 3.2 are shown the corresponding values of Z_k, γ_k, and the thickness d_k mm of every fifth layer for an absorber with 50 layers. Once these parameters

Table 3.2. The wave impedance Z_k, propagation constant γ_k and layer thickness d_k at every fifth layer throughout the hypothetical multilayer absorber considered in Section 3.4.2.1

Layer	Impedance Z_k	Propagation const. γ_k	Thickness d_k (mm)
1	286.175 + 77.29j	0.285 + 25.67j	738.858
5	26.638 + 25.806j	149.921 + 152.731j	13.217
9	6.045 + 5.996j	657.422 + 658.122j	3.041
13	2.405 + 2.393j	1649.259 + 1649.551j	1.213
17	1.228 + 1.224j	3225.379 + 3225.533j	0.62
21	0.724 + 0.722j	5470.328 + 5470.422j	0.366
25	0.468 + 0.467j	8458.966 + 8459.028j	0.236
29	0.323 + 0.322j	12 259.227 + 12 259.27j	0.163
33	0.234 + 0.233j	16 933.756 + 16 933.788j	0.118
37	0.175 + 0.175j	22 540.987 + 22 541.011j	0.089
41	0.136 + 0.135j	29 135.894 + 29 135.913j	0.069
45	0.108 + 0.107j	36 770.544 + 36 770.559j	0.054
49	0.087 + 0.087j	45 494.51 + 45 494.523j	0.044

have been chosen, the recipe given in Section 3.3 can be followed and the reflection coefficient calculated.

The 50 layer absorber considered in this model has a reflection coefficient $\rho = 0.178$ (14.97 dB) at 1 GHz, which means 96.8% of the incident energy is absorbed. The total thickness of this absorber (found by summing the thickness of all the individual layers) is 0.99 m.

Careful examination of Table 3.2 will show that almost all of the thickness of this hypothetical absorber is due to layer 1, which is 738 mm thick. Forcing this layer to be only half as thick as this, the total thickness decreases to 0.62 m but the reflection coefficient only increases to 0.212, which still means 95.5% of the incident energy is absorbed. It is left as an exercise to the reader to experiment with their own computer models to investigate the effects of varying the number and thickness of layers, the distribution of σ and ε_r throughout the material, and produce their own designs for better, and cheaper, absorbers.

3.4.3 Performance of absorbers at different frequencies

It is clearly of great practical importance to know how a given absorber, with its parameters fixed, performs at different frequencies. By running the computer

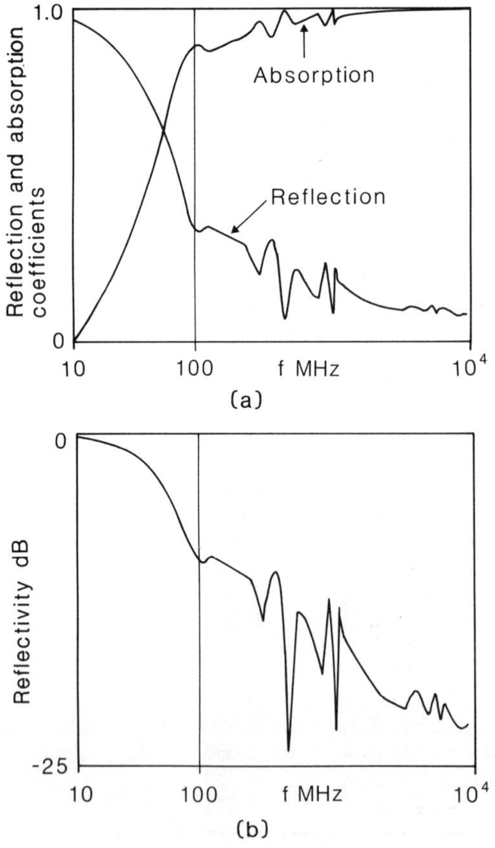

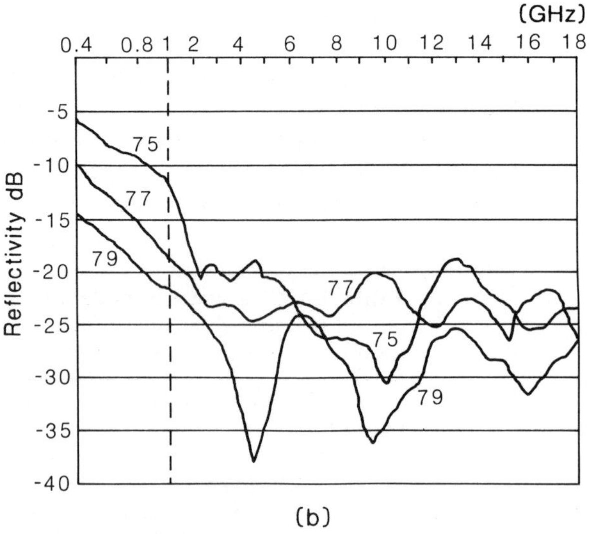

Fig. 3.7 (*a*) The calculated reflection and absorption coefficients of a hypothetical absorber as a function of frequency. (*b*) Reflectivity (i.e. Reflection coefficient in dB) at normal incidence of a hypothetical absorber as a function of frequency compared to good quality flexible foam sheet commercial absorbers, ECCOSORB AN75, AN77 and AN79. (Reproduced by permission of Emerson & Cuming (UK))

model constructed in Section 3.4.2.1 for a whole range of frequencies, Figure 3.7(a) can be calculated. This shows the reflection coefficient $|\rho|$ and the fraction of incident energy dissipated (defined as $1-|\rho|^2$) by the hypothetical absorber between 10 MHz and 10 GHz. It is clear that below about 100 MHz the performance this planar absorber is considerably reduced, but at higher frequencies it improves until at 10 GHz nearly all the incident energy is absorbed. Notice that there is resonant behaviour around certain wavelengths where the absorption becomes particularly good. In Figure 3.7(b) is shown the reflection coefficient in dB as a function of frequency, compared to a good quality lightweight flexible foam sheet commercial absorber. The two show the same general behaviour with the performance improving with frequency, but with some structure at certain frequencies.

3.4.4 Examples of real absorbers

Absorbers have been made since the mid 1930s, with the first patent being granted in 1936. A short but useful history, together with references, of this early period up to the end of the 1960s can be found in [7]; one early absorber, which is still used, consists of carbon black sprayed onto animal hair! Today, absorbers are supplied on a commercial basis by a number of companies. Typical absorbers made to be mounted on the walls of anechoic rooms are shown in Figure 3.8. They are pyramid shaped, with variable conductivity

Fig. 3.8 Photograph of typical absorbers which are mounted on the walls of electromagnetically anechoic rooms. (Reproduced by permission of Emerson and Cuming Microwave Products — Grace Electronic Materials)

throughout, and have a size which depends on the longest wavelength that is required to be absorbed. Three reasons for choosing these shapes are:

(1) specular reflection from the side of an absorber will direct the reflected component further into the absorbing structure rather than back into the environment;
(2) there is more surface area available;
(3) in the vicinity of the sharp point, averaged over the whole wavefront the impedance will be almost the same as the incident wave (there is almost no material), which is advantageous.

The main use of large wideband absorbers is in the building of electromagnetically anechoic and screened rooms for experimental purposes. Here the aim is to limit resonances but the physical size of the absorbers can be very large if the effectiveness is to be extended to low frequencies (50–100 MHz). Marvin [8] has experimented with placing absorbers at specific locations within such screen enclosures, either on the walls or at central points where known field maxima occur, thus selectively damping the resonance. Whether such techniques can ever replace the almost total coverage of all the walls by absorbers required in anechoic rooms has yet to be ascertained. A final example is the present interest in Stealth technology [9] for aircraft, which uses a combination of exotic materials and careful design of shapes to minimise radar reflections.

3.5 OVERVIEW OF TRANSMISSION LINE THEORY

Electromagnetic energy can propagate not only in the form of free electromagnetic waves but as confined waves along various configurations of conductors or dielectrics (in the form of pairs of wires, hollow tubes or fibres etc.) called transmission lines or wave guides. The energy remains locally confined and follows the line or tube whereever it leads. Such devices are clearly of great practical use, but they are of interest in EMC work for a number of reasons.

(*a*) They radiate less than ordinary wires or even twisted pairs, so should be used to make connections which minimise RFI, and they pick up less noise than them.
(*b*) They can occur within complex systems by accident and lead radiation into regions it ought not to go.
(*c*) Electromagnetic waves can couple to transmission lines and in many situations where the electronics is adequately shielded such coupling can be the primary route for interference.
(*d*) Coupling between transmission lines, particularly on circuit boards, can cause EMC problems.

(*e*) There is a direct analogy between the voltage and current distributions along a transmission line and the electric and magnetic fields in a wave. This analogy can be used in shielding calculations in complex geometries by breaking a system up into a number of hypothetical transmission lines, calculating the current and voltage distributions using transmission line theory, which leads to understanding of electromagnetic field distributions. See reference [10] and Section 3.5.7.

This section will review the basic equations needed as well as their most important properties. The properties of transmission lines depend on the frequency range, the way they are terminated, and their geometric properties. There are many books on this subject: the one by Liao [11] is a good companion text for this section. Detailed discussion of their properties can be found in [13, 19].

3.5.1 Basic concepts

Transmission lines may be classified as follows:

(1) TEM-mode type: **E** and **H** entirely transverse. All are two-conductor types, including coaxial lines. Power flows along *between* the two conductors.
(2) Higher-mode type: **E** and **H** or both have components in the direction of transmission. Examples are hollow single conductor waveguides, dielectric rods, and optical fibres. Power flows in the *space* inside the conductor or *in or close to* dielectric rods or fibres.

An equivalent circuit analysis enables the behaviour of current I and the voltage V to be calculated at any point along the line. This approach is found in a number of text books, for example [12]. Normally a transmission line is a two conductor system (two parallel wires, coaxial cable, two flat parallel strips). If the wavelength of the EM disturbance is much less than the length of line then the lumped circuit approximation is valid and ordinary circuit analysis suffices. Once the dimensions of the line becomes comparable to the wavelength, transit time effects become important and transmission line properties must be considered. To discuss the mathematics of transmission lines an equivalent circuit model of basic line types can be constructed, based on the following ideas.

(1) The existence of an electric field between the two conductors suggests a parallel capacitance between the conductors.
(2) Voltage drops down the conductor suggest a series resistance along the line.
(3) The currents mean that magnetic fields exist, so inductive effects occur. This suggests that a series inductance is needed.
(4) There may be leakage across the dielectric insulating the space between the two conductors. This can be modelled as a parallel resistor of conductance G.

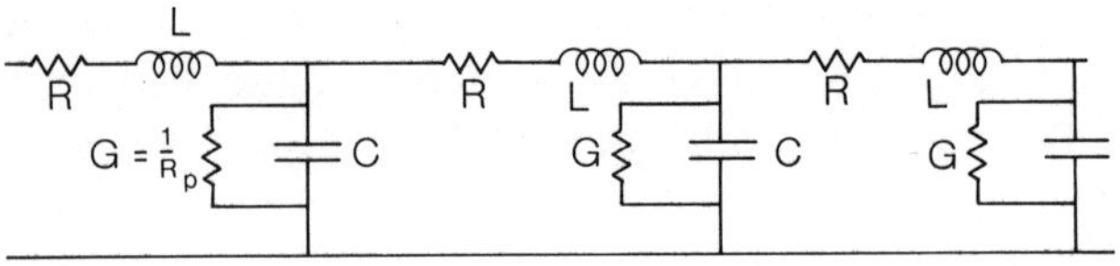

Fig. 3.9 Equivalent circuit model of a transmission line

Putting these ideas together leads to an equivalent circuit model for a transmission line shown in Figure 3.9. The values quoted are ohms, henrys, farads *per unit length* of the line. The construction of an artificial line of a number of discrete R, L and C components is of course possible and such structures do behave as transmission lines, and are useful for carrying signals with long pulse lengths (e.g. $\geqslant 0.5\,\mu s$). Also, in all that follows, it is assumed that these parameters are not functions of frequency.

3.5.2 Impedance and phase shift of an ideal line

If the line has no losses $R = 0$ and $G = 0$ then a simple ladder network of C's and L's results, see Figure 3.10. The section between A and C in Figure 3.10 represents a length d of line. The impedance across AB is $Z_1\,\Omega$, and the impedance looking into CD is $Z_2\,\Omega$. The inductance between A and C is (dL) where L is the inductance per unit length. Using circuit analysis

$$Z_1 = j\omega Ld + \frac{Z_2(1/j\omega dC)}{Z_2 + (1/j\omega dC)}. \tag{3.17}$$

If the line is really infinitely long then the same impedance should be seen looking into AB as into CD, i.e. $Z_1 = Z_2$. Hence making this substitution in (3.17) and rearranging gives

$$Z_1^2 + \frac{Z_1}{j\omega dC} = j\omega LdZ_1 + \frac{L}{C} + \frac{Z_1}{j\omega dC}$$

i.e.

$$Z_1^2 - j\omega LdZ_1 = L/C.$$

If the limit when $d \to 0$ (i.e. model the *continuous* line in terms of more and more

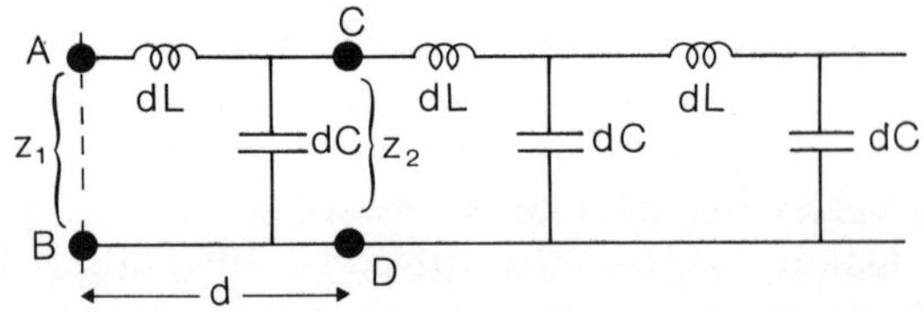

Fig. 3.10 Equivalent circuit model of a lossless transmission line

discrete components), then

$$Z_1 = \surd(L/C) = Z_0\ \Omega. \tag{3.18}$$

This is a pure resistance called the characteristic impedance of the ideal line. If an infinite line were broken at any point and terminated by a resistance Z_0 it would be impossible to detect the difference.

Using the above lossless line model (line with pure L and C, no R) the phase shift per unit length $\beta = \omega\surd(LC)$. This phase shift really represents a wave travelling with velocity

$$V^* = \omega/\beta = 1/\surd(LC)\ \mathrm{m\,s^{-1}}. \tag{3.19}$$

3.5.3 The characteristic impedance of a lossy transmission line

Using the model for the full lossy transmission line (with R, L, C and G per unit length) and assuming the line is excited with sinusoidal signals the voltage V_x and the current i_x at a point x down the line are solutions of the following differential equations [11]:

$$\frac{\mathrm{d}^2 V_x}{\mathrm{d}x^2} = \gamma^2 V_x \tag{3.20}$$

and

$$\frac{\mathrm{d}^2 i_x}{\mathrm{d}x^2} = \gamma^2 i_x \tag{3.21}$$

where

$$\gamma^2 = (R + \mathrm{j}\omega L)(G + \mathrm{j}\omega C). \tag{3.22}$$

The notation $Z = (R + \mathrm{j}\omega L)\ \Omega\,\mathrm{m}^{-1}$ and $Y = (G + \mathrm{j}\omega C)\ \Omega^{-1}\,\mathrm{m}^{-1}$ is often used, so that

$$\gamma = \surd(ZY).$$

These equations have a familiar form (compare them with Equation (2.64)) and have solutions which represent attenuated waves which are

$$V_x = \mathrm{e}^{\mathrm{j}\omega t}(V_\mathrm{A}\mathrm{e}^{-\gamma x} + V_\mathrm{B}\mathrm{e}^{\gamma x}) \tag{3.23}$$

and

$$i_x = \frac{\mathrm{e}^{\mathrm{j}\omega t}}{\surd(Z/Y)}(V_\mathrm{A}\mathrm{e}^{-\gamma x} - V_\mathrm{B}\mathrm{e}^{\gamma x}). \tag{3.24}$$

These solutions represent wave travelling in either direction, and V_A and V_B are constants defined by the boundary conditions, and

$$\surd(Z/Y) = (R + \mathrm{j}\omega L)/\gamma. \tag{3.25}$$

Note the similarity to the equations for **E** and **H** in plane wave propagation in a medium which are the solutions of Equation (2.64) and represent travelling waves, with the real part of γ representing attenuation and the imaginary part causing phase shifts. The first term $V_A e^{j\omega t} e^{-\gamma x}$ represents the forward wave and the second term $V_B e^{j\omega t} e^{\gamma x}$ the backward wave, i.e. waves can go either way, as one might expect! A line is terminated with characteristic impedance $Z_0 = V_x/i_x$ behaves as if it were infinite and no reflected wave is produced, so that $V_B = 0$. Thus using Equations (3.23)–(3.25) gives

$$Z_0 = \surd(Z/Y) = \surd\{(R + j\omega L)/(G + j\omega C)\}\ \Omega. \tag{3.26}$$

This is the general expression for the characteristic impedance of a transmission line.

It is of very great practical importance to make transmission lines with a given impedance. For example, tracks with low characteristic impedance are advantageous for power distribution on circuit boards (see Section 6.6), and impedance matching to a transmission line connected to a device is of great importance. For transmission lines consisting of two conductors separated by a medium, the inductance per unit length depends on the relative geometry of the two conductors, the magnetic properties of the medium and the self inductance of the conductors themselves. The capacitance per unit length depends on the geometry and the dielectric properties of the medium between the conductors. R and G depend on the conductivity of the materials forming the conductors and that of the material between the conductors. The resistance of the conductors increases with frequency due to the skin effect (see Section 2.5.6.3) and dielectric properties are frequency dependent (see Table 2.1). Thus the detailed behaviour of transmission lines as a function of frequency is complex. For an extensive discussion see references [13, 19]. Nevertheless there are a number of simple but useful formulae for the characteristic impedances of transmission lines with different geometries which can be used in circuit board design and some of these have been collected together in Appendix E.

3.5.4 Wave velocity down a lossy line

As discussed in the previous section the propagation constant γ can be written in terms of real and imaginary components as

$$\gamma = \alpha + j\beta.$$

All real transmission lines have some losses which cannot be ignored in practice, although R and G are often very small. By expanding the square root in (3.22) as a series, making the approximations $R \ll \omega L$, $G \ll \omega C$ and equating real and imaginary parts it is readily shown [13] that at high frequencies

$$\alpha = R/(2\surd(L/C)) + G\surd(L/C)/2 \tag{3.27}$$

and

$$\beta = \omega\surd(LC)\{1 + R^2/(8\omega^2L^2) + G^2/(8\omega^2C^2) - RG/(4\omega^2LC)\} \tag{3.28}$$

where $\surd(L/C)$ is the characteristic impedance Z_0 of a lossless line, Equation (3.18).

The wave solution (3.23) can now be written:

$$V_x = e^{j\omega t}(V_A e^{-\alpha x} e^{-j\beta x} + V_B e^{\alpha x} e^{j\beta x})$$
$$= V_A e^{-\alpha x} e^{j(\omega t - \beta x)} + V_B e^{\alpha x} e^{j(\omega t + \beta x)}. \tag{3.23'}$$

A point of constant phase requires $(\omega t - \beta x) = K_1$ or $(\omega t + \beta x) = K_2$. Differentiating these equations with respect to time gives $\omega \pm \beta\, dx/dt = 0$; hence

$$V^* = dx/dt = \pm\omega/\beta$$

where this β is the imaginary part of γ given by Equation (3.28), which is a function of ω, i.e. the velocity can be frequency dependent. This means that the different frequency components in the Fourier expansion of a pulse shape could travel at different velocities, so distorting the pulse shape by the time it reaches the end of the transmission line; this effect is called dispersion. However it is clear from (3.28) that the terms inside the bracket are very small, so that to a first approximation *the speed of a wave along a line with small losses is the same as that for an ideal line,* Equation (3.19). This is an important conclusion.

There is one further significant point. For transmission lines consisting of a pair of conductors of arbitrary cross-section (rather than a collection of discrete components) at frequencies where the internal inductance of the conductors can be ignored, the capacitance and external inductance are reciprocal functions of each other, *so that the product LC, hence the speed of a TEM wave, is independent of the line geometry;* see Appendix 8A in reference [13]. In fact, $LC = \mu_0\mu_r\varepsilon_0\varepsilon_r$, so the velocity is that of a TEM wave propagating through the medium separating the conductors, i.e.

$$V^* = c/\surd(\mu_r\varepsilon_r)$$

where c is the speed of light in free space.

3.5.5 Voltage and current reflection coefficients

For the line shown in Figure 3.11, the currents and voltages V_x and i_x given by (3.23) and (3.24) must satisfy the following boundary conditions (ignoring $e^{j\omega t}$ terms).

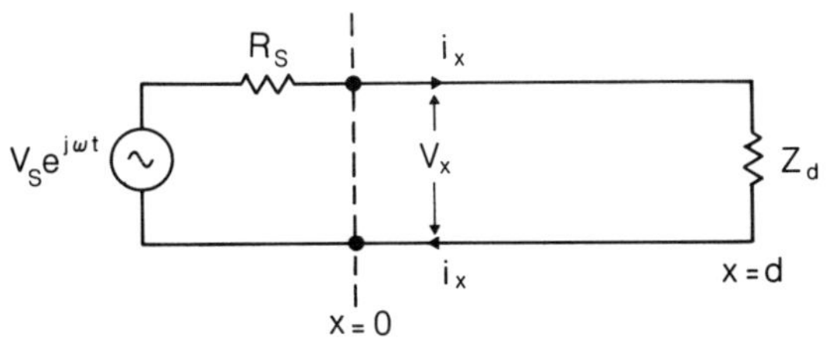

Fig. 3.11 Currents and voltages on a transmission line of length d, terminated by an impedance Z_d ohms

(1) At $x = 0$

$$V_A + V_B = V_S - i_0 R_s$$

(2) At $x = d$

$$Z_d = (V_{x=d})/(i_{x=d})$$

$$= Z_0 \frac{(V_A e^{-\gamma d} + V_B e^{\gamma d})}{(V_A e^{-\gamma d} - V_B e^{\gamma d})} \tag{3.29}$$

where Equations (3.23), (3.24) and (3.26) have been used. If the impedance at the start of the line is a pure resistance $R_s = Z_0$, then at $x = 0$:

$$V_A + V_B = V_S - Z_0(V_A - V_B)/Z_0$$

i.e.

$$V_A = V_S/2. \tag{3.30}$$

Then using (3.30) and manipulating the expression for Z_d/Z_0 obtained from (3.29) it can be readily shown that

$$V_B = (V_S/2)e^{-2\gamma d}(Z_d - Z_0)/(Z_1 + Z_0). \tag{3.31}$$

Thus the voltage at any point x for any transmission line of length d terminated with a pure resistance $R_s = Z_0$ at the start of the line and Z_d at the end can be found by using the general Equations (3.22) and (3.23).

This general solution applies to the situation where the voltage source driving the current is at the $x = 0$ end of the line. Section 3.7 will discuss solutions where the electric field driving the current is caused by the electric vector of an incident electromagnetic wave, giving essentially a voltage source distributed along the line. An expression for the voltage reflection coefficient ρ can be found as follows. At a distance x along the line, the impedance $Z_x = V_x/i_x$, so using (3.23) and (3.24) and (3.26)

$$Z_x = Z_0 \frac{(V_A e^{-\gamma x} + V_B e^{\gamma x})}{(V_A e^{-\gamma x} - V_B e^{\gamma x})}. \tag{3.32}$$

The voltage reflection coefficient ρ is the ratio of the reflected wave to the incident wave so that at the end of the line $x = d$ at the terminating impedance Z_d:

$$\rho = \frac{V_B e^{\gamma d}}{V_A e^{-\gamma d}}$$

$$= \frac{V_B e^{2\gamma d}}{V_A}$$

i.e.

$$V_A = V_B e^{2\gamma d}/\rho.$$

Substituting this in (3.32) at $x = d$ and cancelling the $e^{\gamma d}$ terms gives:

$$\frac{Z_d}{Z_0} = \frac{1/\rho + 1}{1/\rho - 1}$$

so

$$\rho = \frac{Z_d - Z_0}{Z_d + Z_0}. \tag{3.33}$$

Thus the voltage reflection coefficient depends only on the intrinsic impedance of the line and the terminating impedance. This is a specific example of the general point made in Section 2.5.3 that waves are reflected at changes of impedance, and the form of this equation should by now be very familiar. The current reflection coefficient ρ_i is given by

$$\rho_i = -\rho \tag{3.34}$$

(note the $-$ sign) and when the load is another transmission line (as it often is) the voltage transmission coefficient τ given by:

$$\tau = \rho + 1 \tag{3.35}$$

and the current transmission coefficient is

$$\tau_i = 1 + \rho_i. \tag{3.36}$$

The mathematical description and behaviour of plane waves in media and TEM waves on transmission lines are identical. When plane waves meet a medium change the reflection coefficient is calculated characteristic impedances of the media on either side of the boundary. For waves on transmission lines the characteristic impedance of the line and the termination impedance are used in Equation (3.33). The same analogy enables (3.5) to be adapted to calculate the effective impedance at any point a distance L from the end of a transmission line terminated with an impedance Z, i.e.

$$Z_L = Z_0 \frac{(Z + jZ_0 \tan(\beta L))}{(Z_0 + jZ \tan(\beta L))}. \tag{3.37}$$

3.5.6 Input impedance of short transmission lines

For a transmission line with length $L = \lambda/4$ terminated with a *short circuit* $Z = 0$ the input impedance $Z_{\lambda/4}$ is found from (3.37) to be

$$Z_{\lambda/4} = jZ_0 \tan(\pi/2) = \infty \tag{3.38a}$$

and for an *open circuit* termination, where $Z = \infty$,

$$Z_{\lambda/4} = -jZ_0 \cot(\pi/2) = 0. \tag{3.38b}$$

These notable properties of transmission lines have important applications. At frequencies of 100 MHz these effects would occur with line lengths of only 78 cm.

Finally, the input impedance Z_{in} of a very short line ($\beta L \ll 1$) with an open circuit termination is

$$\begin{aligned} Z_{in} &= -jZ_0 \cot(\beta L) \\ &\approx -jZ_0/\beta L. \end{aligned} \tag{3.39}$$

This result will be use in Section 5.4.3.2. Also, in Section 6.3.2 on earthing it is recommended that the earthing leads have a length $L \leqslant \lambda/20$. From (3.37) an earth line with length L and terminating impedance $Z \approx 0$ will have an input impedance $\approx jZ_0 \tan(\beta L)$, which becomes $\approx jZ_0 \tan(\pi/20)$ for $L = \lambda/20$. This is always small, so the danger of accidentally making a high impedance earth connection (see Equation (3.38*a*)) is guarded against.

3.5.7 Concluding remarks

A few final remarks about the properties of transmission lines and the role they play in EMC work are appropriate.

The most important parameter of transmission lines, the characteristic impedance, depends primarily on their geometrical features, i.e. the size and spacing between the conductors, together with the electrical properties of the medium between the conductors. This enables standard lines, often in the form of coaxial cables, to be manufactured with a given Z_0. A very commonly used coaxial cable has $Z_0 = 50\ \Omega$, and this is chosen because, for commonly used dielectrics, the attenuation constant of the line given by Equation (3.27) is a minimum; see Appendix E and reference [19]. However, as has been stressed, the characteristic impedance of a line is a function of frequency, due to the skin effect in the conductors and the frequency dependence of the dielectric between the conductors. Also, changing the geometry by crushing or severely bending cables will change the impedance. Even apparently innocent operations such as soldering together two lengths of cable will yield a short length of line with a different impedance, which may cause troublesome reflections.

An interesting role for transmission line theory depends on the fact that the equations describing the propagation of electromagnetic waves through an arbitrary medium (Equations (2.64)) and along a transmission line (Equations (3.20), (3.21)) are mathematically identical. There is a direct equivalence between the **E** and **H** fields in an electromagnetic wave and the voltage V and current I along a transmission line. This isomorphism leads to a powerful method of calculating field distributions in complex geometries by modelling the space as a regular network of orthogonal intersecting transmission lines composed of lumped circuit elements [10]. Boundary conditions such as electric and magnetic walls can be modelled with lines having short or open circuit terminations. The electrical properties at each intersection can be calculated, and the evolution of a pulse applied at any point on the network can be followed using matrix arithmetic, ideally suited to digital computers. At the end of the calculation the direct equivalence between the voltage and current in each line and the electric and magnetic fields at the same spatial location enables the field distributions to be predicted. Hoefer [10] suggests that the principal advantage of the method is that even complicated structures can be analysed easily using one, rather short, general purpose computer program, within which arbitrary problems can be defined by varying only the input parameters and not the code itself.

3.6 COUPLING BETWEEN TRANSMISSION LINES

Adjacent transmission lines may couple in some way, such that the currents and voltages on one line may cause currents to flow on the other. The effectiveness of the coupling will depend on the type of transmission line, with coupling between coaxial lines being very small compared with that between lines made from parallel pairs of conductors such as tracks on a circuit board. The general problem of two arbitrary coupled lines of different lengths, impedances and terminations is of great mathematical complexity and the study of such situations is not appropriate in this book. However some important lessons can be learned by examining some special cases. In particular one can start from the study of situations where coupling between two lines is made deliberately, in order to make directional couplers. This approach to the subject is the one used by Matick [13], who also discusses the general problem of two coupled lines. The mathematical way in which these examples are tackled depends on whether the lumped circuit approximation can be used. If the coupled lines under consideration are short compared to the wavelength, the coupling can be modelled by using lumped circuit elements distributed along the lines, whereas in the case of 'long' lines the full differential equations for coupled lines must be used. The simplest case to examine is when the coupling between the lines is inductive, such that the induced current in one line is a simple, but small, multiple of the driving current on the other.

3.6.1 Inductively coupled directional couplers

Consider the coupled lines shown in Figure 3.12, where the coupling is small enough not to perturb significantly the behaviour of line 1. The coupling

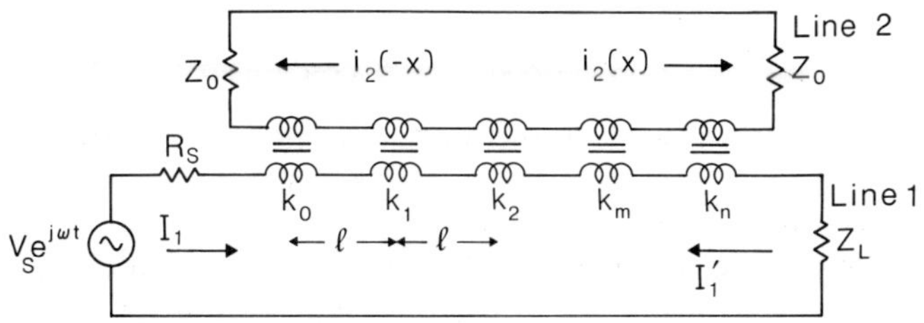

Fig. 3.12 Equivalent circuit model of two inductively coupled transmission lines. The coupling causes currents $i_2(x)$ and $i_2(-x)$ to appear on line 2, adapted from reference [13]. (Reproduced by permission of McGraw-Hill)

mechanism has been modelled as distributed lumped circuit element transformers. Line 1 is being driven by a voltage source $Ve^{j\omega t}$ and is terminated in some arbitrary impedance Z_L. The driven line (sometimes called the victim line), line 2, is terminated at both ends with its characteristic impedance Z_0. It will be noticed that the coupling transformers have been shown distributed in a regular manner; this is intentional as will become clear shortly. The magnitude of the coupling is given by a small coupling constant k, so that in general the current at a point x travelling in the $+x$ direction on line 2, $i_2(x)$ is related to that in line 1 by:

$$i_2(x) = ki_1(x)$$

and in the $-x$ direction by

$$i_2(-x) = ki_1(-x).$$

The problem is to find these currents on line 2 given that the currents on line 1 can be found using the general transmission line theory outlined in Section 3.5.3. The solution to this problem depends on three factors:

(*a*) the values of the coupling constants k_m;
(*b*) the fact that *each* wave with current $i_1(x)$ and $i_1(-x)$ on line 1 induces two waves travelling in opposite directions on line 2;
(*c*) the spatial arrangement of the points where the lines are coupled.

By an appropriate choice of the spacing of the coupling points, the waves travelling in the same direction on the second line can be made to cancel. Thus the second line can be the basis of a device, called a directional coupler, which couples to only one of the two waves travelling on line 1, so that it can be measured independently of the reflected wave on line 1. Matick shows [13] that in the special case where the coupling coefficient of the mth inductive coupler is given by

$$k_m = k_0 n!/m!(n-m)!$$

in a line with $n+1$ coupling points (k_0 is the coupling at the first, $m=0$, point) then the magnitudes of the currents $I_2(-x)$ and $I_2(x)$ flowing in the left and right hand terminating resistances on the victim line 2 are

$$I_2(+x) = 2^n k_0(I_1 + I_1' \cos^n(\beta 1)) \tag{3.40}$$

$$I_2(-x) = 2^n k_0(I_1' + I_1 \cos^n(\beta 1)) \tag{3.41}$$

where I_1' and I_1 are the reflected and incident currents in line 1, and $\beta = 2\pi/\lambda$.

By a suitable choice of ($\beta 1$), where 1 is the distance between the coupling points, the second term in each expression can be removed. When $\cos^n(\beta 1) \approx 0$ then $\beta 1$ is close to $\pi/2$, i.e. the spacing between each coupling point is $1 \approx \lambda/4$. This condition reduces the magnitude of each current, but it also means that the currents at each end of the driven line are caused either by the incident wave or by the reflected wave on the driving line, line 1. This is why a directional coupler works.

For a circuit designer, these two currents are unwanted noise, and this example shows that they can be reduced by terminating properly line 1 so that there is not reflected wave I_1', removing a contribution to the noise current produced in the victim line.

It goes without saying that if the coupling coefficients k_m are made as small as possible by shielding or careful design troublesome currents are suppressed!

3.6.2 Coupling over short line lengths

A second special case of coupled transmission lines which yields some useful lessons is the case when a line carrying a sinusoidal wave is coupled both capacitively and inductively to very short line, such that they can both be treated as being lossless lines; see Figure 3.13. The two lines shown in Figure 3.13 can have different characteristics, but note that line 1 is terminated with its characteristic impedance, so that there is no reflected current on it. An important point is that the capacitance C and inductance L per unit length which each line has when coupled is different from those values which can be attributed to each line when considered in isolation. Following Matick [13], the notation C_{11}, L_{11}, C_{22}, L_{22} refer to the capacitance and inductance per unit length on lines 1 and 2 in the coupled situation. The coupling between the lines is modelled as a capacitance C_{12} and a mutual inductance M_{12}. In many situations an upper limit for the mutual inductance between two inductors with self inductance L and L' is $\approx\surd(LL')$, thus the coupling between these lines will be written as

$$M_{12} = k\surd(L_{11}L_{22}) \qquad (3.42)$$

where k is a small frequency independent coupling constant. Similarly assume that

$$C_{12} = k\surd(C_{11}C_{22}). \qquad (3.43)$$

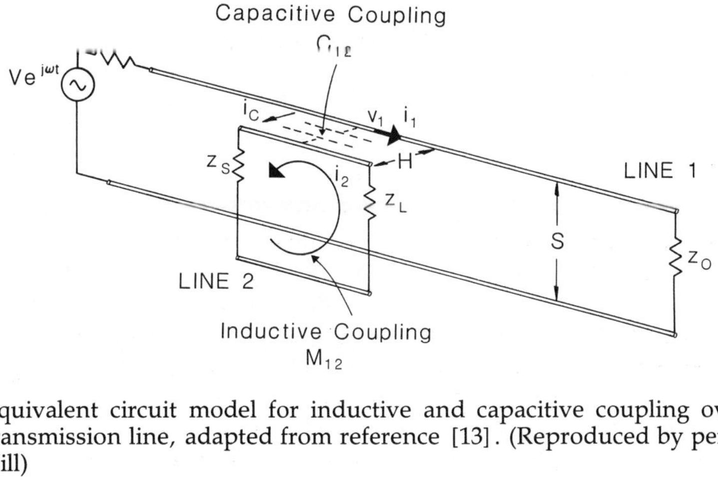

Fig. 3.13 Equivalent circuit model for inductive and capacitive coupling over a short length of transmission line, adapted from reference [13]. (Reproduced by permission of McGraw-Hill)

The inductive coupling between the lines will generate an inductive emf around the closed loop of line 2 given by

$$v_{\text{loop}} = M_{12} \frac{di_1}{dt} \tag{3.44}$$

where i_1 is the current on line 1; this emf will cause a current i_L to flow around the loop in such a way that the voltages generated across Z_s and Z_L are of opposite polarity. Clearly

$$v_{\text{loop}} = i_L(Z_s + Z_L). \tag{3.45}$$

On the other hand the capacitive coupling will produce a current i_c which will flow across the coupling capacitance C_{12} between the lines, given by

$$i_c = C_{12} \frac{dv_1}{dt}. \tag{3.46}$$

This current will divide and flow in the same direction through Z_s and Z_L producing voltages of the same polarity. As far as a designer is concerned the currents generated in line 2 by this coupling are unwanted noise and it is of interest to find if the noise voltages generated across the terminating loads Z_s and Z_L on the short driven line 2 can be reduced.

By circuit analysis it is clear that the voltage v_L generated across the load Z_L is the difference between the two voltages

$$v_L = Z_L\{i_c Z_s/(Z_s + Z_L) - v_{\text{loop}}/(Z_s + Z_L)\}$$

which by using (3.44) and (3.46) can be written

$$v_L = Z_L/(Z_s + Z_L)\{Z_s C_{12}\, dv_1/dt - M_{12}\, di_1/dt\}.$$

This is zero when

$$Z_s C_{12}\, dv_1/dt - M_{12}\, di_1/dt = 0$$

i.e. when

$$Z_s = (M_{12}/C_{12})\, di_1/dv_1$$

and since

$$di_1/dv_1 = \surd(C_{11}/L_{11})$$

$$Z_s = \surd(L_{22}/C_{22}) \tag{3.47}$$

i.e. when Z_s equals the characteristic impedance of line 2 *in the coupled situation* there is no noise voltage generated across the load Z_L! Furthermore, this result is independent of the magnitude of Z_L and frequency of the wave on line 1. All the voltage produced by the coupling between the lines appears across Z_s and none across Z_L. This fundamentally important result could be used to advantage to reduce noise, but the result derived here depends on the fact that line 1 is properly terminated, line 2 is short and that the coupling coefficient between the lines is small. Matick shows [13] that in fact the same result is

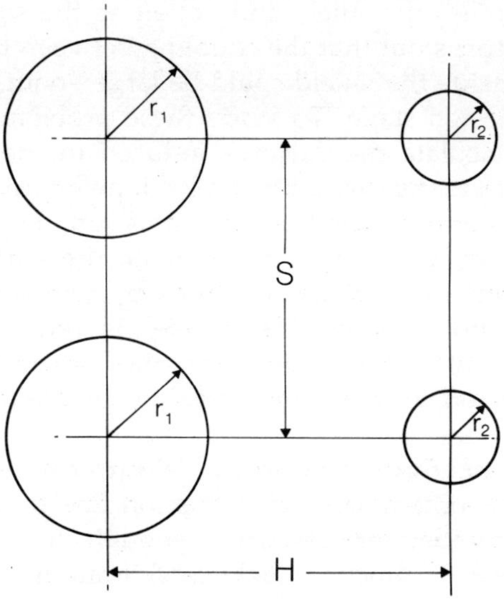

Fig. 3.14 Geometry of two parallel transmission lines each of width S separated by distance H made from wires of different radii

obtained for an arbitrary termination of line 1, Z_1, and large coupling between the lines if the condition

$$Z_L Z_1 = \sqrt{(L_{11}L_{22}/C_{11}C_{22})} \tag{3.48}$$

is satisfied, and this result is also independent of frequency, and so applies when pulses are propagating along line 1!

A final point is that the coupling coefficient k used in the above example can be derived from the relative geometry and wire radii of the two transmission lines. When a pair of lines made from wires with radii r_1 and r_2 metres separated by S metres are a distance H metres apart and carry TEM waves (see Figure 3.14) then the capacitive and inductive coupling coefficients are equal and given by ($r_1, r_2 \ll S, H$)

$$k = \frac{\log_e(\sqrt{(1+(S/H)^2)})}{\sqrt{(\log_e(S/r_1)\log_e(S/r_2))}}. \tag{3.49}$$

This shows that k can be made smaller by decreasing the radii of the wires and the ratio S/H. The reader is referred to Chapter 8 in [13] for further discussion of coupling coefficients between transmission lines.

3.7 COUPLING TO TRANSMISSION LINES

A major source of EMC problems can be traced to radiation being picked up by cables. To overcome these problems many such cables are protected by surrounding them with a shield of some sort, in the hope that the unwanted currents will flow through the shield and not along the signal wires. However in many situations it turns out that the coupling between the currents in shield and the conductors inside the shield could be large enough to warrant serious investigation at the design stage. To study these problems it is first necessary to understand and calculate the currents induced in the cable shield by the impinging radiation. In many cases this can be done by treating the cable shield in conjunction with its environment as a transmission line. This enables the full transmission line theory to be used to calculate the currents. An important reference for these sorts of studies is the book by Vance [14], which contains extensive examples and to which the reader is referred for assistance in performing realistic calculations. He list the advantages of using the transmission line method to treat cable coupling problems to be as follows.

(a) It turns a two- or three-dimensional electromagnetic boundary value problem into a one-dimensional transmission line problem.
(b) The solutions provided are accurate enough for making engineering decisions for almost all practical cable configurations.

This section will firstly discuss one method for finding the currents flowing in shields by modelling the situation as the interaction of a transmission line with incident electromagnetic radiation. A distributed source of emf is generated rather than a voltage source at one end, as is normal. Then, the mechanisms which enable current flowing on the shield to produce noise on the signal cables inside it will be outlined. In realistic situations it must be remembered that additional currents to those caused by the impinging radiation may be injected onto the shields from different sources, and their contribution to the total current flowing in the shield should not be overlooked.

3.7.1 Mathematical framework

Consider the current I and voltage V in a short length dz of transmission line produced by an impinging field $E_z e^{j\omega t}$. This can be considered to be a distributed voltage source on the line, and the equivalent circuit model is shown in Figure 3.15. Applying circuit analysis yields the following pair of equations

$$dV/dz = E_z - I(R + j\omega L) \tag{3.50}$$

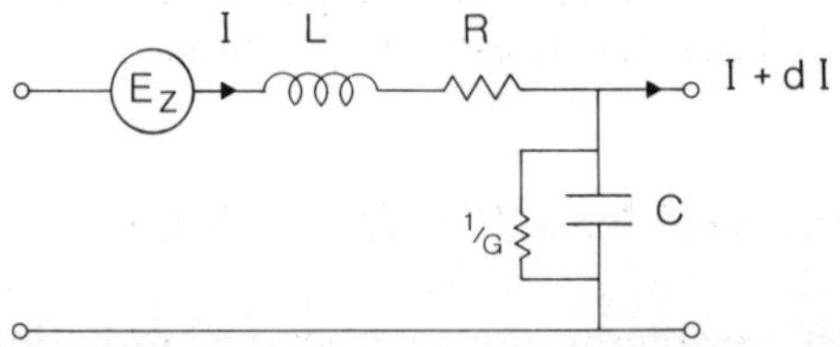

Fig. 3.15 Equivalent circuit model of a transmission line with a distributed voltage source

and

$$dI/dz = -V(G + j\omega C). \quad (3.51)$$

By differentiating and substituting these give:

$$d^2V/dz^2 = \gamma^2 V + dE_z/dz \quad (3.52)$$

and

$$d^2I/dz^2 = \gamma^2 I - E_z(G + j\omega C) \quad (3.53)$$

where $\gamma^2 = (R + j\omega L)(G + j\omega C)$. These are very similar to Equations (3.20) and (3.21) quoted for the transmission line driven by the voltage source at one end, except for the terms involving E_z. The quantity $Z = (R + j\omega L)\ \Omega^{-1}\,m^{-1}$ is the impedance per unit length and $Y = (G + j\omega C)\ \Omega^{-1}\,m^{-1}$ is the admittance per unit length. The characteristic impedance of the line is $Z_0 = \surd(Z/Y)\ \Omega$. For the transmission line shown in Figure 3.16 of length $L = (z_2 - z_1)$ m, terminated with impedances Z_1 and $Z_2\ \Omega$, Vance [14] quotes the following solution for the currents and voltages as functions of z, the distance along the wire from the Z_1 end:

$$I(z) = [K_1 + P(z)]e^{-\gamma z} + [K_2 + Q(z)]e^{\gamma z} \quad (3.54)$$

$$V(z) = Z_0\{[K_1 + P(z)]e^{-\gamma z} - [K_2 + Q(z)]e^{\gamma z}\} \quad (3.55)$$

where the terms in square brackets are currents, given by the following four equations:

$$P(z) = 1/(2Z_0)\int_{z_1}^{z} e^{\gamma\nu}E_z\, d\nu \quad (3.56)$$

$$Q(z) = 1/(2Z_0)\int_{z}^{z_2} e^{-\gamma\nu}E_z\, d\nu \quad (3.57)$$

where ν is a dummy variable (note the limits on these integrals) and

$$K_1 = \rho_1 e^{\gamma z}\,\frac{\rho_2 P(z_2)e^{-\gamma z_2} - Q(z_1)e^{\gamma z_2}}{e^{\gamma L} - \rho_1\rho_2 e^{-\gamma L}} \quad (3.58)$$

$$K_2 = \rho_2 e^{-\gamma z}\,\frac{\rho_1 Q(z_1)e^{-\gamma z_1} - P(z_2)e^{-\gamma z_1}}{e^{\gamma L} - \rho_1\rho_2 e^{-\gamma L}} \quad (3.59)$$

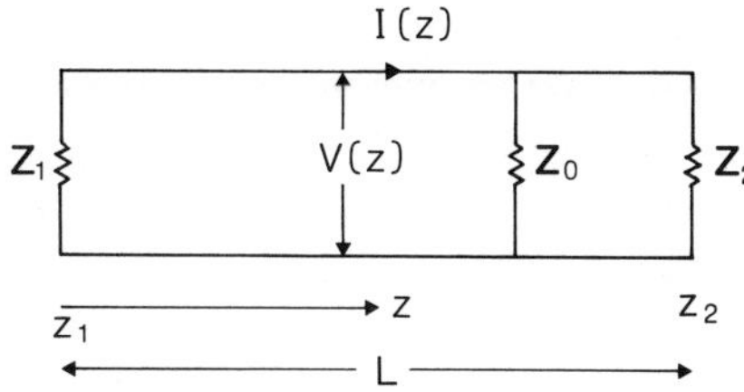

Fig. 3.16 Voltage $V(z)$ and current $I(z)$ produced on a transmission line of a characteristic impedance Z_0 terminated with impedances Z_1 and Z_2 when driven with a distributed voltage source

where ρ_1 and ρ_2 are the reflection coefficients at each end of the line; note that K_1 and K_2 are constants for a given line and given E_z, and if the line is properly terminated, so that ρ_1 and/or ρ_2 are zero, these expressions can be simplified considerably.

Written in this form the physical basis behind these solutions may not be readily apparent, but this mathematics has the following interpretation. At any point z on the transmission line the field is E_z and this will launch a current wave of magnitude $E_z/(2Z_0)$ in both directions along the line. Thus the total current along the line will be the integrals of all such current elements from each end of the line up to the point z, with proper account being taken of the effects of propagation delays, phase shifts, and absorption within the lines. All these effects are accounted for correctly in the expressions for the currents $P(z)$ and $Q(z)$, being contained in the γ term in Equations (3.56) and (3.57).

Careful examination of Figure 3.16 will reveal the similarity between it and situation that has been seen before; a region with characteristic impedance Z_0 and length L is situated between two different impedances Z_1 and Z_2. The analogy between this situation and what happens to waves inside a slab of material, discussed in Section 3.2.2 where re-reflection effects were seen to be important, is striking. If Equation (3.58) is multiplied above and below by $e^{-\gamma L}$ it becomes

$$K_1 = \rho_1 e^{\gamma z} \frac{(\rho_2 P(z_2)e^{-\gamma z_2} - Q(z_1)e^{\gamma z_2})e^{-\gamma L}}{1 - \rho_1\rho_2 e^{-2\gamma L}} \tag{3.60}$$

making the term $1 - \rho_1\rho_2 e^{-2\gamma L}$, which was associated with multiple reflections (see Equations (3.10)), apparent. These currents can be identified with multiple reflections up and down the line, and as such they can be eliminated entirely by properly terminating the line.

Equations (3.54)–(3.59) are the most general set which can be applied in a host of practical situations, some of which have been listed by Vance [14] as

(1) transmission lies formed by cables running above a conducting ground plane, where the source E_z is the resultant electric field at the line height;
(2) buried cables and conduits, where the source is the electric field in the soil;
(3) shielded cables where the source is the internal voltage per unit length *induced by currents in the shield;*
(4) vertical elements such as service conduits and feeds to antennas, where the source is the vertical electric field.

In all these cases it is necessary to find the electric field E_z due to the incident radiation at the position of the transmission line, with proper account being taken of any reflection and absorption effects.

3.7.2 Coupling of shield currents to the signal wires

Consider a current flowing on a cable shield which may be continuous, such as a metal tube, or have small holes or gaps, as in braid or wound conducting tape, so that it is 'leaky'. In all these situations the current on the shield may couple to the internal signal wires via the following mechanisms:

(a) The fields diffuse through the conductor forming the shield (see Section 2.5.9).
(b) Mutual inductance between the shield and the signal wire caused by the circumferential magnetic field generated by the shield currents penetrating through the holes.
(c) The electric field generated by the potential difference between the shield and its surrounding environment penetrates through the holes in the shield to the signal wires and induces charges on them.

Once again it must be stressed that the importance of each of these effects will vary with frequency. The voltage per unit length, dV/dz, induced in the signal wire and the current I_0 flowing on the shield are related by

$$E_z = dV/dz = I_0 Z_T \tag{3.61}$$

where Z_T *is a frequency dependent property of the shield* called the transfer impedance. Good shielding corresponds to having a small transfer impedance. It can be written as

$$Z_T = Z_d + j\omega M_{12} \tag{3.62}$$

where the term Z_d accounts for the diffusion of the fields through the shield and the term in M_{12}, the mutual inductance between the shield and the signal wire, is due to magnetic leakage through holes in the shield. These terms can be calculated in some situations, and several examples for different types of cylindrical shields are discussed by Vance [14]. Often the transfer impedance is measured and tabulated at different frequencies.

By way of illustration, consider the transfer impedance of a shield made from a continuous thin-walled cylinder of radius a, thickness $T \ll a$, and conductivity σ. This was derived by Schelkunoff [15] to be

$$Z_T = \frac{1}{2\pi a\sigma T}\,\frac{(1+j)T/\partial}{\sinh(1+j)T/\partial} \tag{3.63}$$

where ∂ is the skin depth and the tube radius a is assumed to be much smaller than the smallest wavelength being considered; there is no contribution from the term in M_{12} in Equation (3.62) because there are no holes in the cylinder, thus $Z_T = Z_d$. At low frequencies ($T \ll \partial$) (3.63) becomes $Z_T = 1/(2\pi a\sigma T)$, which is just the ohmic resistance R_0 of one metre of the cylinder. In Figure 3.17 the ratio of $|Z_T|/R_0$ is plotted against T/∂, and shows that for a solid walled tube at high frequencies where $T > \partial$ the transfer impedance becomes smaller than the ohmic resistance, so the shielding effectiveness increases. Also shown is the schematic behaviour of a braided shield with holes. Its shielding effectiveness decreases with frequency due to the increasing importance of the $j\omega M_{12}$ term in (3.62). Further discussion of this topic can be seen in Section 4.6.4.

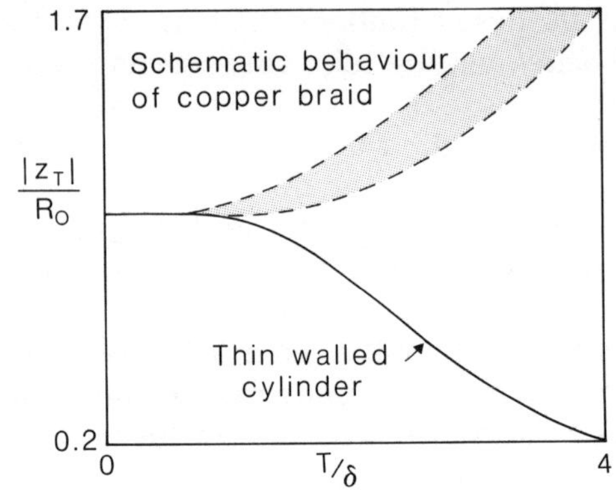

Fig. 3.17 Normalised transfer impedance as a function of frequency for a thin-walled tube. For comparison the shaded area shows, in schematic form, the general behaviour of the transfer impedance of a shield with holes in it, i.e. braided copper

Using $\partial = 1/\surd(\omega\sigma\mu_0/2)$ and recalling that the diffusion time constant for a metal of thickness T is $\tau_s = \sigma\mu_0 T^2$ (see Equation (2.84)), a little algebra will show that $(1+\mathrm{j})T/\partial = \surd(\mathrm{j}\omega\tau_s)$, so substituting in (3.63) yields

$$Z_T = \frac{R_0\surd(\mathrm{j}\omega\tau_s)}{\sinh\surd(\mathrm{j}\omega\tau_s)}. \tag{3.64}$$

This illustrates that in the case of a solid cylinder the transfer impedance Z_T is associated with diffusion processes.

To summarise, the effect of external electromagnetic fields coupling to the signal wires inside shielded cables is calculated in three stages.

(**a**) The shield of the cable and the surrounding environment is considered to be a transmission line, and the currents on the shield are calculated with the theory outlined in Section 3.7.1, using the appropriate models to obtain estimates for E_z, Z_0, γ etc.

(*b*) The field E_z generated inside the cable by these currents is calculated using the calculated or measured transfer impedance.

(*c*) This internal field is used as the driving field E_z for a further application of the theory of a source distributed along a transmission line to calculate the noise currents flowing on the signal wires, this time using the cables' own characteristic values for Z_0 and γ.

Further discussion of these topics can be found in Section 4.6.3, where the transfer admittance, which describes the coupling of external electric fields to the signal wires through holes in the shield, will be introduced.

3.8 WAVEGUIDES AND CAVITIES

3.8.1 Waveguides

A waveguide is a long hollow structure made of conducting materials within which electromagnetic waves propagate and follow the meanderings of the structure (other waveguides such as single wires or surfaces are not discussed in this section). They can be considered to be the limiting case of a coaxial transmission line where the central conductor has been reduced to zero thickness. The cross sectional shape, as well as the electrical properties of the walls and the material filling the guide, influence such parameters as the velocity of propagation, the rate at which energy is absorbed from the wave by currents created in the walls and, very importantly, the cut-off frequency. This cut-off frequency is the lower limit of frequency below which wave propagation is prevented, or, put the other way, it defines the *maximum* wavelength which can be transported by the guide. A mnemonic to remember which way round this rule applies is to recall that light can be seen down a hollow tube so there cannot be a high frequency cut-off! All sorts of everyday items made of metal, such as water pipes, holes drilled through metals, conduits for wire runs, service and heating ducts, etc. could behave as waveguides *unintentionally* so it is necessary to be aware of their properties even if they are not used as active components in the system being studied. A proper understanding of the Crawford cell, described in Section 5.5.2, also requires familiarity with waveguide concepts.

There are two properties of electromagnetic waves which are fundamental to understanding waveguides. The first is that they are reflected with little energy loss from metals. The second is that electromagnetic fields are linear, so that any wave can be considered to be the resultant of an arbitrary number of different plane waves which propagate independently of each other. A single wave propagating along a waveguide can be considered to be the sum of two independent waves, each reflected from the walls in such a way that they recombine in phase, recreating their starting condition, so that the cycle can be repeated. Clearly to do this there must be some restrictions on the allowable combinations of the two waves, so that there are only certain modes and frequencies for which these rather severe restrictions will actually work. If the waveguide is too small compared with the wavelengths of the waves attempting to propagate, such that they do not have the room to complete their cycle of reflection and recombination to get back into the same state as when they started and repeat the cycle all over again, then there will be a cut off maximum wavelength above which waves just cannot propagate; Feynman [4] discusses these ideas and uses them to show how the cut-off wavelength is related to the waveguide dimensions.

This book will not consider the mathematics of waveguides in the same detail as that devoted to transmission lines. However, it is worth writing down and discussing some of the equations given by theory with a view to obtaining a feeling for orders of magnitude and the importance of different parameters. Liao [11] discusses the mathematical details of rectangular and circular waveguides with great clarity.

3.8.2 Cut-off frequency and attenuation constant

In both rectangular and circular waveguides the two principal modes of propagation of the waves are called TM and TE modes (transverse magnetic/electric), where the M and E refer to the magnetic or electric component which is *normal to the axis* of the waveguide. Thus in the TM mode there is an electric component along the axis whereas in the TE mode it is the magnetic component which is so orientated. These waves have a very different structure from free space TEM waves and arise because of the different boundary conditions which have to satisfied at the surrounding conductor. Allowable waves are characterised by pairs of integers m, n which appear in the equations for the different components. For example, in rectangular guides with cross sectional dimensions a and b, terms in $\cos(m\pi x/a)$ and $\sin(n\pi y/b)$ arise in the expression for the field components. Each allowable mode is written with the notation TM_{mn}, TM_{nm}, TE_{mn} or TE_{nm} mode and in general each mode has a different cut-off frequency. In rectangular waveguides the cut-off frequency is given by

$$f_c = \frac{1}{2\sqrt{(\mu\varepsilon)}} \sqrt{\left(\frac{m^2}{a^2} + \frac{n^2}{b^2}\right)} \tag{3.65}$$

where m and n are integers. The formula for f_c in circular waveguides is a little more complicated and involves the arguments of the Bessel function (but also depend on the choice of a pair of integers), but in has the same dependence on $1/\sqrt{(\mu\varepsilon)}$. There are tables of cut-off frequencies for typical rectangular and circular guides in reference [11], and a small selection are shown in Table 3.3 for the TE_{10} (lowest frequency) mode, called the dominant mode, in an air-filled rectangular guide.

The fact that the dimensions of these standard waveguides are in the ratio 2:1 is to ensure that only mode, the dominant mode, propagates when it is used to transport energy [11].

Table 3.3. A selection of lowest cut-off frequencies in standard rectangular waveguides [11]

Width a (cm)	Height b (cm)	f_c (GHz)
58.24	29.21	0.257
29.21	14.605	0.514
12.954	6.477	1.158
3.485	1.58	4.304
1.067	0.432	14.058
0.109	0.056	137.615

3.8.3 Shielding effectiveness of apertures

It must not be thought that if the frequency of a wave at the start of a waveguide is less than the critical frequency f_c then the electromagnetic fields associated with it are prevented absolutely from penetrating into the guide. What actually happens [11] is that when $f < f_c$ the initial field amplitudes are attenuated in the guide by a factor $e^{-\alpha z}$ where

$$\alpha = \omega\sqrt{(\mu\varepsilon)}\sqrt{\{(f_c/f)^2 - 1\}} \tag{3.66}$$

and z is the distance along the waveguide. From (3.65), the TE_{10} mode cut-off frequency in an air-filled rectangular guide where $a > b$ is

$$f_c = c/2a \tag{3.67}$$

where $c = 1/\sqrt{(\mu\varepsilon)}$. For frequencies well below the cut-off frequency Equation (3.66) can be approximated by

$$\alpha \approx \omega\sqrt{(\mu\varepsilon)}(f_c/f) \tag{3.68}$$

so that using the definitions introduced in Section 3.2.3 the shielding effectiveness of a narrow slot of length a m and depth z m which is acting as a waveguide below the cut-off frequency can be written as

$$\begin{aligned} SE &\approx 8.686\,|\alpha z|\ \text{dB} \\ &\approx 8.868\omega\sqrt{\mu\varepsilon}(f_c/f)z\ \text{dB} \\ &\approx 8.868\pi z/a\ \text{dB} \\ &\approx 27.2z/a\ \text{dB}. \end{aligned} \tag{3.69}$$

This is a remarkable little formula as it is independent of frequency, and shows that a rectangular hole in a metal shield which acts as a waveguide (i.e. where $z > a$) can allow fields below the cut-off frequency to penetrate into the shielded region. For the dominant TE_{11} mode propagating in a hole acting as a circular waveguide of radius r metres, the cut-off frequency of an air-filled waveguide is given by [11] $f_c = 1.841c/2\pi r$ so that using (3.68) for the attenuation at a distance z m within the circular hole for frequencies below f_c gives

$$SE \approx 8.686(1.841z/r)\ \text{dB} \approx 15.991z/r\ \text{dB}. \tag{3.70}$$

Ott [16] quotes this as $SE \approx 32z/2r$ dB, and these formulae have been derived by Quine [17]. Equation (3.70) is also independent of frequency below the cut-off frequency, but again its use is only justified when the conductor is thick enough to allow $z \geqslant 2r$, so that the hole can act as a waveguide. The relevance of these two formulae (3.69) and (3.70) for the manufacture of shields and screened rooms is discussed in Section 4.6.1. In cases where the depth of the slot or hole is much less than its width, the aperture should be treated as a slot antenna rather than a waveguide and these will be discussed in Chapter 5.

A final point to be noted is that, as can be seen from Equation (3.65), the cut-off frequency depends on $1/\sqrt{\varepsilon_r}$. This means that if the waveguide is filled with a dielectric, the cut-off frequency is reduced by $1/\sqrt{\varepsilon_r}$ compared to the case where it is air-filled. This follows from the fact the wavelength in the dielectric is smaller by the same factor $1/\sqrt{\varepsilon_r}$ compared with that in free space. In some situations, such as when a dielectric rod (e.g. an insulating spindle to control some component) is inserted into a hole drilled through a metal shield, the corresponding reduction in the cut-off frequency for that sized hole acting as a waveguide may have EMC consequences that should be guarded against by suitable construction techniques; see Section 4.6.

3.8.4 Cavities and cavity resonance

Cavities of any shape made from conducting materials can act as containers for electromagnetic waves, confining them until their energy is dissipated by currents flowing in the resistive walls of the container. Even if a structure has holes in it, like the passenger compartment of an automobile, at certain frequencies there will be resonant behaviour that will cause field amplitudes to become very large in some locations within the cavity.

One way to think about the behaviour of a cavity is to imagine a waveguide with the open ends being closed by metal plates. These will act as good reflectors, resulting in the allowed modes within the waveguide being reflected back and forth within the enclosure. Feynman [4] has given a different picture which illuminates the relationship between the fields around a resonant circuit and those within a cavity. In Figure 3.18(*a*) is shown an LC circuit making up a resonant circuit. This will have a natural frequency $f = 1/\sqrt{(LC)}$ Hz. If C is made smaller by increasing the separation of the plates and L also reduced by decreasing the number of turns, see Figure 3.18(*b*), the resonant frequency increases. The limit of this modification is reached when the inductance becomes a single strand of wire. However a further reduction in the inductance can be achieved by putting two inductances L in parallel, see Figure 3.18(*c*), for then the combined inductive impedance L' is given by $1/\mathrm{j}\omega L' = 1/\mathrm{j}\omega L + 1/\mathrm{j}\omega L$, i.e. $L' = L/2$ which increases the resonant frequency still further. This process can be continued until the additional parallel single loops forms a complete wall, as in Figure 3.18(*d*). The resonant circuit has been transformed into a resonant cavity. In fact, for a rectangular cavity with dimensions, *a*, *b*, *d* metres containing a lossless dielectric the resonant frequencies for the TE and TM

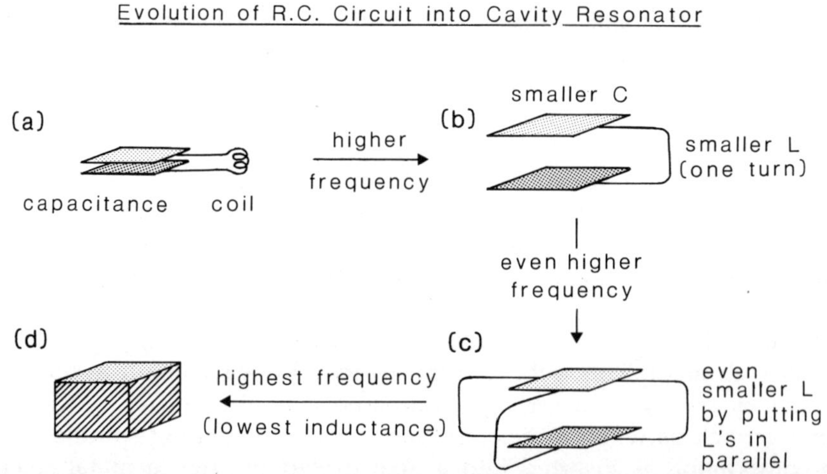

Fig. 3.18 Evolution of a resonant circuit into a resonant cavity. (Adapted from Richard P. Feynman, *The Feynman Lectures on Physics*, volume IIc (1964), California Institute of Technology. Reprinted with permission of Addison-Wesley Publishing Company, Inc.)

modes are [5]

$$f_r = \frac{1}{2\sqrt{(\mu\varepsilon)}} \sqrt{\{(m/a)^2 + (n/b)^2 + (p/d)^2\}} \tag{3.71}$$

where m, n and p are positive integers, one of which can be zero (formula number 5-7-4 in reference [11] for the resonant frequencies is misprinted). At these frequencies the boundary conditions are that the tangential electric fields are zero at all six walls, a fact which, perhaps, could be used when positioning and orienting sensitive components within a screened enclosure.

By way of illustration, Table 3.4 contains resonant frequencies between 50 and 200 MHz for an air-filled room of $2.3 \times 2.3 \times 4.6$ m, calculated with Equation (3.71). Figure 3.19 shows the measured frequency response of a screened room of about this size excited by a small magnetic loop [8]. There is very nice overall agreement. However some differences can be seen, where frequencies which should be degenerate (the same) are very close. This can be explained by the facts that the real room had structural ridges inside and a wooden floor, which acts as a dielectric slab, neither of which were included in the calculation [8].

Table 3.4. Some frequencies of resonant modes in an air-filled cavity with dimensions $a \times b \times d = 2.3 \times 2.3 \times 4.6$ m, calculated using Equation (3.71). These can be compared with those shown in Figure 3.19

m	n	p	F(MHz)
1	0	1	72.9
0	1	1	72.9
1	1	0	92.2
0	1	2	92.2
1	1	1	97.8
1	1	2	113.0
0	1	3	117.6
1	1	3	134.4
0	2	1	134.4
2	1	0	145.8
0	1	4	145.8
0	2	2	145.8
1	2	1	149.4
1	1	4	159.7
1	2	2	159.7
0	2	3	163.0
1	2	3	175.6
0	1	5	175.6
0	2	4	184.5
2	2	0	184.5
1	1	5	187.3
2	2	1	187.3
2	2	2	195.7
1	2	4	195.7
0	3	1	198.4

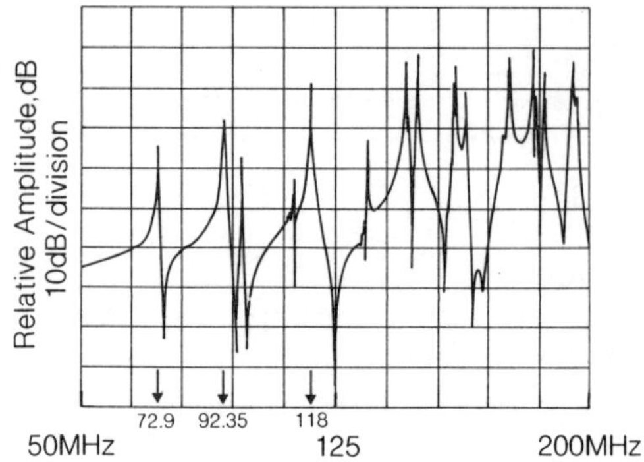

Fig. 3.19 Frequency response of a room excited by a small magnetic loop antenna, adapted from reference [8]

As a very first step in the study of a potential EMC problems within cavities it may be worth tabulating the resonant frequencies to see if they are near frequencies, or harmonics, of active circuit elements, such as clock pulses, used in electronics situated within the cavity.

The danger of cavity resonances in screened rooms is that they might invalidate EMC measurements by generating large fields in the vicinity of probes and antennas being used in a test. Section 3.4 discussed how absorbers can be used effectively to dampen such resonances. As well as this problem, metal shields around equipment will form resonant cavities at some frequencies, and any holes or apertures in the shield could act as transmission lines or waveguides, allowing extraneous radiation into the cavity, which could feed resonances. This is one reason behind the warning in Chapter 1 that the correct treatment of connections through shields is of paramount importance, and suitable methods of making such connections will be reviewed in Chapter 6.

A number of ways have been used to overcome cavity resonances, for example the Crawford Cell, GTEM cells and Reverberating Chambers; see Section 5.5.

Finally, recent work suggests that by using asymmetric, non-rectangular, screened rooms a significant economic improvement can be made to chamber performance [18].

REFERENCES

[1] Simon, B. and Wilson, R. M. (1988) Supercalculators on the PC *Notices of the American Mathematical Society* **35** 978–1001.

[2] *MATHCAD*, Mathsoft Inc., 201 Broadway, Cambridge, Massachusetts, 02139, USA.

[3] Schultz, R. B., Plantz, V. C. and Bush, D. R. (1988) Shielding Theory and Practice *IEEE Trans. EMC* **EMC-30** 187–201.

[4] Feynman, R. P. (1964) *Lectures on Physics* vol. 2, Addison-Wesley Publishing Corp. Inc., Reading, MA.
[5] Liao, S. Y. (1985) *Microwave Devices and Circuits* 2nd edition, Englewood Cliffs, NJ, Prentice Hall. See also Klein, C. A. (1990) 'Microwave Shielding Effectiveness of EC-Coated Dielectric Slabs' *IEEE Trans. Microwave Theory and Techniques* **MTT-38** 321–324.
[6] Lorrain, P., Corson, D. P. and Lorrain, F. (1988) *Electromagnetic Fields and Waves* 3rd edition, W. H. Freeman and Co., New York.
[7] Emerson, W. H. (1973) Electromagnetic Wave Absorbers and Anechoic Chambers through the Years *IEEE Trans. Antennas and Propagation,* **AP21** 484–490.
[8] Dawson, L. and Marvin, A. C. (1988) Alternative methods of Damping Resonances in a Screened Room in the frequency range of 30 to 200 MHz *6th Int. Conf. on EMC, University of York,* IERE Conference Publication No. 81, 217–223.
[9] Adam, J. A. (1988) How to design 'invisible' aircraft *IEEE Spectrum* **25** 26–31.
[10] Hoefer, W. J. R. (1985) The Transmission-Line Matrix Method-Theory and Applications *IEEE Trans. Microwave Theory and Techniques* **MTT-33** 882–893.
[11] Liao, S. Y. (1988) *Engineering Applications of Electromagnetic Theory* West Publishing Co., St. Paul.
[12] Johnk, C. T. A. (1988) *Engineering Electromagnetic Fields and Waves* 2nd edition, Wiley, New York.
[13] Matick, R. E. (1969) *Transmission Lines for Digital and Communication Networks* McGraw-Hill, New York.
[14] Vance, E. F. (1978) *Coupling to Shielded Cables* Wiley, New York. See also Smith, A. A. (1977) *Coupling of External Electromagnetic Fields to Transmission Lines* Wiley, New York.
[15] Schelkunoff, S. A. (1938) *Bell System Techn. J.* **17** 17–48.
[16] Ott, H. W. (1988) *Noise Reduction Techniques in Electronic Systems* 2nd edition, Wiley, New York.
[17] Quine J. P. (1957) Theoretical Formulas for Calculating the Shielding Effectiveness of Perforated Sheets and Wire Mesh Screens *Proc. 3rd Conf. on Radio Interference Reduction,* Armour Research Foundation pp 315–329.
[18] Edwards, D. J. and Burbridge, R. F. (1990) An Experimental Investigation into the Performance of Asymmetric Screened Rooms for Electromagnetic Measurements *7th Int. Conf. on Electromagnetic Compatibility, York, August 1990* IEE Conference Publication 326 114–118.
[19] Jackson, W. (1945) *High Frequency Transmission Lines* Wiley, New York.

4 Shielding Theory and Practice

4.1 BASIC PHYSICAL IDEAS

The shielding of a source of electrical disturbance from a potential receiver is a powerful technique in EMC control. In this context a shield is any enclosure (partial or complete) that is used to lower the electromagnetic field inside (or outside) the space enclosed. As such, shields can be as large and rigid as whole steel lined buildings (to test military equipment, to shield sensitive computer systems) or as small and flexible as the braids on coaxial cables. The understanding of the principles involved however is not entirely obvious except in a few idealised cases where the use of a particular approximation is justified. It is always possible in principle to solve Maxwell's equations with the appropriate boundary conditions to obtain an accurate solution. This procedure can seldom be applied in the real world. Analytic solutions can only be obtained for a few idealised shapes (spherical, cylindrical shells, etc.) composed of known homogeneous materials excited by a well defined field. The results of these rigorous calculations will be presented in this chapter wherever necessary so as to highlight the accuracy or otherwise of the approximations to be discussed. The range of shapes can be extended using numerical techniques but the solution for any engineering shield composed of real materials, its associated jointing and excited by an arbitrary source having non-ideal characteristics is currently prohibitively expensive in computer requirements. It is for this reason that the various approximations are so important both in obtaining an estimate of a particular shields effectiveness and in building up a physical picture of how the shield functions. This physical picture enables a designer to deviate more safely from published design tables for homogeneous shields [1] when involved in the development of a specific practical shield.

It is possible to identify three approximations which give good results and a simple physical picture when used in their respective areas of applicability. These are as follows.

(1) The 'ducting' model which applies for static (or quasi-static) electric and magnetic fields.
(2) The 'equivalent circuit' model which applies to situations where the shield dimensions are considerably smaller than the wavelength of the impinging fields. This model clearly shows the difference between electric and magnetic field screening, highlighting the importance of the induced current

flow in the total shield structure. Both high and low frequency estimates can be obtained, including skin depth effects.

(3) The 'plane wave or transmission line' model which applies for shields considerably larger than the wavelength of the impinging fields and when these fields can be treated as plane waves. This implies that the shield should be remote from the source. Use is made of the transmission, reflection and skin depth equations developed for plane waves in the last two chapters.

Many design formulae are developments of (3) allowing for the effects of source type and proximity of the shield [1–4]. These latter formulae will be discussed towards the end of this chapter after the three well defined approximations have been considered. The performance of a shield is usually discussed in terms of a figure of merit called the shielding effectiveness SE(f) which is a function of frequency f. For low frequency or quasti-static systems the electric and magnetic can be considered separately.

Thus the shielding effectiveness is defined in decibels as

$$\mathrm{SE}(f) = 20\ \log_{10} [H1/H2]$$

or

$$\mathrm{SE}(f) = 20\ \log_{10} [E1/E2]$$

where

$H1$ = the magnetic field in the absence of the shield,
$H2$ = the magnetic field with the shield,
$E1$ = the electric field in the absence of the shield,
$E2$ = the electric field with the shield.

At higher frequencies E and H cannot be considered separately so that SE is defined in terms of power flow,

$$\mathrm{SE}(f) = 10\ \log_{10} [P1/P2]$$

where

$P1$ = the power flow in the absence of the shield,
$P2$ = the power flow with the shield.

In every case above the source is in one fixed position and the various parameters ($H1$, $H2$, $E1$, $E2$, $P1$, $P2$) refer to values measurable at another fixed location. Other forms of definition are possible [5, 6] but they will only be introduced as and when necessary.

Before starting to discuss the various approximate theories of shielding it is important to remember that most practical shields are not limited by the conductivity, permeability or thickness of the material but by defects (joints, apertures, etc.) in the overall construction. The effects of these will be considered at the end of this chapter.

4.2 STATIC (OR QUASI-STATIC) FIELD SHIELDING

Solving Maxwell's equations for the static case reduces to using Laplace's equation with the appropriate boundary conditions to determine the scalar potential (V or V_M). In the case of magnetostatics, in order to define a scalar magnetic potential V_M, it is necessary for the region within the boundaries to contain no current. If the problem can be made to fit a number of simple geometries an exact solution may be possible using the separation of variables technique [7]. In order to understand the basic principles however, consideration of an infinite cylindrical shell in a previously uniform transverse field is all that is required. Some other useful formulae will be quoted without proof.

4.2.1 Magnetostatic shielding

The infinite cylinder of magnetic material shown in Figure 4.1 has an outer radius b and an inner radius a. It has a relative permeability μ and is orientated at right angles to a previously uniform magnetic field H_0. The problem is to calculate the field within the shell. Such a problem requires the cylindrical form of the Laplace equation.

$$\nabla^2 V_m = 0 \qquad \text{or} \qquad \frac{1}{r}\frac{\partial}{\partial r}\left(\frac{r\partial V_m}{\partial r}\right) + \frac{1}{r^2}\frac{\partial^2 V_m}{\partial \theta^2} + \frac{\partial^2 V_m}{\partial z^2} = 0. \tag{4.1}$$

Since there is no z variation this reduces to:

$$\frac{\partial}{\partial r}\left(r\frac{\partial V_m}{\partial r}\right) = \frac{1}{r}\frac{\partial^2 V_m}{\partial \theta^2} = 0. \tag{4.2}$$

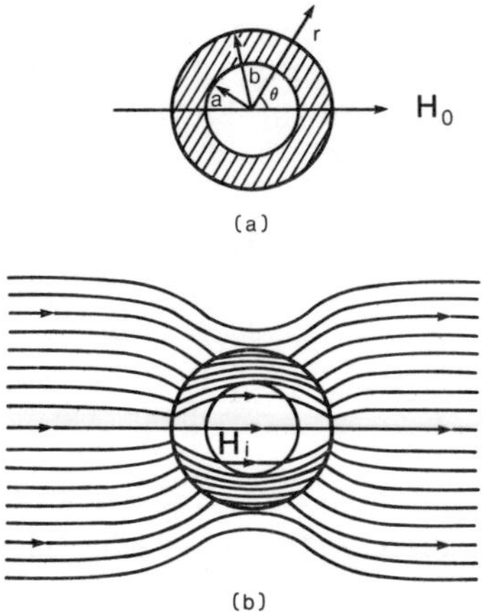

Fig. 4.1 Definition of the coordinates and an illustration of the shielding effect of a magnetic cylinder in a previously uniform static magnetic field, adapted from reference [7]. The ducting of the lines of **H** through the walls of the magnetic shield is clearly shown. (Reproduced by permission of McGraw-Hill)

For the problem concerned the boundary conditions are

(1) $\mathbf{H}_0$ is uniform remote from the cylinder;
(2) V_m is constant across the interface at $r = b$ and $r = a$;
(3) the components of $\mathbf{B}_i$ and $\mathbf{B}_0$ normal to the surface of the cylinder are continuous across the interfaces at $r = b$ and $r = a$.

Three solutions for V are required corresponding to the three regions of the problem, which are V_1 for $r > b$ (outside the cylinder), V_2 for $b > r > a$ (in the material of the cylinder) and V_3 for $r < a$ (in the cavity inside the cylinder) since

$$\mathbf{H} = -\partial V_m / \partial r \tag{4.3}$$

so that

$$\mathbf{B} = -\mu_r \mu_0 \, \partial V_m / \partial r. \tag{4.4}$$

Using the separation of variables technique, where the solution for V_m is assumed to have the general form $V_m = R(r)F(\theta)$ it can be shown that the most general form for V_m in cylindrical coordinates is given by [7]:

$$V_m = A_0 + B_0 \log_e r + \sum_{n=1}^{\infty} \times \{r^n (A'_n \sin(n\theta) + A_n \cos(n\theta)] + r^{-n} [B'_n \sin(n\theta) + B_n \cos(n\theta)]\} \tag{4.5}$$

where $A_0 \ldots A_n$ and $B_0 \ldots B_n$ are arbitrary constants. Applying the boundary conditions given earlier and with some manipulation [7] it can be shown that:

$$H_i = \frac{4H_0 \mu_r p}{(\mu_r^2 + 1)(p - 1) + 2\mu_r(p + 1)} \tag{4.6}$$

where $p = b^2/a^2$ and H_i is the modulus of a uniform magnetic field within the bore of the cylinder. If $\mu_r \gg 1$ this can be approximated to

$$H_i = \frac{4H_0 p}{\mu_r(p - 1) + 2(p + 1)} \tag{4.7}$$

i.e.

$$\frac{H_0}{H_i} = \frac{\mu_r(1 - 1/p) + 2(1 + 1/p)}{4}. \tag{4.8}$$

If the thickness of the shell $t = (b - a)$ and $R = (b + a)/2$ is the mean radius then for a large radius, thin-walled, cylinder (such that $a^2 \approx b^2 \approx R^2$) this can be written:

$$\frac{H_0}{H_i} \approx 1 + \frac{\mu_r t}{2R}. \tag{4.9}$$

This is the result obtained by Thomas [8] who also quotes a similar formula for a spherical shell:

$$\frac{H_0}{H_i} \approx 1 + \frac{2\mu_r t}{3R}. \tag{4.10}$$

From these formulae it is possible to deduce some important properties of magnetostatic shells.

(*a*) Equation (4.6) indicates that when $\mu_r = 1$ no shielding occurs because $H_i = H_0$. Thus a conducting but non-magnetic shield such as copper has no effect on a static magnetic field.

(*b*) The shielding action occurs because flux lines have been diverted or 'ducted' into the magnetic material as shown in Figure 4.1(*b*).

(*c*) As the thickness t of the shield increases so does the general formula for which SE is obtained by using Equation (4.6) to give

$$\mathrm{SE} = 20 \log_{10}(H_0/H_i) = 20 \log_{10}\left[\frac{(\mu_r^2+1)(p-1)+2\mu_r(p+1)}{4\mu_r p}\right] \text{ dB} \tag{4.11}$$

which for a large radius thin-walled cylinder becomes (using Equation (4.9))

$$\mathrm{SE} \approx 20 \log_{10}\left[1 + \frac{\mu_r t}{2R}\right] \text{ dB}. \tag{4.12}$$

Nesting a series of shields one within another (provided that the spacing between the shields is adequate) can produce a much better solution [8]. This is because nesting produces a multiplicative effect as opposed to an additive effect of just increasing the thickness.

(*d*) The ducted flux increases the flux density within the shield walls leading to the danger of saturation with a consequent reduction in μ_r and loss of shielding effectiveness. The outer shield of a nested structure has therefore to be made thick enough to stop this happening and so this tends to limit the range of $\mathbf{H}_0$ and size of structures that can be shielded [8].

All magnetic material exhibits hysterisis and a tendency for the material properties to vary with their exact treatment prior to use. Some typical values of μ_r are given in Table 4.1.

Details of manufacturing limits are given in reference [8] by Thomas. For low frequency fields the shielding effectiveness is altered from that estimated above

Table 4.1. Some properties of magnetic materials

	Saturation flux ($\mathrm{T\,m^{-2}}$)	Initial value of relative permeability μ_r	Conductivity $\sigma(\mathrm{S\,m^{-1}})$
Steel	2.1	500–1500	1.8×10^6
78 Permalloy	1.07	3000	6.25×10^6
Maganese–zinc ferrite		800–10 000	1–0.05

because

(1) Eddy currents start to flow in the shield structure. This effect cancels internal fields far more effectively than ducting and depends on the conductivity of the material.
(2) The effective value of μ_r drops, see reference [1, 8].

However, ducting shield designs are important for static and VLF magnetic field screens. The effect of the eddy currents will be considered in Section 4.3.

4.2.2 Superconducting shields

A superconductor exposed to a static magnetic field has currents flowing in it that exactly cancel the magnetic field inside, so that the field is expelled; this property is called the Meissner effect [38]. However this effect vanishes if the field is too high and exceeds a critical value, and the temperature of the material must remain below a critical temperature, T_c, which is normally below about 15 K. This property has been used for magnetic shielding in some situations, such as transporting a beam of charged particles through a region of magnetic field, by sending it along the bore of a superconducting tube running through the field. Pure niobium is often used as the superconductor but unfortunately it has $T_c = 9.25$ K, which means using expensive cryogenic systems to keep it this cold.

The discovery in 1987 of ceramic materials which were superconducting at liquid nitrogen temperatures, the so-called high T_c superconductors, opens the possibility that the Meissner effect in similar materials could be used more widely for magnetic shielding. So far there are problems in producing the new materials, as the superconducting regions appear to be confined to small grains within the bulk material, and the critical field values are low. However if it becomes possible to make sheets or long wires of this superconductor which could tolerate high fields, it would then be possible to use this material to obtain very effective magnetic screening.

4.2.3 Electrostatic shielding

It is easy to make very effective screens against electric fields with complete conducting shields. Usually however electrostatic shields are not complete so that some of the external electric field can enter the shielded enclosure. This situation can often be analysed in terms of the capacitance distribution. Capacitance is an electrostatic concept but provided the current flow on the relevant conductors is small enough the capacitance describes the shielding up to reasonably high frequencies (unlike the case for magnetic fields). In order to illustrate some of the practical problems dealt with by considering capacitance distribution the following simple examples are discussed.

Consider the case of two parallel conductors A and B, as shown in Figure 4.2. They will have capacitance between each other, and each will have capacitance

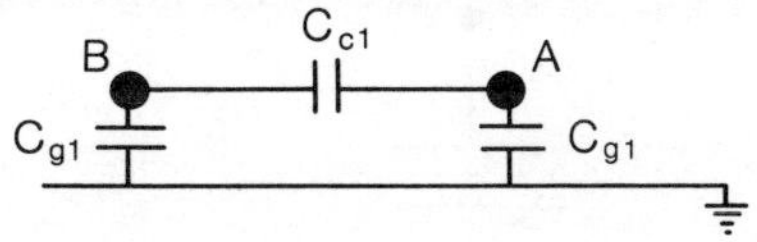

Fig. 4.2 Equivalent circuit model for capacitive coupling between two conductors A and B and ground

to ground. A pulse voltage V on conductor A will produce a noise voltage V_n on conductor B given by:

$$V_n = \frac{VC_{c1}}{C_{g1} + C_{c1}}. \tag{4.13}$$

This is because the capacitors C_{g1} and C_{c1} act as a potential divider for the pulse V. To reduce V_n it is necessary to increase C_{g1} and reduce C_{c1}. One way to do this is by inserting a conducting sheet between A and B. When the sheet is grounded the resulting situation is shown in Figure 4.3. Here it is assumed that $C_{c2} \approx 2C_{c1}$, where C_{c1} is the capacitance between the conductors as shown in Figure 4.2. Because the sheet is grounded the effective capacitance C_{c2} between A and ground is rendered unimportant for transmitting a pulse to B since such a pulse travelling across this is shorted to ground, so that the equivalent circuit becomes that shown in Figure 4.4. The small residual stray capacitance C_{s1} between A and B is the route by which noise is transmitted. In

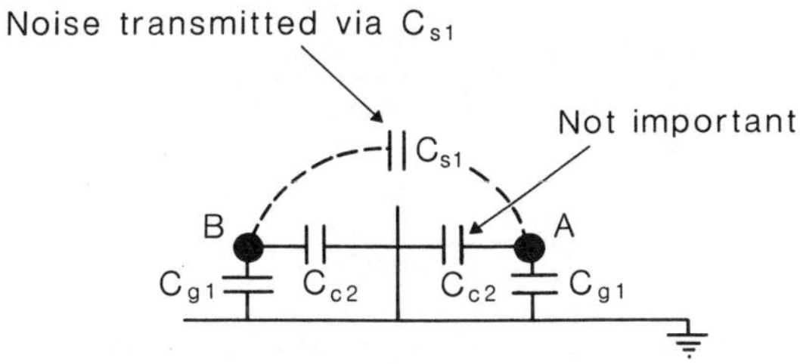

Fig. 4.3 Capacitances present when two conductors are separated by a grounded plane

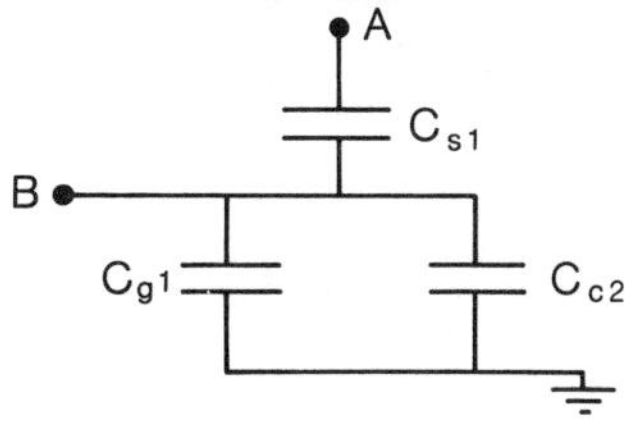

Fig. 4.4 Equivalent circuit model for Figure 4.3

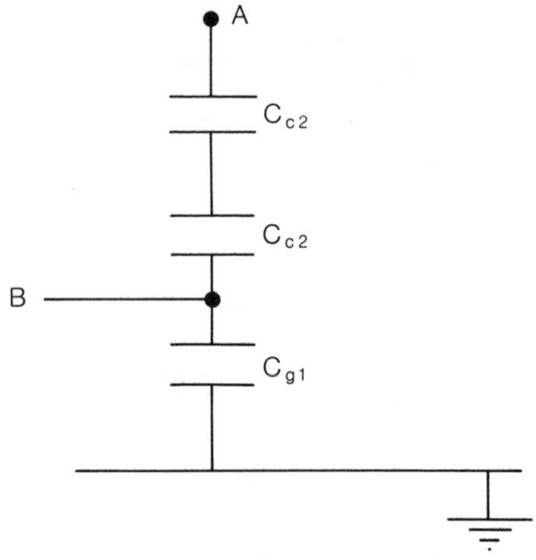

Fig. 4.5 Equivalent circuit model for two conductors separated by an ungrounded sheet

this case the noise voltage at B is given by:

$$V_n = \frac{VC_{s1}}{(C_{g1} + C_{c2}) + C_{s1}}. \tag{4.14}$$

This is much smaller than (4.13) because the denominator in (4.14) is bigger than C_{c1} and the numerator C_{s1} is smaller than C_{c1}. If the central plane is not grounded but floating then the equivalent circuit becomes that shown in Figure 4.5. To obtain Figure 4.5 the capacitance C_{s1} has been neglected (because $C_{s1} \ll C_{c2}$) and for this circuit the noise voltage at point B is

$$V_n = \frac{V(C_{c2})/2}{C_{g1} + (C_{c2})/2} \tag{4.15}$$

i.e.

$$V_n = \frac{VC_{c1}}{C_{g1} + C_{c1}} \tag{4.16}$$

i.e. the same as (4.13) since $C_{c2} \approx 2C_{c1}$, so that putting the ungrounded sheet between the lines has not had much effect. However this conclusion is not true if there is an appreciable capacitance from the plate to ground. In this case the pulse circuit approximates that shown in Figure 4.6. In Figure 4.6 the noise voltage at the point B can be obtained by considering this to circuit be a two stage potential divider, dividing the voltage V between the point A and the point P, then between P and the point B, such that

$$V_P = \frac{VC_{c2}}{C_{c2} + C_{plate}}$$

and

$$V_B = \frac{V_P C_{c2}}{C_{c2} + C_{g1}}$$

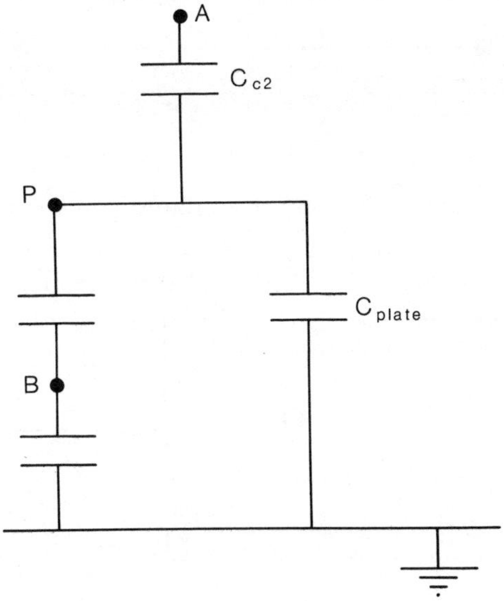

Fig. 4.6 Equivalent circuit for two conductors separated by an ungrounded sheet which has large capacitance to ground

so that at point B the noise voltage is given by:

$$V_n = \frac{V(C_{c2})^2}{(C_{c2} + C_{plate})(C_{g1} + C_{c2})} \tag{4.17}$$

which is much smaller than (4.13) when C_{plate} is large. Thus the noise is reduced if an ungrounded sheet with large capacitance to ground is inserted between the lines.

If the conductors are close to the ground plane, then the capacitance to ground and the capacitance between the conductors changes significantly from the values C_{g1} and C_{c1} obtained from the geometrical setup shown in Figure 4.2. This is because the presence of the plate changes the structure of the electric field lines significantly, so changing the capacitance. In Figure 4.7 the capacitances C_{g2} and C_{c3} are such that $C_{g2} \gg C_{g1}$ and $C_{c3} \ll C_{c1}$ (see Figure 4.2). Thus at point B in Figure 4.7.

$$V_n = V\frac{C_{c3}}{C_{g2} + C_{c3}} \tag{4.18}$$

which is much less than that given by (4.13), or (4.15) so that the noise is reduced compared to the arrangement shown in Figure 4.3 where the grounded plate is put between the wires.

If the plate is not grounded and remote capacitively from ground, point A becomes closely coupled to B and large noise voltage will result. Thus when trying to improve electrostatic shielding it is sensible to ground shields but this is not always essential if there is adequate capacitance between the shield and

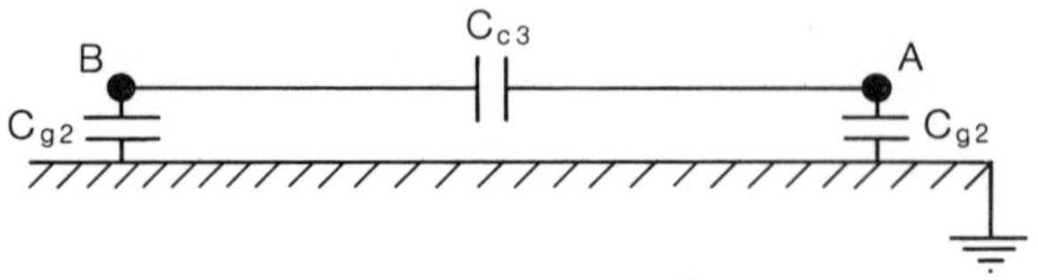

Fig. 4.7 Capacitance distribution for a ground plane close to conductors

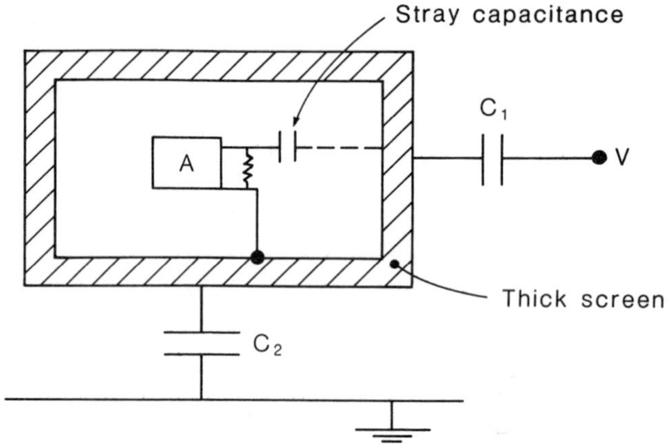

Fig. 4.8 Schematic diagram illustrating the coupling via stray capacitances of an amplifier inside a screened enclosure

ground. The case where it is not necessary to ground the shield (beware safety requirements here) is when it completely surrounds a circuit.

In Figure 4.8, an amplifier A is situated in a screened container, and there is stray capacitance between the amplifier input and the interior of the shield. If the screen is continuous the electric field inside it is zero and hence no induced voltages are possible. Voltages will only appear on the amplifier input if current flows on the interior of the shield generating voltage drops with corresponding internal electric fields, which will couple to the amplifier. Note in the case shown in Figure 4.8 the screening container itself will reach a pulse voltage level of order $VC_1/(C_2+C_1)$, even though this voltage will not appear on the amplifier input.

All the above examples are often called electrostatic screening in that they neglect the effect of current flow in the shield. In the next section the equivalent circuit model will be compared with accurate calculations and the neglect of current flow will no longer be possible.

4.3 EQUIVALENT CIRCUIT MODELS OF SHIELDING

This approach to understanding and calculating shielding behaviour has recently been reviewed by Bridges [9]. This review contains an extensive list of useful references. For a situation where it is possible to consider E and H field effects separately the circuit model gives a clear physical picture of the mechanisms by which shields work and how they are degraded by faults in their construction.

4.3.1 Electric field screening

Consider first of all the response of a box shield to an originally uniform electric field. The situation is illustrated in Figure 4.9. When the E field is static, induced charges are set up as shown so reducing the internal field to zero. This charge will concentrate on the surface and near the corners of the box. If the field is now allowed to oscillate (initially at low frequency) then the charges have to continually re-adjust themselves for each new field value, so causing currents to flow on the box walls. At low frequencies these currents will be distributed throughout the thickness of the conducting walls. The current on the internal surfaces will generate E (and H) fields within the shielded enclosure. Since the current will increase with frequency it might be expected that the shielding effectiveness would decrease as the frequency increases. However, the skin effect (see Section 2.5.5) at higher frequencies causes the current to become more and more localised on the outside surfaces of the box exposed to the exciting field. The internal currents are reduced, and so are any internal fields. Thus the shielding effectiveness starts to increase once the skin effect starts to dominate. It is possible to derive a simple formula for the shielding effectiveness by considering the induced charges on a spherical conducting shell, as shown in Figure 4.10. The time derivative of the charge on one hemisphere (assuming the charge is always the static value at any given field) enables the current flowing through the equatorial plane to be calculated [9, 10], giving the result

$$I_0(t) = \mathrm{j}\omega 3\pi\varepsilon_0 a^2 E_0 \mathrm{e}^{\mathrm{j}\omega t}. \tag{4.19}$$

Thus the voltage drop across an equatorial loop of thickness d, radius a and

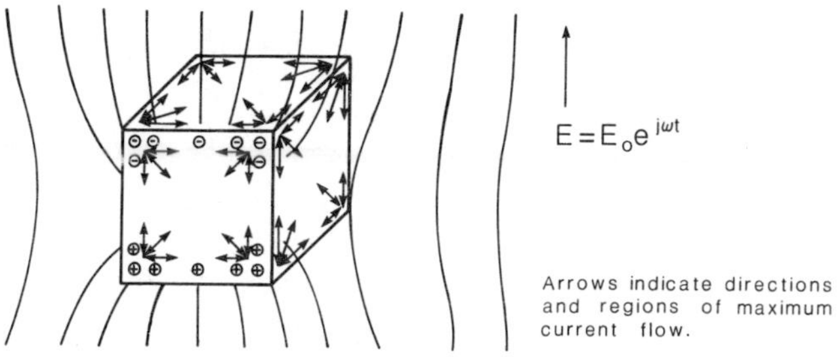

Fig. 4.9 Electric field and surface charge distribution for a rectangular enclosure situated in a quasi-static electric field [9]. (Reproduced by permission of IEEE)

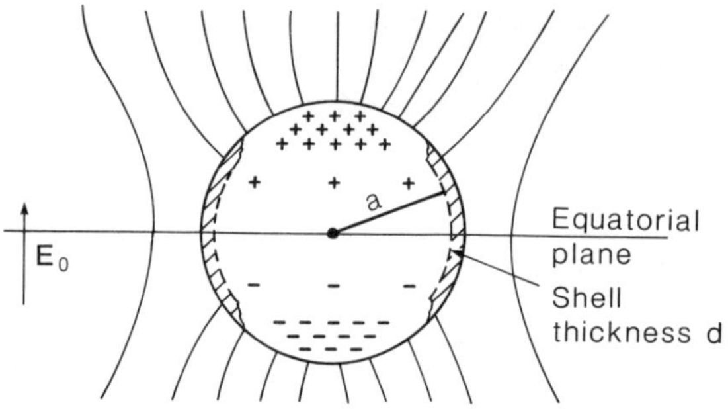

Fig. 4.10 Charge distribution on a spherical conducting shell situated in an originally uniform electric field [9]. (Reproduced by permission of IEEE)

height y is given by

$$V(t) = \frac{j\omega 3\varepsilon_0 a E_0}{2\sigma d}\, e^{j\omega t} y \tag{4.20}$$

where σ is the conductivity of the sphere. The internal surface field is not easily calculated and this is taken to be equal to the field E_i at the centre of the sphere [9], which is

$$E_i = \frac{j\omega 3\varepsilon_0 a E_0}{2\sigma d}\, e^{j\omega t}. \tag{4.21}$$

Equation (4.21) assumes that the skin depth $\partial \gg d$. If the frequency is high enough so that $\partial \ll d$ then the bulk of the current flows on the outer surface and sees a much higher impedance. Allowing for the skin depth depletion of the current on the internal surface it is possible to show [9] that E_i is given by

$$E_i = \frac{j\omega 3\sqrt{2}\varepsilon_0 a E_0 e^{j\omega t} e^{-d/\partial}}{\sigma\partial}. \tag{4.22}$$

Hence it is possible to derive two formulae for the electric field shielding effectiveness, SE (see reference [9] for details):

$$\text{for } \partial \gg d \qquad \text{SE} = 20 \log_{10}\{2\sigma d/\omega 3\varepsilon_0 a\} \text{ dB} \tag{4.23}$$

$$\text{for } \partial \ll d \qquad \text{SE} = 20 \log_{10}\left(\frac{\sigma\partial}{\omega 3\sqrt{2}\varepsilon_0 a e^{-d/\partial}}\right) \text{ dB} \tag{4.24}$$

Rigorous scattering calculations have been carried out for a number of aluminium spheres (with $\sigma = 3.54 \times 10^7$ S m^{-1}) by Harrison and Papas [11]. Their results are given, in a slightly modified form, in Figure 4.11 and it can be shown that (4.23) and (4.24) give almost the same values in appropriate regions of validity [9].

In order to develop these ideas into a general technique for dealing with spheres, spheroids, long thin rods and related structures the current flowing perpendicular to an equipotential plane of symmetry due to a field parallel to

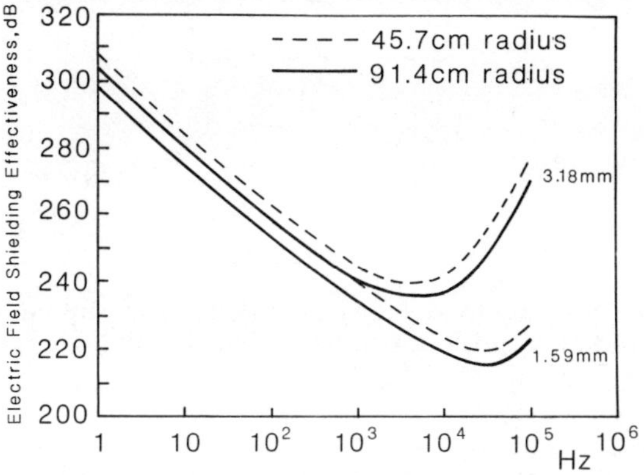

Fig. 4.11 Steady state electric field shielding effectiveness of thin-walled aluminium spheres of thickness 3.18 and 1.59 mm, adapted from reference [11]. (Reproduced by permission of IEEE)

the major axis must be determined [9]. The approximation used is to regard the structure as a 'black box' that relates the uniform exciting field E to the open circuit voltage, V_{oc}, of the network. This open circuit voltage is equal to the applied field intensity E multiplied by the effective height of the structure, h. Next, the network source impedance is determined and used to calculate the mid-plane current. For high conductivity barriers this current is essentially the short circuit current of the network. This current, when multiplied by the surface impedance, gives the interior fields. Thus for the conducting sphere example considered above, h is taken to be the radius a. The open circuit voltage is then:

$$V_{oc}(t) = aE_0 e^{j\omega t}. \tag{4.25}$$

Rewriting (4.19) gives

$$\begin{aligned} I_0(t) &= \frac{aE_0 e^{j\omega t}}{-j/(3\pi\varepsilon_0 a\omega)} \\ &= \frac{aE_0 e^{j\omega t}}{-j/(C\omega)} \\ &= V_{oc}/Z_{source}. \end{aligned} \tag{4.26}$$

This is because for the effective height chosen, $C = 3\pi\varepsilon_0 a$, and so it is possible to use the equivalent circuit shown in Figure 4.12 to calculate the internal field E_i using $E_i = I_0(t)R_s$.

It has been claimed that this technique can be used to calculate the approximate shielding effectiveness of nearly cubical structures, and that long structures can be characterised as spheroids or thick antennas [9]. The antenna approach that is suggested in this reference is useful because the effective height and input impedance have been developed for a number of antenna

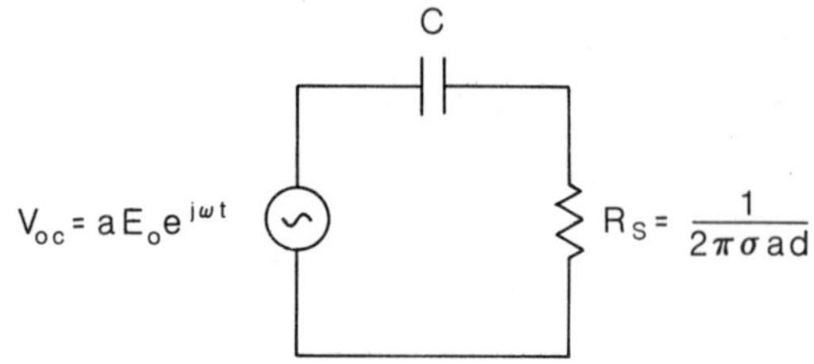

Fig. 4.12 Low frequency circuit model for E field penetration into a conducting spherical shell of thickness d, radius a and conductivity σ. This is valid when the skin depth $\partial \gg d$ and $2\sigma d/3\varepsilon_0 > \omega$

shapes. Bridges [9] has also used the technique for analysing the effect of an applied electric field having a gaussian pulse shape. Once again the results agreed well with scattering theory.

4.3.2 Quasi-static magnetic field screening

Consider the response of a box shaped shield to an originally uniform magnetic field applied along one axis of the box, as shown in Figure 4.13. In Section 4.2 it has been shown that for a static magnetic field, no ducting occurs with non-magnetic materials and thus the static **H** field penetrates the whole box structure. As the field is allowed to oscillate (initially at low frequency), induced currents begin to flow in the conducting material, the direction of flow being such as to generate new magnetic fields which oppose changes to the applied field. Because the total flux linkage increases as one moves out radially from the centre of the box (indicated by O in Figure 4.13) the currents tend to increase at the edges and around the walls, as indicated. This means that the circuit behaviour of each system can be visualised as a shorted loop having a resistance R_S and an inductance L_S. This time varying magnetic field induces a voltage in the shorted turn that is proportional to the frequency of the applied magnetic field. The circuit analogue is shown in Figure 4.14.

At low frequencies the loop current will be proportional to the induced voltage divided by the resistance of the loop. Note however that this will mean

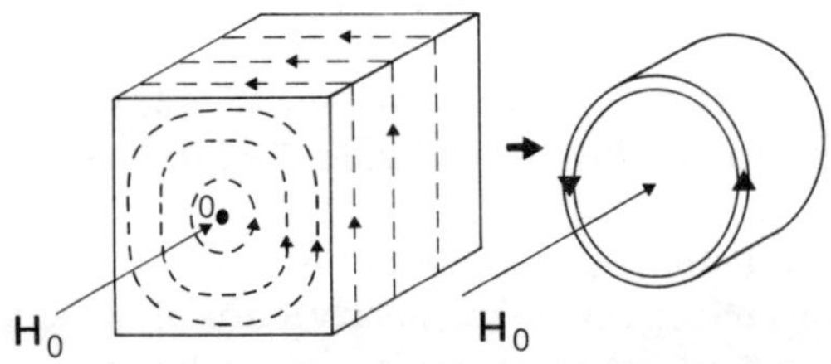

Fig. 4.13 Schematic diagram showing how the currents induced in a box generated by an alternating magnetic field can be modelled as a single shorted turn

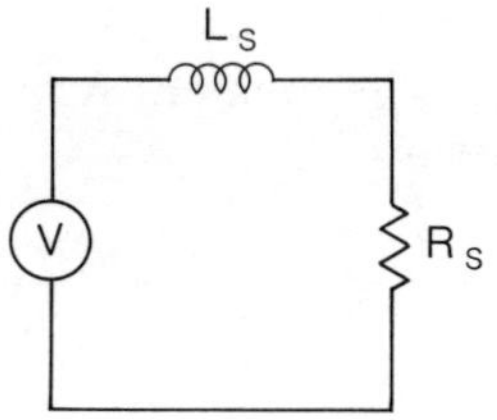

Fig. 4.14 Circuit analogue for magnetic screening, where L_s and R_s are the effective inductance and resistance of the shorted turn

that the current (and its opposing field) will be 90° out of phase with the applied field. Therefore one can expect little cancellation of the internal field. As the frequency increases the loop impedance becomes dominated by the inductance so that the current will come into phase with the applied field. Thus some cancellation of the field within the shorted turn occurs, together with an increase outside.

If the frequency is increased further then significant attenuation of the current flow on the inside of the shield occurs due to the skin effect. This reduces the inside fields even further. The circuit model prediction for the magnetic shielding effectiveness SE_H can be derived using the following argument (see reference [14] for a similar discussion) which is only valid for very low frequencies.

For a loop of area A the induced emf, V_{oc}, due to the applied field $H_0 e^{j\omega t}$ is

$$V_{oc} = -\partial/\partial t \int \mu_0 \mathbf{H} \cdot \mathbf{dS}$$

$$= -j\omega\mu_0 H_0 A.$$

Hence the loop current is

$$i(t) = V_{oc}/Z_{loop} = V_{oc}/(R_s + j\omega L_s). \tag{4.27}$$

Now inside the shell the field is assumed to be H_i so that the induced emf (due to $\partial H_i/\partial t$) around the inside of the shell, V_i, is given by

$$V_i = -j\omega\mu_0 H_i A.$$

But this must be equal to $i(t)R_s$ at the boundary so that using (4.27)

$$-j\omega\mu_0 H_0 A = V_{oc} R_s/(R_s + j\omega L_s)$$
$$= -j\omega\mu_0 H_0 A R_s/(R_s + j\omega L_s).$$

Hence

$$H_i/H_0 = R_s/(R_s + j\omega L_s) \tag{4.28}$$

as given by Bridges [9]. For a spherical coil Wheeler [12] gives

$$R_s = 2\pi n^2/3d\sigma \qquad \text{and} \qquad L_s = 2\pi a\mu_0 n^2/9 \tag{4.29}$$

where d = wall thickness, σ = conductivity, a = the outer radius of the sphere, b = inner radius of the sphere, n = equivalent number of turns, so that $d = a - b$ and for a thin shell $a \approx b$. Hence

$$H_i/H_0 = \frac{(1/(3d\sigma))}{(j\omega\mu_0 a/9 + 1/(3d\sigma))}$$

which gives

$$SE_H = 20 \log_{10}(H_i/H_0) = 20 \log_{10}(1/(1 + j\omega\mu_0 a\sigma d/3)) \text{ dB} \tag{4.30}$$

which is identical to the very low frequency limit of the rigorous formula due to King [13]. Similar arguments concerning skin depth effects lead to another limit where $d \gg \delta$ and $j\omega L_s \gg R_s$ giving [9],

$$SE_H = 20 \log_{10}\left[\frac{3\sqrt{2}\delta}{be^{d/\delta}}\right] \tag{4.31}$$

with δ = the skin depth.

This formula can also be shown to correspond to a high frequency approximation to the general results of King [13]. Equations (4.30) and (4.31) also agree well in the appropriate regimes with the rigorous scattering calculations of Harrison and Papas [11] which are given in Figure 4.15. All these results are for non-magnetic materials. In the case of magnetic materials the rigorous results of King can be represented by relatively simple formulae for the two cases $d \ll \delta$ and $d \gg \delta$. These formulae are given in Table 4.2 and have been confirmed by equivalent circuit calculations (ignoring ducting effects) given by Bridges [9], Shenfeld [15] and Cooley [14]. The formulae in Table 4.2 for the low frequency limit contain terms allowing for the effect of ducting. As $f \to 0$ these equations reduce to the forms given earlier (Equations (4.9) and (4.10)) for the static shielding case.

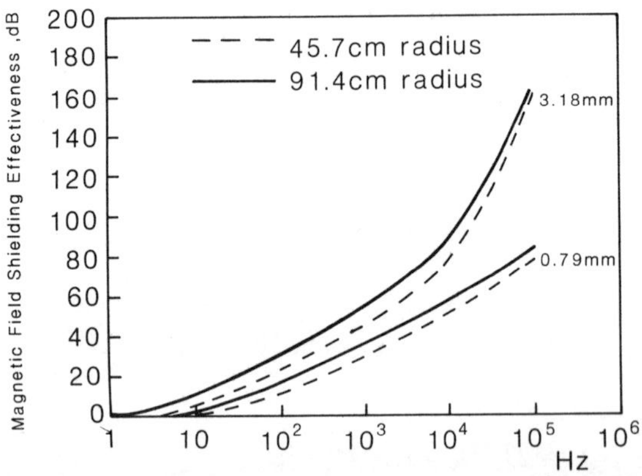

Fig. 4.15 Steady state magnetic field shielding effectiveness for thin-walled aluminium spheres having radii of 45.7 cm and 91.4 cm and of thickness 3.18 mm and 0.79 mm, adapted from reference [11]. (Reproduced by permission of IEEE)

Table 4.2. Formulae for the magnetic shielding effectiveness for thin cylindrical and spherical magnetic shells of thickness d, relative permeability μ_r, and $a \approx b$ [13]

Sphere	Cylinder (transverse H_0)
Low frequency $d \ll \delta$	
$20 \log_{10}\left[\dfrac{1}{1+\dfrac{2(\mu_r-1)^2 d}{3\mu_r b}+\dfrac{j\omega\mu_0\sigma b d}{3}}\right]$	$20 \log_{10}\left[\dfrac{1}{1+\dfrac{(\mu_r-1)^2 d}{2\mu_r a}+\dfrac{j\omega\mu_0\sigma a d}{2}}\right]$
High frequency $d \gg \delta$	
$20 \log_{10}\left[\dfrac{3\sqrt{2}\mu_r\delta}{a e^{d/\delta}}\right]$	$20 \log_{10}\left[\dfrac{2\sqrt{2}\mu_r\delta}{a e^{d/\delta}}\right]$

The equivalent circuit approach has highlighted the importance of the exact shape of the shield and how much current flows on its inner surface. This latter point will become even more relevant when it is shown in Section 6.3.4 how the interior of a shield may be excited by an incorrectly routed ground wire rather than by some external field. The importance of shield shape has been clearly outlined in the work of Lee and Bedrosian [16] on the penetration of EMP induced H waves into sealed enclosures.

In discussing this problem they suggest that four crucial times define the situation, namely t_1 the pulse width, t_2 the time it takes the pulse to pass the enclosure ($t_2 = a/c$ where a is the system size and c is the velocity of light), t_3 the diffusion time (see Section 2.5.9), and finally t_4 the characteristic L/R time for current decay in the enclosure walls. Typically $t_1 \approx$ 100's ns, $t_2 \approx$ tens of nanoseconds, $t_3 \approx$ tens of microseconds, and $t_4 \approx$ tens of milliseconds. These time parameters are well separated and it is possible to describe the event in the following way. If a pulse strikes the enclosure at $t = 0$ then for $t < t_3$ no internal fields appear. For $t \simeq t_3$ the internal fields reach a maximum but by this time the exciting pulse has long passed the enclosure. The wall currents now decay with a time constant t_4 ($\simeq \mu_0\sigma a d/3$ for a spherical shell). Lee and Bedrosian present rigorous transfer functions for a number of shell shapes and show that there exist vast differences at low frequencies between planar sheets and spherical enclosures. For example they show that at 1 kHz a planar approximation to a shielding enclosure overestimates the shielding effectiveness by 100 dB.

To extend the equivalent circuit model to other geometries than the spherical is not always easy though some guidance has been given by Bridges [9] and Franceschetti [17].

4.3.3 Conclusions

The circuit model has only been discussed here for the spherical shell case (though some results for other simple shapes have been mentioned) and the obvious drawback of the method lies in obtaining the correct circuit to use. It does however give good results for some cases and highlights the fact that the current flow on the interior surface of the enclosure coupled with the effective surface impedance dominate the interior fields. Such an approach naturally leads to explaining why poor bonds and seams cause difficulties [9, 18]. Essentially any increase in local impedance will increase the internal fields. Such an explanation is not so easily given by the plane wave model discussed in the next section.

4.4 PLANE WAVE OR TRANSMISSION LINE MODELS OF SHIELDING

In Chapters 2 and 3 it was shown how to treat reflection and transmission of plane waves at various planar media change boundaries by means of the reflection and transmission coefficients calculated from the characteristic impedance of the media concerned. The attenuation of these waves in conducting media was discussed and various aspects of the skin effect developed. Finally it was shown how reflecting surfaces develop standing wave patterns leading to the concept of an effective impedance at different distances in front of the reflecting (partial or complete) boundary. Similar equations were developed for TEM waves on transmission lines. This similarity between plane waves and waves on transmission lines leads to the obvious title for the model to be discussed in this section. Consider the planar shielding arrangement shown in Figure 4.16. The three regions are shown to have in general three different characteristic impedances Z_1, Z_2, and Z_3. Within the shield it is obvious that secondary reflections and a standing wave pattern will occur unless the shield length L is appreciably longer than the skin depth. It is also clear that this model of shielding can only apply in real situations when the shield is remote from a real source and has an extent greater than many wavelengths of the incident radiation [4, 19]. The incident and exit field intensities are shown as E_i, H_i and E_t, H_t respectively. Note that (as for the

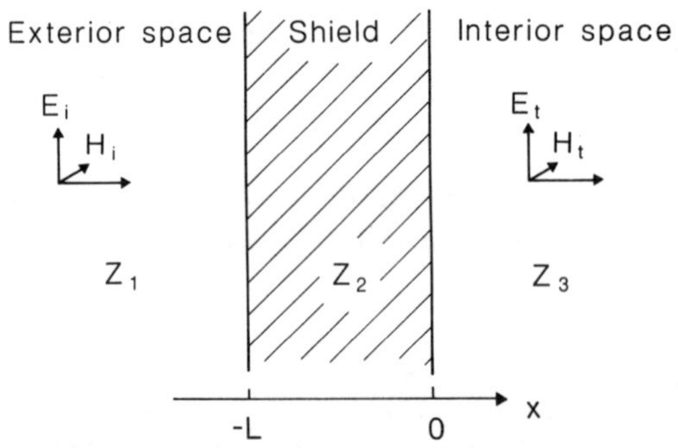

Fig. 4.16 Plane wave model of shielding

other models considered in this chapter) resonances exterior to the shield, within the shield or within the space enclosed are not considered. Because of the need to consider multiple reflections in the shield we make use of the formulae developed in the last chapter for the general case of wave propagation into layered structures [21]. It is then possible to show fairly easily [4] that the transmission coefficient for the magnetic field T_H is given by (see Section 3.2.2, Equation (3.10*a*))

$$T_H = p_H(1 - q_H e^{-2\gamma L})^{-1} e^{-\gamma L} \tag{4.32}$$

where

$$p_H = \frac{4Z_1Z_2}{(Z_1 + Z_2)(Z_3 + Z_2)}$$

and

$$q_H = \frac{(Z_1 - Z_2)(Z_3 - Z_2)}{(Z_1 + Z_2)(Z_3 + Z_2)}. \tag{4.33}$$

Usually in the planar case $Z_1 = Z_3$, then $T_H = T_E = T$ the total transmission coefficient.

Putting $Z_1/Z_2 = k$ with $p = 4k/(k+1)$ and $q = (k-1)/(k+1)$ so that

$$T = p(1 - qe^{-2\gamma L})^{-1} e^{-\gamma L}. \tag{4.34}$$

The total shielding effectiveness is then

$$\begin{aligned} \text{SE} &= -20 \log_{10} |T| \\ &= 20 \log_{10} |e^{\gamma L}| - 20 \log_{10} |p| + 20 \log_{10} |1 - qe^{-2\gamma L}| \\ \text{SE} &= A + R + B \text{ dB} \end{aligned} \tag{4.35}$$

where

A = the absorbtion or penetration loss in the conducting shield = 20 $\log_{10} |e^{\gamma L}|$,
B = re-reflection term (always negative or nearly zero) = 20 $\log_{10} |1 - qe^{-2\gamma L}|$,
R = the reflection loss due to the two boundaries = 20 $\log_{10} |1/p|$.

When re-reflection is negligible (as is usual for plane waves and reasonably thick shields) one obtains the standard formula

$$S = 20 \log_{10} |\tau_{12}\tau_{23} e^{-\gamma L}|^{-1} \tag{4.36}$$

where τ_{12} = the transmission coefficient at the first boundary and τ_{23} = the transmission coefficient at the second boundary with $e^{-\gamma L}$ = the absorbtion due to a single pass through the shield.

Notice that $p = (\tau_{12}\tau_{23})^{-1}$ and so (4.36) is (4.35) with B set to zero.

Figures 4.17 and 4.18 show some results calculated from (4.35) for a copper shield. It is clear that for a thin shield ($L = 10^{-6}$ m) that reflection and re-reflection dominate the shielding effectiveness. These terms dominate also for the even thinner conducting films used to produce windows that are opaque to

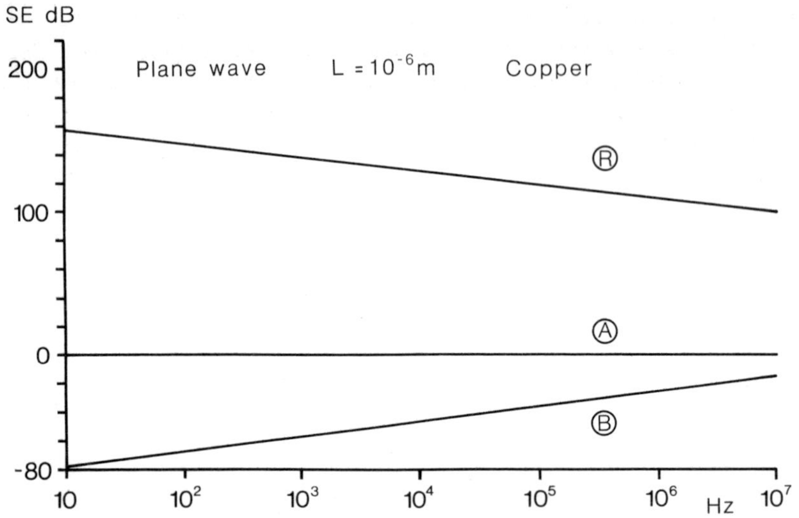

Fig. 4.17 Relative contributions as a function of frequency of absorbtion, A, reflection, R, and secondary reflection, B, to the plane wave shielding effectiveness for a copper shield of thickness 10^{-6} m

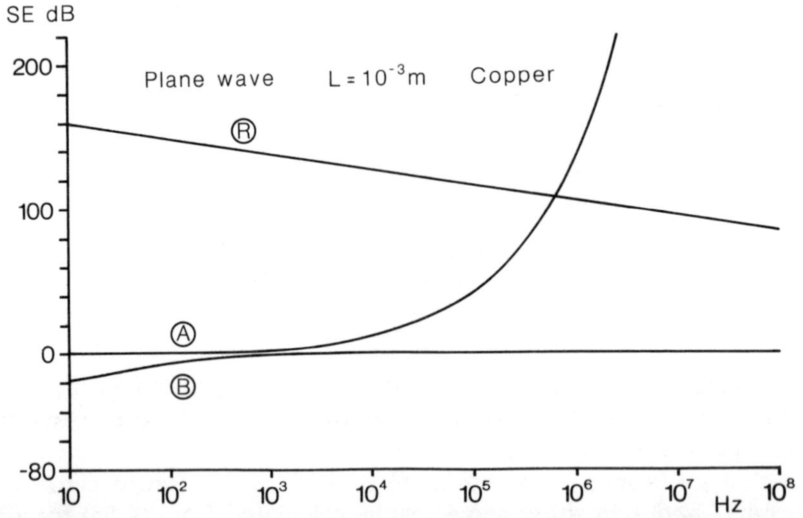

Fig. 4.18 Relative contributions as a function of frequency of absorbtion A, reflection, R, and secondary reflection, B, to the plane wave shielding effectiveness for a copper shield of thickness 10^{-3} m

microwaves but allow visible light to be transmitted [20]. On the other hand, for thicker shields (such as might be used for screened rooms ($L = 10^{-3}$ m)) the re-reflection term B becomes unimportant as is shown in Figure 4.18. It is clear also that the reflection term remains the same but that the absorption term dominates at the higher frequencies ($f > 10^6$ Hz). This simple plane wave model of shielding can be used to investigate the properties of multi-media and multiple shields [4]. In some circumstances two shields with a spacing between them can produce an appreciable improvement over a single shield having the same total thickness of metal [4]. However, this is not always the case and in general one needs to examine the worth of multiple shields fairly carefully. Finally it is essential to note that the plane wave model predicts shielding values considerably larger than seen with practical shielded rooms even when they are exposed to plane wave illumination. When the dominant fields are magnetic (H) the model is even more in error predicting SE as much as 160 dB too large at 10 Hz. In real world situations it is virtually impossible to experience the conditions required by the plane wave model at the lower frequencies (unless deliberately set up for experimental purposes [22, 23]). In order to study these problems the ideal theory of this section has been extended to non-planar waves with some loss of rigour.

4.5 EXTENSIONS OF PLANE WAVE THEORY INTO NON-IDEAL SITUATIONS

In Chapter 2 it was shown that the energy flow from a small dipole was given by the time averaged Poynting vector which only involved E_θ, H_ϕ terms (E_r not contributing). It was also shown that E_θ / H_ϕ could be evaluated at all distances r from a small dipole and a small loop (see the full derivation in Appendix C) leading to the concept of high and low impedance fields. These results are for the situation where the dipole or the loop are small compared with r. If a spherical shield is located with its centre at the dipole (loop) then E_θ and H_ϕ will be tangential to the surface at whatever distance r the shield is located. We can use the results of the last section to calculate the shielding effectiveness SE utilising the following assumptions.

(1) The input impedance of the spherical wave which arrives from the source and impinges normally on the shield is given by the value of E_θ / H_ϕ at that point. This is constant across the whole surface of the shield.

(2) As the spherical wave enters the shield material it reverts to the plane wave impedance characteristic of that material. For conducting shields this can be justified since we have seen that in this case the wavelength is given by (2.73) ($\lambda = 2\pi\delta$) which means that it is considerably reduced over the free space value. Thus for most practical shields the shield radius is much larger than wavelength in the shield (except perhaps at the lowest frequencies).

(3) On leaving the shield the final transmitted wave reverts to the wave impedance calculated for the distance r from the unshielded source as defined in (1) above. Thus it is assumed that the shield thickness is much smaller than r.

(4) The various reflection and transmission coefficients can be calculated using (2.44) and (2.45) but now using the requisite characteristic impedances as defined above.

The equations developed in Section 4.4 can now be used to calculate the shielding effectiveness SE of the spherical shield.

Note that $k = Z_1/Z_2$ as before but that now for a dipole source

$$Z_1 = 377c/2\pi fr \quad \text{for } r \leqslant c/2\pi f \tag{4.37}$$

whereas for the loop source

$$Z_1 = 377(2\pi rf/c) \quad \text{for } r \leqslant c/2\pi f. \tag{4.38}$$

In both cases $Z_1 = 377$ for $r > c/2\pi f$ whereas at all distances

$$Z_2 = (1+\mathrm{j})\sqrt{\pi\mu f/\sigma} = (1+\mathrm{j})/\sigma\delta \quad \text{(see (2.79))}.$$

With these changes Equation (4.35) can be used to calculate SE. The relative importance of reflection R, absorbtion A, and re-reflection B is illustrated for a 1 mm thick, 2 m diameter, copper sphere in Figure 4.19(*a*). It is clear that, at low f, R dominates whereas, at high f, A has most importance.

The wave impedance plays a crucial role at low frequencies which explains why it is so difficult to shield against the effect of low frequency magnetic fields (low impedance waves (or H waves)). The changes caused by altering the source–shield distance are illustrated in Figure 4.19(*b*). The main practical inference that can be drawn is that it is worthwhile situating the shield as far away as is practical in the case where magnetic shielding is desired. Both sets of curves show that high impedance waves are easier to screen than plane waves which in turn are easier than low impedance waves. Figure 4.20 gives calculated curves for SE as a function of frequency for $r = 1$ m. The three thicknesses of copper chosen correspond approximately to those used in

(1) coating optically transparent windows, with $L \approx 10^{-8}$ m;
(2) coating plastics to provide some shielding capability; with $L \approx 10^{-6}$ m
(3) construction of shielded cabinets and rooms, $L \approx 10^{-3}$ m.

Figure 4.21 gives calculated results for a material similar to iron. Similar but more extensive curves have been presented by Violette *et al.* [1]. From these types of results the following general points can be made.

(1) Re-reflection corrections are always important when reflection is high and absorbtion is small.
(2) For copper absorbtion becomes important ($A \geqslant 10$ dB) when the product $f^{1/2}L > 0.1$. Thus for most thin coatings reflection is the dominant mechanism in shielding.
(3) Any reasonable thickness ($L > 10\ \mu$m) of metal screens plane and E waves adequately. (SE $\geqslant 100$ dB at all frequencies).

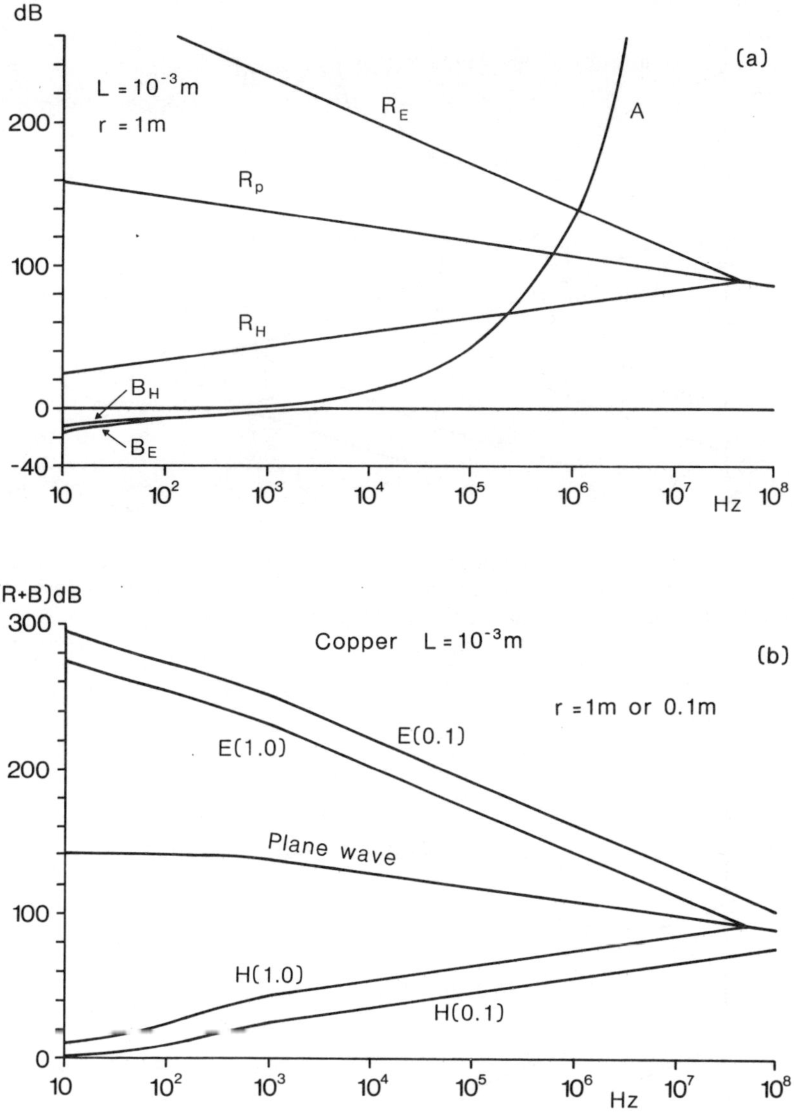

Fig. 4.19 (*a*) Relative contributions as a function of frequency of absorbtion, *A*, reflection, *R*, and secondary reflection, *B*, to the shielding effectiveness as affected by the wave impedance. In the figure R_E and B_E are the reflection and secondary reflection contributions for high impedance *E* fields, and R_H and B_H are for low impedance *H* fields. The reflection term R_P for plane waves is included for comparison. The absorption *A* is not affected by the wave impedance. All the results are calculated for a spherical copper shield 1 mm thick, 2 m in diameter, with a source at its centre. (*b*) Net reflection $(R + B)$ from the inner surface of a 1 mm thick copper sphere as a function of frequency, showing the effect of changing the source–shield distance from 0.1 m to 1 m. In the figure the curves labelled $E(0.1)$ and $E(1.0)$ refer to the net reflection for a high impedance *E* wave for the two source distances, and the curves labelled $H(0.1)$ and $H(1.0)$ to the net reflection for a low impedance *H* wave

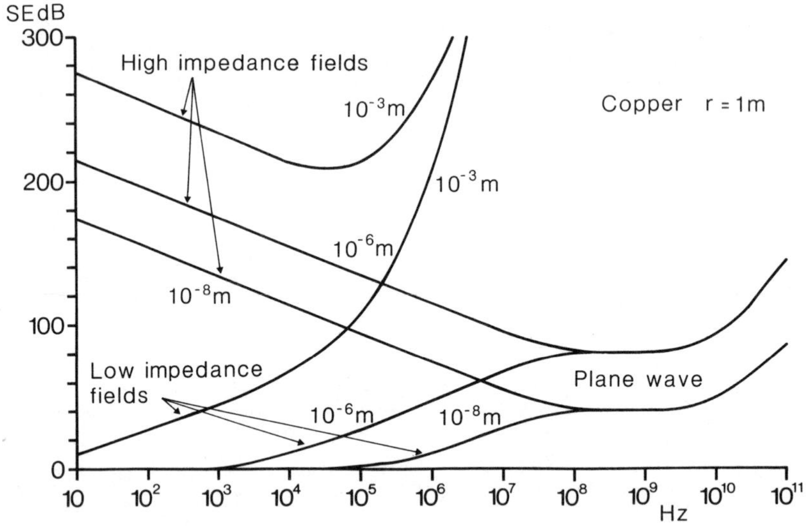

Fig. 4.20 Total shielding effectiveness SE dB as a function of frequency for three thicknesses of copper with the source at the centre of a 1 m radius shield. The lines corresponding to high and low impedance fields are indicated

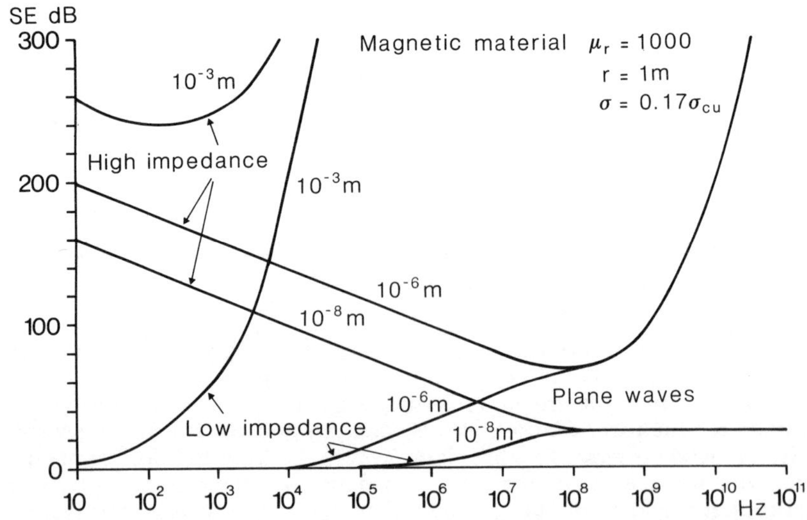

Fig. 4.21 Total shielding effectiveness SE dB as a function of frequency for three thicknesses of an iron-like material (with $\mu_r = 1000$, $\sigma = 0.17\sigma$ of copper) with the source at the centre of the 1 m radius shield. The lines corresponding to high and low impedance fields are indicated

(4) H waves are very difficult to shield against at low frequencies unless thick layers of magnetic material are used. It pays to keep the source as far away as possible from the shield in this circumstance. The curves in Figure 4.21 assume that the permeability is constant with frequency but this is not true in general the permeability decreasing as the frequency increases [4]. Thus the theoretical difference between copper and iron is not so marked at frequencies in excess of 500–1000 MHz. On top of this it must be remembered that the permeability depends on the material's history and the applied field level, saturation often being a problem for large thin shields. Moreover in practice once the theoretical value of SE exceeds 150–200 dB (such high values are never achieved in real systems) defects in any shield system dominate over material properties. Hence as will be seen later, once 1–10 MHz is exceeded there is usually little difference between iron and copper.

(5) Thin layers ($L = 0.01–1\ \mu$m) of conducting material have useful shielding properties provided a shielding effectiveness of between 20–100 dB is all that is required. These calculations assume that the copper films have constant properties at frequencies less than 10 GHz. A further discussion of their behaviour at very high frequencies is given in Section 3.2.4.

As has been pointed out, these shielding estimates are for a small dipole or loop situated a relatively long way from a spherical shield. By analogy a similar approach can be used for a cylindrical shield enclosing two parallel current filaments located at $r = 0$ [4]. Note however that all the systems modelled using the plane wave approach are unlike any seen in practice except in special circumstances. For example no circuit will have the high impedance associated with the dipole nor the low impedance associated with the small loop but will have some intermediate value. The question then is how to define the external field. It was shown in Section 2.4.5 that one can associated low impedance waves with low impedance circuits and vice versa for high impedance waves. It has been suggested [1] that one can make some allowance for the impedance of radiating systems by making the following changes to the calculation procedure outlined above. Let the circuit impedance be Z_c, then Equation (4.37) is modified to read

$$Z_1 = Z_c \qquad \text{for } r < \frac{377c}{2\pi f Z_c} \tag{4.37a}$$

$$Z_1 = \frac{377c}{2\pi f r} \qquad \text{for } \frac{377c}{2\pi f Z_c} < r \leqslant \frac{c}{2\pi f}$$

and

$$Z_1 = 377 \qquad \text{for } r > c/2\pi f$$

whereas Equation (4.38) becomes

$$Z_1 = Z_c \qquad \text{for } r < cZ_c/(377(2\pi f)) \tag{4.38a}$$

$$Z_1 = 377(2\pi r f/c) \qquad \text{for } cZ_c/(377(2\pi f)) < r \leqslant c/(2\pi f)$$

and

$$Z_1 = 377 \qquad \text{for } r > c/(2\pi f).$$

Recent work [24] has measured the wave impedance at different distances away from a variety of loaded loops. The results show that far from the loop the impedance approaches 377 Ω and that near the loops, low impedance loops produce low wave impedance and vice versa. Unfortunately the results show that the equations above are not well obeyed and depend on the loops' orientation. When this is added to the fact that in any real circuit the loops are not simple and are at differing distances from the shield, one can only expect inaccurate estimates based on the above procedure.

At low frequencies a number of phenomena become apparent if the results obtained from the extended plane wave model are examined carefully.

(*a*) For magnetic materials ($\mu_r > 1$) a steady residual value of SE is reached no matter how low the frequency is taken. This can be identified with ducting phenomena discussed in Section 4.2. Indeed a suitable manipulation of (4.35) for small k, γ, δ leads to an expression close to the ducting result (4.10) for the spherical case [1] even though the two field geometries are not identical. This phenomena is illustrated in Figure 4.22 for a spherical shield situated 0.1 m away from a small loop placed at its centre. If the shield is

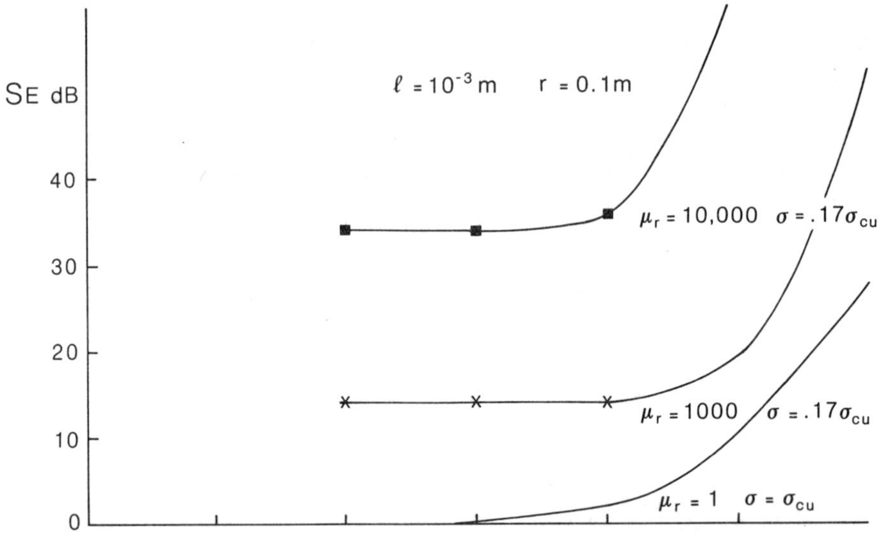

Fig. 4.22 The total magnetic shielding effectiveness SE dB as a function of frequency, calculated with the extended plane wave model, of a 0.1 m radius sphere made from 1 mm thick material with different electrical properties. The sphere has a low impedance source at its centre. For magnetic materials ($\mu_r > 1$) residual shielding is obtained even at the lowest frequencies

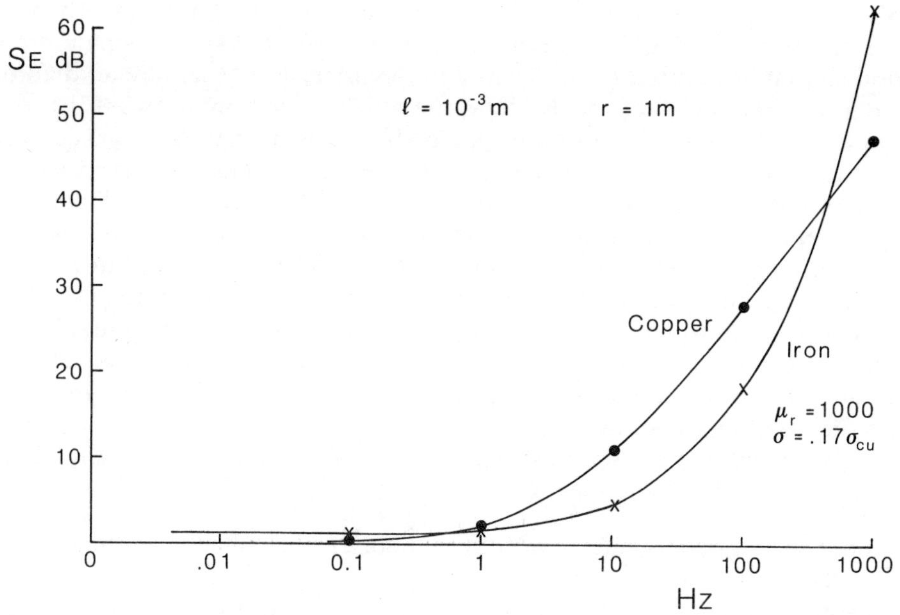

Fig. 4.23 Comparison of the low frequency magnetic shielding effectiveness SE dB as a function of frequency, calculated with the extended plane wave model, for thin (1 mm), spherical (1 m radius), copper and iron shields. The calculation is done assuming the loop source is at the centre of the shield

made of copper then SE approaches zero whereas if it is made of iron then, depending on the value of μ_r, SE reaches a variety of residual values.

(*b*) Reflection loss becomes important at low frequencies so that in some circumstances copper can make a better shield material than iron (over a restricted range of frequencies). This is illustrated in Figure 4.23. In some circumstances a conducting coating (e.g. copper) is used on magnetic shields (on the side facing the impinging field) so as to improve their low frequency performance [25].

4.6 THE RELATIONSHIP OF SHIELDING THEORIES TO PRACTICAL SHIELDS

It has been shown that shielding theories give good agreement with experiments provided the field geometries are essentially solutions of Maxwell's equations for specific boundary conditions. These solutions usually have a restricted range of applicability and have been arrived at using some approximations. In spite of this under the right circumstances the theories give good agreement with experiments provided the field geometries are appropriate in both cases. Of the theories presented the extended plane wave model is probably the most generally useful. It is important however to realise that it is hardly even representative of a real industrial shielding problem. Thus the source of any disturbing field cannot usually be represented by a single dipole or loop so that one would have to allow for a variety waves all having different impedances and angles of incidence at the shield. Such an approach is rarely appropriate. The theories presented have shown even for the idealised cases that a huge range of SE is possible and this would suggest that even slightly different boundary conditions can lead to markedly different results. Figure 4.24 compares the theoretical results of Harrison and Papas [11] for the case of a spherical Al shield immersed in a uniform magnetic field with those for an identical shield surrounding a small current loop, the last case being calculated using the extended plane wave model. At frequencies exceeding 10 kHz the results of the two approaches agree well. This is because skin depth effects then dominate. At lower frequencies however there are considerable differences due to the different field geometries.

The main results of the various theories can be summarised in the following way.

(1) In the appropriate regime, each of the theories is capable of producing good results for specific geometries.
(2) In some cases approximation to one theory can lead to similar equations to those derived by another [1, 2].

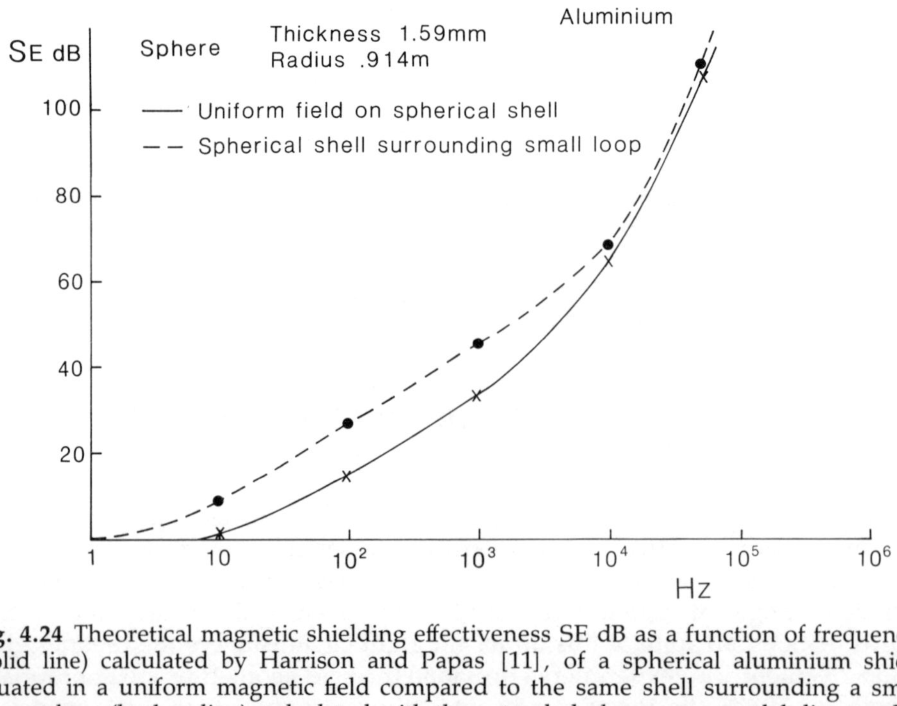

Fig. 4.24 Theoretical magnetic shielding effectiveness SE dB as a function of frequency, (solid line) calculated by Harrison and Papas [11], of a spherical aluminium shield situated in a uniform magnetic field compared to the same shell surrounding a small current loop (broken line), calculated with the extended plane wave model discussed in the text

(3) For shields thick enough to be self supporting ($L \approx 10^{-3}$ m) and for frequencies greater than 10^4 Hz the calculated values of SE are very high (>200 dB). Such values exceed by a long way any practically achievable shielding performance.
(4) Low impedance (H waves) waves are the most difficult to shield at low frequencies and as such are usually taken to be the performance determining environment at these frequencies.
(5) Thin shields can produce useful attenuation factors in a number of practical situations.
(6) The equivalent circuit model indicates in a very clear way the importance of resistive sections (joints, corners, holes) in any shield. Such defects allow fields to penetrate so degrading the shielding behaviour.

Figure 4.25 gives the shielding performance of a typical well constructed commercial screened room. These results given in curve A are for a modular screened room system which uses two 0.56 mm steel panels separated by plywood as the basic shielding material. If one compares the results for SE given in Figure 4.21 for 1 mm of iron with the ones in Figure 4.25 it is clear that there is a very large difference for $f > 10$ kHz. This is due to the inevitable imperfections in any real system, however well constructed. These imperfections are caused by three (usually unavoidable) departures from the concept of a perfectly closed screen, namely:

(1) Apertures. These may be doors, access panels, gaps in joints, air vents, or entry holes for cables and necessary mechanical drives.

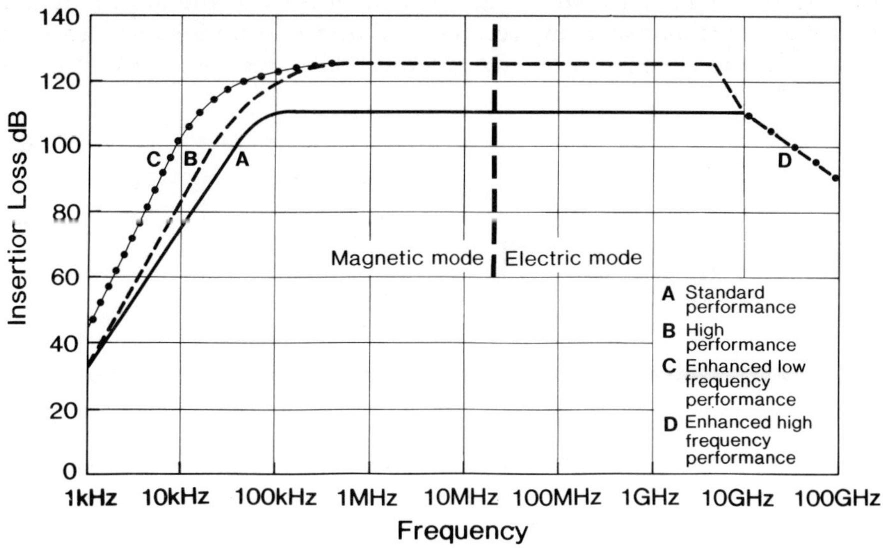

Fig. 4.25 Shielding performance of a well constructed commercial modular screened room system, courtesy of Ray Proof Ltd. Curve A is for standard rooms; curves B, C, D are for high performance versions. (Reproduced by permission of Ray Proof Ltd)

(2) Regions of high resistance in the shield. These are usually associated with joints in the structure. They will also occur when partially conducting windows are used to provide visual access.
(3) Cables entering the screened enclosure. Such cabling can carry unwanted signals into the inner space so lowering the perceived shielding effectiveness.

These defects mean that shielding theory cannot be easily used to design practical shields. Such imperfections also have a bearing on the use of standard measurement techniques to quantify shielding performance as an alternative route for practical designs. This has been discussed by Vance *et al.* [6, 26]. A standard insertion loss measurement often used to quantify shielding effectiveness is that specified by the United States MIL-STD-285 which is illustrated in Figure 4.26. In order to measure the shielding properties the following procedure is carried out. In Figure 4.26(*a*) the source is set to fixed output voltage V_s and the attenuator adjusted until the receiver produces a signal voltage V_1. This attenuation A_s is noted. The shield is then removed as indicated in Figure 4.26(*b*). The source and receiving antenna remain the same distance apart. The attenuator is then adjusted so that for the same source voltage the receiver signal remains at the level V_1 given in the first test. The new value of the attenuation A_0 is noted. Thus $A_0 - A_s$ gives the insertion loss of the system. It is important to recognise that even for a perfectly closed shield this measurement contains factors that depend on both the measurement system as well as on the properties of the shield. This is because, as is clear from shield theory, the penetration of the fields depends on the exact nature of the source together with its distance and orientation from the shield. It is also obvious that the response of any receiver will depend on the exact location and orientation of the effective receiving antenna. It is therefore unlikely that one

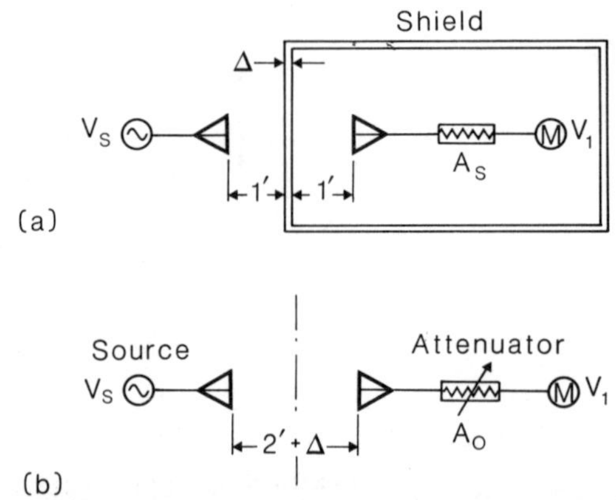

Fig. 4.26 Principle of shielding effectiveness measurement in MIL-STD-285, adapted from reference [6]; (*a*) coupling with shield present, (*b*) reference coupling. (Reproduced by permission of IEEE)

can use data obtained from MIL-STD-285 to design specific shields for arbitrary circuits located anywhere within the shield when the shield is excited by an unspecified source. The results will only be of direct use when the practical system closely matches the test system. This general problem is made worse when defects in the shield are considered. Some examples of situations that could arise are shown in Figure 4.27 [6].

It is clear for the situation depicted in Figure 4.27(a) that the low frequency coupling of the source S' is likely to be large for the receiver R because it is dominated by the power line configuration. Thus it is in no way related to the coupling experienced in the standard insertion loss measurement. Similarly Figure 4.27(b) indicates what might happen in a high frequency case involving, say, a mesh shield. When the source and receiver are in the position for a standard insertion loss measurement a number of different coupling paths could determine the response of the receiver. It might be that destructive interference occurs so indicating an apparently low coupling. On the other hand with the receiver at R and the source at S' constructive interference might occur so leading to a much larger coupling than would be expected from the shielding effectiveness measurements.

When the possibility of internal (and external) resonances is added to all the above problems it is clear that any shielding calculation or experiment has to be considered very carefully before applying it to a specific design. In particular such designs must concentrate on the correct estimation of the effects of the various shield imperfections once the shield is thick enough to provide adequate attenuation of any fields penetrating directly through the material [6].

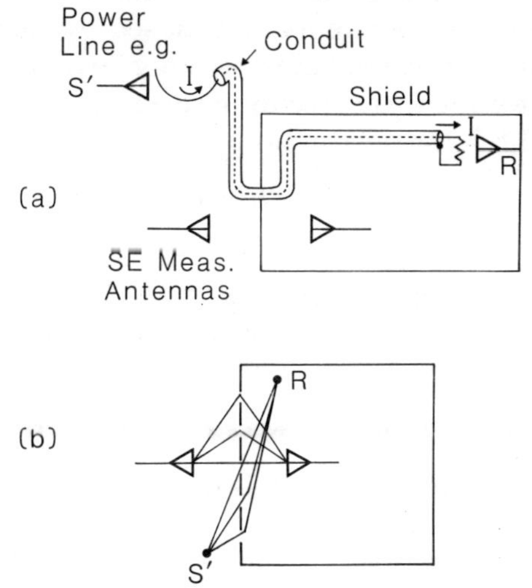

Fig. 4.27 Hypothetical situations where the actual coupling between an antenna R inside a shield and a source S' outside the shield are different from the assumed standard SE measurement situation, adapted from reference [6]. (a) S' couples to R via a power cable carried in a conduit. (b) transmission through different apertures in a mesh shield interfere. (Reproduced by permission of IEEE)

4.6.1 Apertures

The problem of calculating the fields penetrating an aperture in a conducting plate is usually dealt with by computing the equivalent magnetic and electric dipoles which when situated in the aperture generate the required fields. These dipoles depend on the shape of the aperture and the nature of the exciting fields [27–29]. Detailed studies [28] indicate that the in most situations magnetic field penetration is the dominant process creating the far field generated by the excited aperture. It is possible to estimate the effect of holes in an otherwise impermeable conducting shield by using the standard formula for the gain G (see Section 5.3.3) of an aperture illuminated by a plane wave [29].

For a circular aperture of radius a:

$$G = (2\pi a/\lambda)^2 \tag{4.39}$$

then the shielding effectiveness SE is

$$\mathrm{SE} = 10 \log_{10}(P_1/P_2) \text{ dB}$$

with P_1 = the power received with no shield and P_2 = the power received with the shield plus hole in place. Since $G = P_2/P_1$

$$\mathrm{SE} = 20 \log_{10}(\lambda/2\pi a) \text{ dB}. \tag{4.40}$$

Since G is proportional to area we can write for n apertures

$$S = 20 \log(n^{-1/2}\lambda/2\pi a) \text{ dB} \tag{4.41}$$

where λ = the wavelength of the radiation.

In order to allow for the fact that real shields have finite thickness it is necessary to calculate the extra attenuation caused by each aperture when it is considered to be a length of waveguide operating beyond cut-off. Ott [27] gives a convenient formula for the extra shielding effectiveness due to a circular guide namely (see Section 3.8.3 for derivation):

$$\mathrm{SE} = 32t/2a \text{ dB} \tag{4.42}$$

where t = length of the waveguide and a = the radius of the waveguide.

This formula only really applies for guides of length $t > a$. Thus the total shielding effectiveness for a thick sheet with n apertures can be approximately written

$$\mathrm{SE} = 20 \log(n^{-1/2}\lambda/2\pi a) + 32t/2a \text{ dB}. \tag{4.43}$$

The first term can be considered to represent the reflection loss and the second term the absorption loss of the aperture [1, 31]. Figure 4.28 shows this function (for $t = 1$ mm) superimposed on a SE curve for a typical screened room. The curve for $n = 2.5 \times 10^5$ simulates what might be expected for a room with mesh walls ($a = 0.5$ mm, holes on a 2 mm pitch, wall size 1 m × 1 m). Here it is assumed that n is set by the number of apertures in one side only (the side facing the impinging radiation). Results are also given in Figure 4.28 for an infinite wire mesh screen calculated using an equivalent circuit theory for the

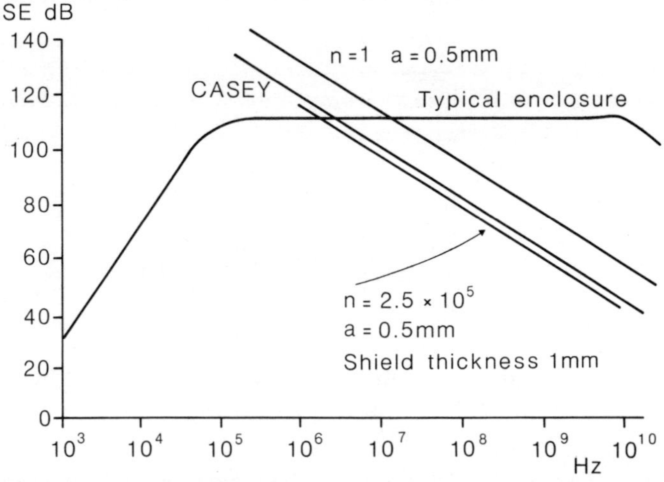

Fig. 4.28 Effect of a single $n = 1$, or an array, $n = 2.5 \times 10^5$, circular apertures of 0.5 mm radius on the shielding effectiveness of a typical (screened) enclosure, calculated using Equation 4.43. The line labelled CASEY refers to results for an infinite wire mesh screen of 1 mm wires having a 2 mm spacing calculated using an equivalent circuit theory, reference [30]

shielding effectiveness due to Casey [30]. In this case it is assumed that the wires are 1 mm in diameter and that their axes are 2 mm apart. Reasonable agreement is apparent between the results obtained using the two different methods demonstrating that it is important to allow for the shields thickness when calculating the effect of apertures. At higher frequencies than the limit given by the SE = 0 condition various resonances occur since the apertures act as efficient antennas [5]. These results suggest that mesh walls could be safely used in low frequency protection schemes. Unfortunately one always has to be on guard that frequencies outside the obvious limits of the system do not effect its operation for example by introducing level shifts due to unexpected rectification effects.

Any large aperture (e.g. for admitting a mechanical drive, providing ventilation etc.) can be made to have a higher screening effectiveness by turning it into a length of waveguide operating beyond cut-off. Such a system is shown in Figure 4.29. Note that the predictions of (4.42) are for an open guide only. If metal is inserted (cables, drive rods, armatures etc.) the situation could be better or worse than indicated by (4.42) depending on the exact way the penetration is treated.

So far only circular apertures have been considered. An important type of aperture in practise is a slot (length $l \gg$ width w). Simple arguments show that a slot is much more important than a circular aperture of the same area [27]. Following Ott [27] and realising that magnetic leakage is usually most important one can use the circuit model of shielding to picture what is happening. For effective magnetic shielding the induced shield currents have to

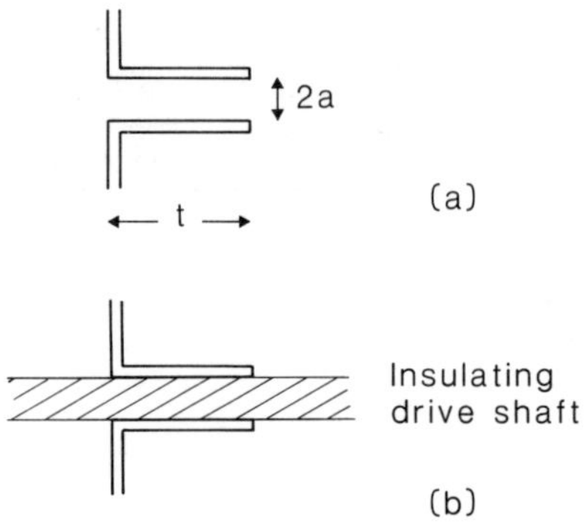

Fig. 4.29 Use of a waveguide operating beyond cut-off to introduce an insulating mechanical drive. Conducting drives need to be treated differently

flow in precisely the correct place in order to cancel out the impinging field (see (4.32)). Thus as is shown schematically in Figure 4.30, a slot causes far more current deviation than a single aperture or even a row of apertures. The orientation of the slot with respect to the current flow is also important as is clear from comparing Figure 4.30(*c*) with 4.30(*d*). At high frequencies the slot can act as an efficient slot antenna (see Section 5.3.3). For the least favourable orientation (the one needed for worst case design) Ott [27] gives a formula for

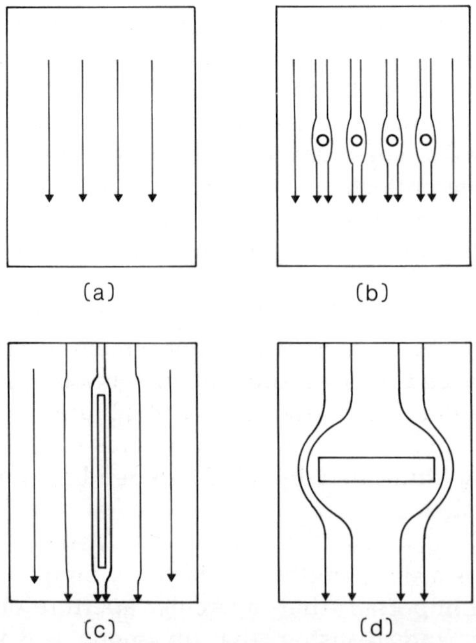

Fig. 4.30 Schematic diagram showing the effects of slots and holes on shield surface currents

the shielding effectiveness due to the presence of a slot in an impervious shield which is similar to (4.43) above but only involving the slot length rather than the area,

$$S = 20\ \log(\lambda/2l)\ \text{dB} \tag{4.44}$$

where l = the slot length. Once again this formula only applies to slots in thin shields. In order to calculate the extra shielding effectiveness due to a real shields thickness one can make use of the attenuation formula for a waveguide operating beyond cut-off. In this case a rectangular guide is the required shape and Ott [27] gives the following formula for the resulting value of SE (see Section 3.8.3)

$$\text{SE} = 27.2t/l\ \text{dB} \qquad \text{for } t > 1 \tag{4.45}$$

where t = the thickness of the shield and l = the maximum transverse dimension of the guide (i.e. the slot length).

Thus for an impervious shield of finite thickness containing a slot the total shielding effectiveness is given by;

$$\text{SE} = 20\ \log(\lambda/2l) + 27.2t/l\ \text{dB}. \tag{4.46}$$

Once again the first term represents a reflection loss (for a slot in a thin sheet) and the second the absorption loss. Similar but slightly more complex formula to (4.43) and (4.46) are to be found in EMC handbooks [1, 31]. Slots are often found in practice when two pieces of metal have to be joined and welding (or continuous soldering) is impracticable. One can use (4.46) to examine the behaviour of some joint designs that might be met in real situations. Figure 4.31 shows two idealised designs, the first of which in Figure 4.31(a) consists of just butting two thin sheets (t = 1 mm) together and holding them in that position by screw fixings every so often along the seam. Even if the sheets are very carefully made the regions away from the fixing points will tend to separate as shown (much exaggerated) in Figure 4.31(b). Such small separations constitute slots. The total length of such slots will depend on the quality of construction and the frequency of fixing locations. Figure 4.32 shows calculated values of SE for such a joint design with different assumed final slot lengths compared with the typical screened enclosure shielding effectiveness. It is clear that slot apertures have to be kept to a minimum if a reasonable performance is required (it is usually recommended that l be kept to less than $\lambda/50$). Fortunately the problem can be helped to a large extent by creating thicker regions near the joint by, for example, bending the sheet through a right angle before bolting or screwing through the overlapping region. This is equivalent to increasing t as is shown in Figure 4.31(c). The effect of doing this on the calculated values of SE is shown in Figure 4.32 for t = 2.5 cm. It is clear that the overlapping joint has improved the expected values of SE for lower slot lengths. This technique of overlapping joints is commonly used in many practical designs though the exact procedure can vary considerably (see for example Figure 4.35(d) below and reference [1]). Overlapping also can be considered beneficial when it is realised that a poorly conducting joint could be modelled as a capacitor. Thus at high frequencies much lower joint impedances are obtained when overlapping is carried out.

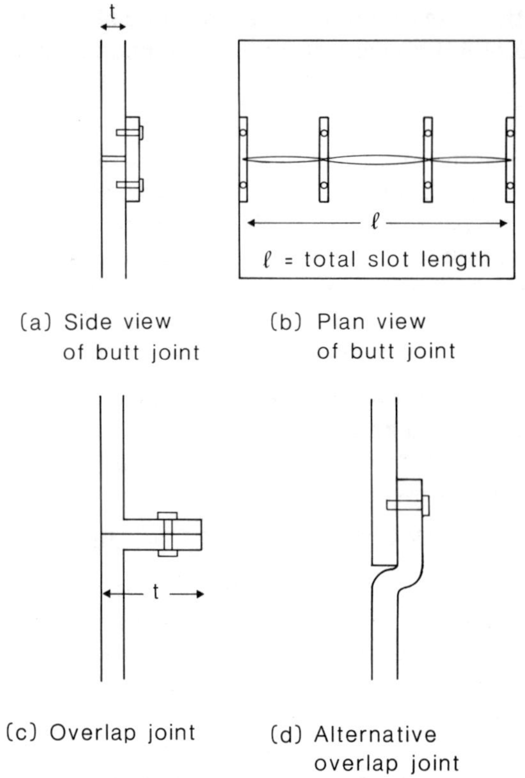

Fig. 4.31 Critical dimensions of some joint designs

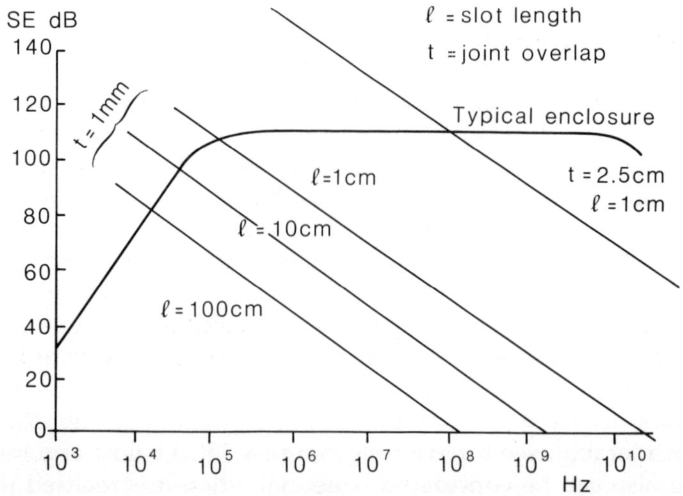

Fig. 4.32 Effects of various different length slots of length l with overlap t on the Shielding Effectiveness SE dB as a function of frequency of a typical (screened) enclosure, calculated using Equation (4.46)

Besides overlapping a number of other techniques have to be employed if successful joints are to be constructed. Increasing the frequency of bolting points will decrease l and some guidelines for good design are available [1]. It is also clear that smooth flat mating surfaces also lower the effective value of l but as these are difficult to manufacture it is usual to consider the use of gaskets to make the most reliable joints. A gasket is a deformable conducting interface between the two materials to be joined. The gasket may be a metallic mesh or a conducting elastomer [1, 27]. Such a deformable interface must have a low enough resistivity such that if it fills the aperture no electromagnetic penetration can occur. The joint, gasket and its fixing must be capable of withstanding the life cycle of vibration and handling the structure is expected to encounter without changing its initial performance. The final problem to be tackled either in the case of a plain or a gasketed joint is the correct treatment of any possible corrosion effects on the conductivity of the joint. It is obvious that any high resistance region behaves similarly to an aperture in that it will impede current flow. Thus oxidation (this is a major problem with untreated aluminium structures) and the corrosion that occurs between incorrectly chosen pairs of metals has to be avoided or minimised. Full details of the various procedures available (e.g. surface treatments, correct choice of gaskets/metals), whilst outside the scope of this book, are to be found in a number of EMC handbooks [1, 27]. It is clear from the preceding discussion that the best type of joint would be to weld the metal parts together and this is in fact the procedure adopted for the most secure screened enclosures. Finally it should be noted that the formulae given in this section (e.g. (4.43) and (4.46)) are approximate, apply only to plane wave excitation and give results for the far field of the aperture. Therefore their use for other types of excitation or in the near field could result in serious error.

4.6.2 Windows and thin conducting films

Whenever visual access is required into a screened enclosure some sort of window has to be employed. This window must permit visible electromagnetic wavelengths to pass but prevent the passage of the longer wavelengths important in determining the screening behaviour of the system. Such an aperture or window can be constructed in two ways:

(1) A fine metal mesh can be sandwiched between two transparent sheets (glass or suitable plastic).
(2) A thin conducting layer can be applied to glass or a transparent plastic. For example a metal layer about 10^{-8} m thick will suffice (see reference [20] for a discussion of the optical transmitting properties of thin metal films). As before a sandwich is usually preferred to prevent corrosion and protect the thin metal layer. Such layers are often specified in terms of their surface resistivities and values between 5 and 20 $\Omega\,\square^{-1}$ are typical. As the metal layer is reduced in thickness so the surface resistivity and light transmission increase but the shielding effectiveness is reduced. Fortunately at some optimum thickness a reasonable compromise can be reached [1, 20].

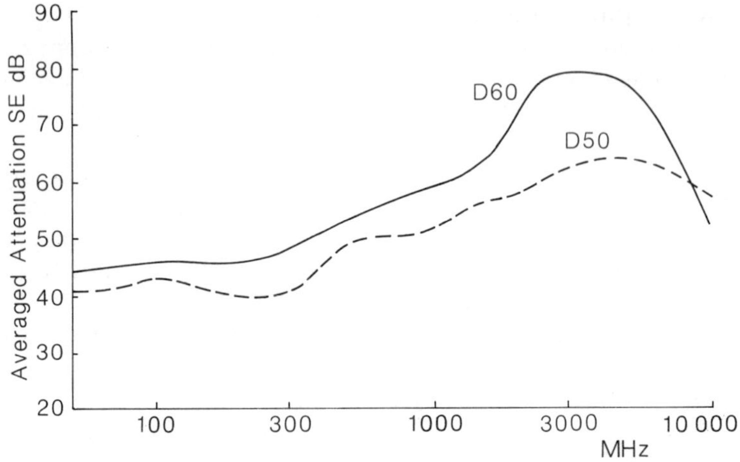

Fig. 4.33 Averaged attenuation measurements as a function of frequency for two commercial double glazed conducting metal layer windows, tradename 'DATASTOP'. (Reproduced by permission of Pilkington Glass Ltd.)

In both cases good contacts at the edge between the rest of the screened enclosure and the conducting region of the window are required. Typical results for some commercial double glazed units are those given in Figure 4.33. For the results given conducting coatings are on three of the four possible surfaces. It should be noted that these measurements (as is usual) are taken using a variation of the MIL-STD-285 insertion loss technique with all its attendant problems (see Section 4.6). Nowadays [28] a variety of more advanced measurement techniques are used to define the shielding properties of materials. These are designed so that the properties of the measurement system do not intrude into the overall result.

It is generally accepted that for small total areas mesh screens are able to achieve higher values of SE than single thin metal layer designs. Typical values are given in Table 4.3. On the other hand the visual access and appearance of the transparent film designs are superior so that final choice depends on the exact needs of a particular system. Estimates of the expected values of SE can be made (for plane wave situations) using the formulae given in Section 4.6.1 for the mesh and by using the results given in Figure 4.21 for thin continuous metal films. Further information on the installation and choice of windows can be found in a number of handbooks and commercial guides [1, 31].

Table 4.3. Window shielding effectiveness

	f (MHz)	SE (dB) Conductive coating	SE (dB) Range of wire meshes
E wave	10	70–80	60–110
Plane wave	1000	30–40	20–70

The metal films used for windows are generally about 10^{-8} m thick. Much thicker films (greater than about 10^{-6} m) can be used to improve the performance of the plastic enclosures widely used by modern industry to produce economic container designs for electrical equipment. A number of different processes can be used to make plastic act as a useful shield. Amongst these are:

(1) conductive paints,
(2) vacuum evaporation,
(3) electrodeless plating (chemical deposition),
(4) flame/arc spray.

It is also possible to stick metal foils onto the plastic or to load the plastic with conducting particles. At the present time however the first four are the most common. Factors which affect the choice in any given situation are:

(a) The thickness and surface resistivity of the deposited layer or layers. As seen in Section 4.5 these factors affect the expected shielding effectiveness.
(b) The stability of the film over the working life of the equipment. Problems here are corrosion resistance, adherence to the substrate and the interaction between the plastic, the film and the processing technique.
(c) The ease with which contacts to the shield (necessary for cable entry, grounding, proper treatment of apertures etc.) can be made. This can be a problem when plastics are loaded with metallic particles.
(d) Economics and aesthetic appearance. This factor tends to rule out the use of stuck on metal foils except in the development process.
(e) The influence of the chosen process on the casing design. Factors here can be limitations on the choice of plastic and allowable shapes.

At one time the dominant processes were flame/arc spraying of metals and conventionally sprayed conductive paints [1]. More recently, however, electrodeless plating with nickel/copper and vacuum evaporation of aluminium have started to more widely used [27]. In some cases moreover special metals are being deposited using vacuum sputtering. Sputtering allows almost any metal to be deposited on any plastic. For good design of metal coated plastic shields the same considerations as with all metal shields apply. That is to say, that once adequate metal thickness is obtained the design must concentrate on dealing with shield defects.

4.6.3 Alternative ways of describing shield quality

It is clear from the preceding sections that shielding effectiveness as determined by measurement or by theory does not necessarily provided a good design basis for practical equipment. As Graf and Vance [6] have pointed out neither theory nor experimental testing can usually mimic the exact electromagnetic environment met in typical systems. It would appear therefore that there is a need to develop a measure of a shields performance that could be shown to be clearly independent of any system used to measure it. Moreover this new measure should enable some bounds to be set on the shields behaviour in practical systems [6]. One possibility they considered is the use of the concept of the transfer impedance (and admittance) which describes cable shielding reasonably adequately. See Section 3.7.2 for an introduction.

When an electromagnetic wave impinges on a coaxial cable it is convenient to split the interaction into three parts so that the relevant features can be highlighted. Firstly the time changing E and H fields will induce a current in the cable screen. In general this current will be made up of the direct field induced value plus any current injected from either end of the cable shield. If the cable is electrically short the current will be constant along its length. The internal values of E and H will be lower than those outside and this of course will be affected by the perfection of the shield. The presence of the shield current causes a voltage drop along the internal surface of the shield (dependent on the skin depth, the shield construction and thickness) which then causes an electrical disturbance to the signal on the central conductor. This direct effect of the shield current is described by the transfer impedance. This impedance will consist of a resistive term dependent on the skin depth and an inductive part which is caused by the presence of any apertures in the shield [5, 32, 33]. Secondly the attenuated E field will couple to the centre conductor and this is described by the transfer admittance. Thirdly any residual value of H could also couple to the centre conductor. However, provided the cable is truly coaxial the flux cutting either side of the centre conductor will produce an equal and opposite induced voltage and so can usually be neglected.

The transfer impedance Z_t for a cable shield that is electrically small in diameter is defined as

$$Z_t = \frac{1}{I} \times \frac{dV}{dx} \tag{4.47}$$

where I = the external current on the shield, dV/dx = the open circuit voltage per unit length (x) produced on the internal surface of the shield, and Z_t = the transfer impedance/unit length.

Hence for a cable of length $L (L < \lambda/4$ [32]) the maximum internal shield voltage is given by:

$$V = Z_t L I. \tag{4.48}$$

In general the current and voltage appearing on the signal wire will depend on the cable length and the terminating impedances Z_E at either end. General formulae for some cases have been given by Harrison [32] and Vance [33] which show that it is usually best to terminate a long cable at both ends with its characteristic impedance Z_0. For a short length similarly terminated the noise current i will be

$$i = Z_t L I / 2 Z_E. \tag{4.49}$$

The effect of the residual internal electric field can be considered to produce a current source between the shield and the centre conductor [5] and this leads to the concept of the transfer admittance Y_t which is defined as:

$$Y_t = \frac{1}{V} \times \frac{dI}{dx} \tag{4.50}$$

where V = the external voltage on the shield which will depend on the external electric field or charge density on the surface, Y_t = the transfer admittance/unit length, and dI/dx = current flowing directly from the external source to the internal conductor.

Once again the effect on the signal will depend on the terminations but usually these can be chosen [5] (L large and Z_t small) so that the dominant overall effect of the impinging E, H field is given by the contribution calculated from the transfer impedance alone. It is this fact that makes Z_t such a good description of the cable shield. The next question to be answered is: can this concept be extended to model the effect of arbitrary shaped shielding enclosures?

In the case of a three-dimensional field problem one can define a transfer impedance for each point on the shields surface in terms of the local field values [6]. Thus

$$Z_t = E_{t_i}/H_{t0} \tag{4.51}$$

where E_{t_i} is the tangential electrical field produced inside the shield by the tangential magnetic field H_{t0} on the outside of the shield. As before this has the units of ohms. Unfortunately Z_t will vary from position to position on the shield and this when combined with the need to calculate the current distribution at every point means that the calculation of the internal surface fields is very difficult. Similarly the calculation of the effect of these internal fields on the circuits within the shield is also going to be complex so that the use of the transfer impedance approach is going to be no easier than that experienced when attempting to use the concept of shielding effectiveness for enclosures of arbitrary shape.

To overcome this problem of assessing a real shield's performance Graf and Vance [6] have suggested the following approach which concentrates on assessing the effect of the defects of the shield. They observed for nearly all screening scenarios (excluding cables, thin shields and situations involving low frequency magnetic fields) that any direct field penetration through the material of the shield is negligible. With this assumption one has only to allow for the effects of the apertures and the cable entries on the design. In the case of an aperture we know that any exterior magnetic or electric fields will penetrate through the opening. It is possible to estimate the maximum effect this penetration might have on any internal circuit by using the idealised situations illustrated in Figures 4.34(*a*) and (*b*). First of all consider the case of magnetic field penetration. The maximum open circuit voltage induced on any internal circuit will be when all the flux entering through the aperture cuts the circuit. A circuit which does this [6] is shown in Figure 4.34(*a*). It is orientated in the direction of the current flow on the shield (which of course is at right angles to the magnetic field near the surface) and for this situation the open circuit voltage V_{oc} is clearly

$$V_{0c} = \mu_0 A_m \, dH/dt. \tag{4.52}$$

Here A_m is an effective area of the aperture which can be defined as being involved in the flux penetration. This voltage can be measured and represents an upper limit of the effect of this aperture. A similar approach can be used for the case of electric field penetration, but in this case the situation is shown in Figure 4.34(*b*) and the quantity estimated is the short circuit current I_{sc} flowing from the interceptor plate to ground. The plate is located so as to intercept all the electric flux penetrating the aperture. This current, which is the maximum the external E field could induce on any internal circuit, is given by

$$I_{sc} = \varepsilon_0 A \, dE/dt. \tag{4.53}$$

In this case A is approximately the area of the aperture.

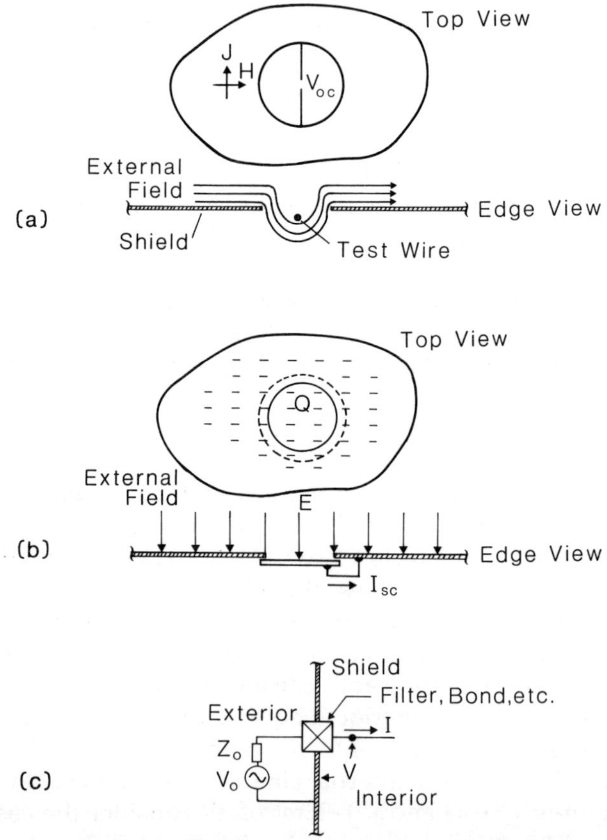

Fig. 4.34 Schematic methods for estimating the maximum effects of different shield defects, adapted from reference [6]. (Reproduced by permission of IEEE)

Of course the limits given by Equations (4.52) and (4.53) are upper bounds and more realistic estimates might be obtained by enforcing exclusion zones for the internal wiring. For example [6] one might require any wiring to be located at least x m from the aperture. In this case the wire and sheet used in Figures 4.34(a) and (b) would be on the surface of an x m radius hemisphere inside the shield, so producing much lower estimates of the disturbing effects of the penetrating fields.

It is also possible to estimate the maximum effect of any penetrating conductor. The transfer characteristic of the conductor can be obtained by driving the external circuit with its Thévenin equivalent source and then measuring the voltage and current that are transmitted through the shield as is schematically illustrated in Figure 4.34(c). The resulting values can be used to construct an equivalent Thévenin (noise) source which could be used to drive the internal circuit so as to estimate any disturbing effects.

This overall approach, whilst simplifying the calculation of internal effects, still leaves the problem of calculating the external fields and equivalent sources. Vance and Graf [6] suggest that this is easier than having to solve the total problem when attempting to use shielding effectiveness or transfer impedance concepts.

If SE values (either measured or calculated) are to be used then care has to be taken in assessing the degree of protection provided. This is not because of any imprecision in the procedures but because the practical electromagnetic boundary conditions are often insufficiently close to those utilised in a particular theory or measurement system.

4.6.4 Cables and connectors

In the last section the importance of the transfer impedance in assessing the response of cables to external fields was emphasised. The transfer impedance can be measured using a system [5] that passes a known current on the outside of the cable. One end of the coaxial cable is shorted and the voltage appearing on the central conductor at the other (open circuited) end is measured. Typical results [32] are shown in Figure 4.35. The results for solid sheath cables show a characteristic drop in Z_t with frequency. The low frequency limit is just the DC resistance per unit length of the cable sheath, R_0. Such results can also be calculated from the classical formula discussed in Section 3.7.2. (see also Vance [33]). When braided cables are measured (see for example curve 3 in Figure 4.35) it becomes clear that above a certain frequency the inductive term starts to dominate and the magnitude of Z_t increases. This makes normal braided cables very poor at frequencies greater than about 10 MHz [5, 32–36]. Vance [33] has presented approximate theoretical expressions for Z_t/R_0 for certain designs of flexible screened cable. Some of the results of these calculations are shown in Figure 4.36 for a variety of optical coverages K and different parameters of the braiding. It is clear that the presence of the holes in any braided design leads to appreciable values of Z_t at high frequencies. See references [37] and [31] for more data.

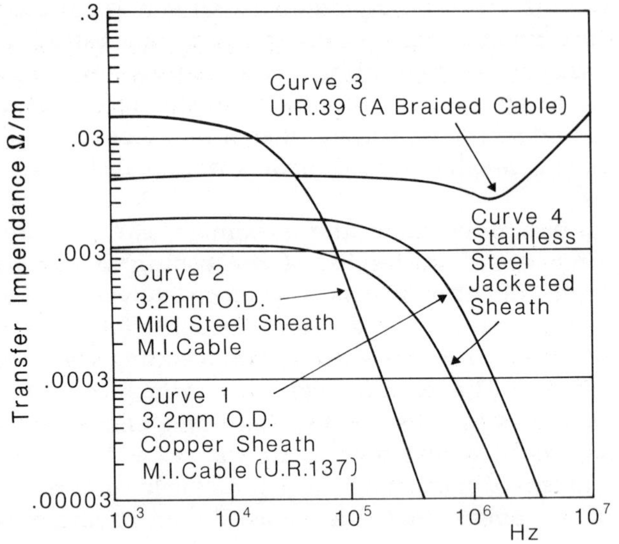

Fig. 4.35 Measured transfer impedance for a number of cable designs, adapted from reference [32]. (Reproduced by permission of IERE)

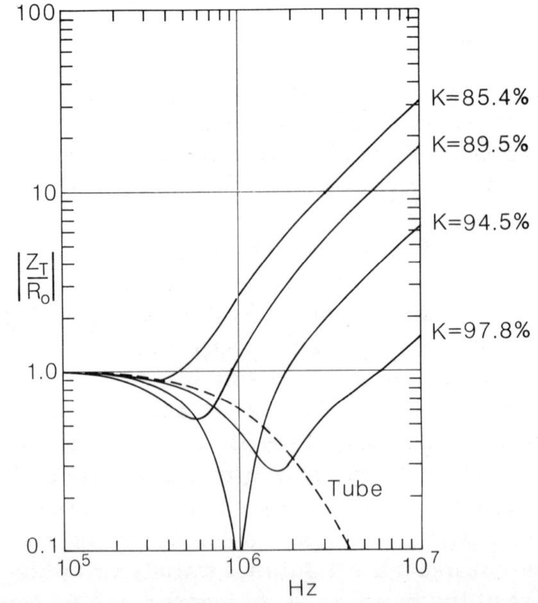

Fig. 4.36 Calculated transfer impedance of various braided-wire shields, after Vance, reference [33]. (Reproduced by permission of Wiley, New York)

If large currents are expected to flow in the screens of any flexible cable then thought must be given to decreasing Z_t from the values measured and calculated above. The effects are likely to be important when Z_tIL is greater than about one tenth the signal voltage. It is obvious that L should be kept as small as possible. Improvements in Z_t can be obtained in a number of ways some of which are listed below.

(1) Increase the optical coverage of the braid.
(2) Use two or three braids (or more) overlaying the primary screen. The use of ferrite beads in between layers can force the current to flow in the outer layers so lowering Z_t.
(3) Lay the cable inside a suitably connected copper tube or other metallic conduit.
(4) Use special superscreen designs [39, 40] which utilise helically wrapped mu-metal tape between two normal braids.

In general most design changes which improve Z_t also make the cable system more expensive and much stiffer.

At each end of any cable connectors must be used to mate the cable to the system it serves. These joints have all the problems outlined for cables plus the extra need to withstand the opening and closing of the joint during the system lifetime. A good connector should have the following attributes:

(1) It must make a low resistance joint and have a good match to the cable characteristic impedance.
(2) The screen should make a 360° connection to the screened enclosure of the next part of the system so as to keep Z_t low and avoid the 'pigtail' effect [27, 41, 42]. This is discussed further in Chapter 6.
(3) The resulting transfer impedance should be low [33].

4.7 SOME CONCLUSIONS AND A COMMENT ABOUT GROUNDING

In this chapter the various theories of shielding have been presented and compared with both ideal experiments and more practical systems. It is possible to make a few general points in what is admittedly a complex area.

(1) Plane and E waves are easy to shield against.
(2) H or low impedance waves are difficult to shield at low frequencies.
(3) Shielding theories give good results where their selected boundary conditions match those of the experimental test system.
(4) Theoretical SE values for $f > 10$ kHz and for reasonable shield thicknesses always exceed those achieved practically.
(5) In this regime shield behaviour is dominated by a number of unavoidable defects. A correct assessment of the shields of performance is only obtained by concentrating on these defects.

(6) Apertures and cable entry are the major reasons for the occurrence of the low values of SE normally achieved.
(7) Cable systems should have a low transfer impedance and be properly connected to any shield they enter or leave. The need for this as well as the requirements for proper filtering are more extensively discussed in Chapter 6.

Finally it should be emphasised that an isolated perfect shield that surrounds a protected system does not in principle require earthing. The potential of the shield may take any value without effecting the circuits inside. However real shields have apertures and are connected to cables (signal, mains etc.) so that unless the shield is earthed noise voltages on the shield can couple to the interior circuits. The safety requirements for mains connection will also require an earth. Earthing and grounding will be discussed further in Chapter 6.

REFERENCES

[1] Violette, J. L. N., White, D. R. J. and Violette, M. F. (1987) *Electromagnetic Compatibility Handbook* Van Nostrand Reinhold, New York.
[2] Moser, J. R. (1967) Low Frequency Shielding of a Circular Loop Electromagnetic Field Source *IEEE Trans. on EMC* **EMC-9** 6–18.
[3] Bannister, P. R. (1968) New Theoretical Expressions for Predicting Shielding Effectiveness for the Plane Shield Case, *IEEE Trans. on EMC* **EMC-10** 2–7.
[4] Schulz, R. B., Plantz, V. C. and Brush, D. R. (1988) Shielding Theory and Practice *IEEE Trans. on EMC* **EMC-30** 187–201.
[5] Ricketts, L. W., Bridges, J. E. and Miletta, J. (1976) *EMP Radiation and Protective Techniques* Wiley-Interscience, New York.
[6] Graf, W. and Vance, E. F. (1988) Shielding Effectiveness and Electromagnetic Protection *IEEE Trans. on EMC* **EMC-30** 289–293.
[7] Duffin, W. J. (1968) *Advanced Electricity and Magnetism,* McGraw Hill, London.
[8] Thomas, A. K. (1968) Magnetic Shield Enclosure Design in the DC and VLF Region *IEEE Trans. on EMC* **EMC-10** 142–152. Magnetostatic shielding by iron was first reported by Gilbert W (1600) 'De Magnete, magneticisque corporibus, et de magno magnete tellure; Physiologia nova, plurimis et argumentis et experimentis demonstrata', *Excudebat Petrus Short,* Book 2, Chapter XVI, Londini.
[9] Bridges, J. E. (1988) An Update on the Circuit Approach to Calculate Shielding Effectiveness *IEEE Trans. on EMC* **EMC-30** 211–221.
[10] Lorrain, P., Corson, D. P. and Lorrain, F. (1988) *Electromagnetic Fields and Waves* 3rd edition, W. H. Freeman and Company, New York.
[11] Harrison, C. W. and Papas, C. H. (1965) On the Attenuation of Transient Fields by Imperfectly Conducting Spherical Shells, *IEEE Trans. on Antennas and Propagation* **AP-13** 960–966.
[12] Wheeler, H. A. (1958) The Spherical Coil as Inductor, Shield or Antenna *Proc. IRE,* **46**, 1595–1602.
[13] King, L. V. (1933) Electromagnetic Shielding at Radio Frequencies *Phil. Mag. and J. Sci.* **15** 201–223.
[14] Cooley, W. W. (1968) Low-Frequency Shielding Effectiveness of Nonuniform Enclosures *IEEE Trans. on EMC* **EMC-10** 34–43.
[15] Shenfeld, S. (1968) Shielding of Cylindrical Tubes *IEEE Trans. on EMC* **EMC-10** 29–34.

[16] Lee, K. S. H. and Bedrosian G. (1979) Diffusive Electromagnetic Penetration into Metallic Enclosures *IEEE Trans. on Antennas and Propagation* **AP-27** 194–198.
[17] Franceschetti, G. (1979) Fundamentals of Steady-State and Transient Electromagnetic Fields in Shielding Enclosures *IEEE Trans. on EMC* **EMC-21** 335–348.
[18] Bridges, J. E. and Miller, D. A. (1968) Correspondence: 'Comparison of Shielding Calculations *IEEE Trans. on EMC* **EMC-10** 175–177.
[19] Schelkunoff, S. A. (1938) *Bell System Tech. J.* **17** 17–48.
[20] Liao, S. Y. (1985) *Microwave Devices and Circuits* 2nd edition, Prentice-Hall, New Jersey. See also Klein, C. A. (1990) Microwave Shielding Effectiveness of EC-coated dielectric Slabs *IEEE Trans. on Microwave Theory and Techniques* **38** 321–324.
[21] Ramo, S., Whinnery, J. R. and Van Duzer, T. (1965) *Fields and Waves in Communication Electronics* Wiley, New York.
[22] Miller, D. A. and Bridges, J. E. (1966) Geometrical Effects on Shielding Effectiveness at Low Frequencies *IEEE Trans. on EMC* **EMC-8** 174–186.
[23] Crawford, M. L. (1974) Generation on Standard EM Fields using TEM Transmission Cells *IEEE Trans. on EMC* **EMC-16** 189–195.
[24] Upton, M. E. G. and Marvin, A. L. (1988) Measurement of Transverse Wave Impedance in the Vicinity of Electrically Small Radiating Structures *Proc. 6th Int. Conf. on Electromagnetic Compatibility, University of York* IERE Conference Publication 81, pp. 163–166.
[25] Rizk, F. A. M. (1977) Low Frequency Shielding Effectiveness of a Double Cylinder Enclosure *IEEE Trans. on EMC* **EMC-19** 14–21.
[26] Vance, E. F., Graf, W. and Nanevicz, J. E. (1981, 1983) Unification of Electromagnetic Specifications and Standards, *Defense Nuclear Agency report DNA 5433F-1, DNA 5433F-2.*
[27] Ott, H. W. (1988) *Noise Reduction Techniques in Electronic Systems*, 2nd edition, Wiley, New York.
[28] Wilson, P. F. and Ma, M. T. (1986) A study of Techniques for Measuring the Electromagnetic Shielding Effectiveness of Materials *National Bureau of Standards report NBS/TN1095* (Available from Superintendent of Documents, US Government Printing Office, Washington, DC 20402, USA).
[29] Ramo, S., Whinnery, J. R. and Van Duzer, T. (1965) *Fields and Waves in Communication Electronics* Wiley, New York, 661–668.
[30] Casey, K. F. (1988) Electromagnetic Shielding Behaviour of Wire-Mesh Screens *IEEE Trans. on EMC* **EMC-30** 298–306.
[31] *EMI Shielding Design Guide* (1984) and *Design Guidelines to EMI Shielding Windows* (1984) (Available from Tecknit® EMI Shielding Products, Cranford, NJ, USA).
[32] Harrison, D. (1965) The Mechanism of Interference Pick-up in Cables and Electronic Equipment with special reference to Nuclear Power Stations, *J. Inst. Electron. Radio Eng.* **29** 149–158.
[33] Vance, E. F. (1978) *Coupling to Shielded Cables*, Wiley, New York.
[34] Vance, E. F. (1975) Shielding Effectiveness of Braid-Wire Shields *IEEE Trans. on EMC* **EMC-17** 71–77.
[35] Casey, K. F. and Vance, E. F. (1978) EMP Coupling Through Cable Shields *IEEE Trans. on EMC* **EMC-20** 100–106.
[36] Hoeft, L. O. and Hofstra, J. S. (1988) Measured Electromagnetic Shielding Performance of Commonly Used Cables and Connectors *IEEE Trans. on EMC* **EMC-30** 260–275.
[37] Krugel, L. (1956) Abschirmwirkung von Außenleitern flexibler Koaxialkabel *Telefunken-Zeitung* **29** 256–266.
[38] Krugel, L. (1957) Mehrfachschirmung flexibler Koaxialkabel *Telefunken-Zeitung* **30** 207–215.
[39] Fowler, E. P. (1979) Superscreened Cables *The Radio and Electronic Engineer* **49** 38–44.

[40] Best, D. A. and Fowler, E. P. (1983) On Screens and Layout for Instrument Circuits *The Radio and Electronic Engineer* **53** 147–152.

[41] Paul, C. R. (1980) Effect of Pigtails on Crosstalk to Braided-Shield Cables *IEEE Trans. on EMC* **EMC-22** 161–172.

[42] Hejase, H. A. N., Adams, A. T., Harrington, R. F. and Sarkar, T. V. (1989) Shielding Effectiveness of 'Pigtail' Connections *IEEE Trans. on EMC* **EMC-31** 63–68. See also Hejase, H. A. N., Adams, A. T. and Harrington, R. F. (1989) A Quasi-Static Technique for Evaluation of Pigtail Connections' *IEEE Trans. on EMC* **EMC-31** 180–183.

[43] Schrieffer, J. R. (1964) *Theory of Superconductivity* Benjamin, New York.

5 Spectral Analysis and Antenna Theory in EMC

This chapter will concentrate on presenting the mathematical tools needed to deal with EMC problems where there is radio frequency disturbance, and discuss some aspect of the general theory of antennas which have a bearing on making EMC measurements. The cornerstone of this chapter is Fourier analysis, which explains how periodic non-sinusoidal waveforms actually contain a whole spectrum of sinusoidal components with frequencies which are multiples of the fundamental frequency. This feature is vitally important, for it has been stressed repeatedly that there are many frequency dependent aspects to consider when tackling an EMC problem, so that before any effective action can be taken it is necessary to be able to identify and measure the spectrum of frequencies involved.

The chapter will be arranged as follows. In the first section the mathematics of Fourier theory will be outlined and used to examine in detail the spectrum generated by a train of rectangular pulses. This is then compared with the spectrum associated with an isolated pulse, and the effect of having a finite rise time on the pulse is examined. In Section 5.2.9 the importance of understanding how the bandwidth of measuring devices has a bearing on the reliability of measurements is discussed. In Section 5.4 transmission and reception of radiation from antennas is discussed. It is important to be familiar with antennas because they are often used in EMC testing to generate the incident field in susceptibility tests, to detect radiated fields in emissions tests, and they can occur within systems *by accident*, radiating or receiving unwanted signals. It is also important to have some understanding of why there are many different types of antennas available. Certain other properties, such as the fact that antennas do not radiate isotropically and the radiation is often polarised, can be used to advantage to minimise unwanted radiative coupling between different systems.

5.1 BASIC CONCEPTS

The most familiar form of a periodic voltage signal $v(t)$ is a sinusoidal like $v(t) = M \sin(2\pi ft + \phi)$, where M is the amplitude, f is the frequency and ϕ is the phase angle. However periodic waveforms need not be pure sinusoidals. Any waveform which repeats itself after P seconds such that $v(t + P) = v(t)$ for all P, is periodic. Graphs of $v(t)$ against time display the function in the 'time domain'. In any given system a number of repetitive signals and isolated pulses are generated. Real signals are rarely pure sinusoidals of a fixed frequency (but are rectangular clock pulses, pulsed RF etc.). Besides the intended complex pulse structure, various distortions to the intended signals can also (and usually do) occur and these are known as 'harmonic distortion' and 'intermodulation distortion'. It is convenient for these to be discussed at this stage.

5.1.1 Harmonic distortion

This is the generation by a non-linear device of output frequencies other than the input. In a non-linear device the current and voltage are not related by a frequency independent constant impedance (Ohm's law) but by a more complex transfer impedance which can be represented by a power series, the number of terms and the coefficients depending on the system, e.g.:

$$I = a_0 + a_1 V + a_2 V^2 + a_3 V^3 \quad \text{etc} \tag{5.1}$$

where I is the instantaneous current, a_0 is a background current (which may be zero), V the voltage, and the coefficients a_i have units of Ω^{-1}.

If $V = A\cos(\omega t)$, then

$$I = a_0 + a_1 A\cos(\omega t) + a_2(A\cos(\omega t)^2 + \cdots . \tag{5.2}$$

This expression can be written in different ways by making use of trigonometrical identities such as

$$\cos^2(\omega t) = 1/2 + (1/2)\cos(2\omega t),\ \cos^4(\omega t) = (3 + 4\cos(2\omega t) + \cos(4\omega t))/8$$

etc. Thus, rearranging, it is clear that (5.2) can be written as

$$I = C_0 + C_1\cos(\omega t) + C_2\cos(2\omega t) + C_3\cos(3\omega t) + \cdots \tag{5.3}$$

i.e. harmonics (*integral* multiples of the basic frequency) are generated. Other forms of distortion can be produced by driving devices into saturation so that their output becomes independent of the input. For example clipping off a sine wave can produce extreme distortion (i.e. an approximate square wave).

5.1.2 Intermodulation distortion or mixing

This is the generation by a non-linear transfer function of output frequencies that are *not integral* multiples of the input frequencies. Consider a signal $v(t)$ consisting of two components with different angular frequencies ω and ω_2 input into the non-linear device described by Equation (5.1):

$$v(t) = A\cos(\omega t) + B\cos(\omega_2 t). \tag{5.4}$$

Using (5.1) the output current will be of the form

$$i(t) = a_0 + a_1(A\cos(\omega t) + B\cos(\omega_2 t)) + a_2(A\cos(\omega t) + B\cos(\omega_2 t))^2 + \cdots .$$

With a fair amount of maths this can be written as

$$i(t) = C_0 + \sum C_k \cos((m\omega \pm n\omega_2)t) \tag{5.5}$$

Table 5.1. Output from two frequencies f_1 and f_2 input to a non-linear device

	f_1 f_2
$f_1 - f_2$	$f_1 + f_2$
$2f_1$	$2f_2$
$2f_1 - f_2$	$2f_1 + f_2$
$f_1 - 2f_2$	$f_1 + 2f_2$
$3f_1$	$3ff_2$
...	...

where m and n are independent integers. Table 5.1 shows the frequencies that are output when two frequencies are input. The exact number of terms in a given situation is decided by the precise form of the transfer function of the non-linear device used. In order to analyse and deal with this variety of signals it is useful to employ spectra analysis techniques.

5.2 SPECTRAL ANALYSIS

The basic idea behind spectral analysis is that a function $v(t)$ in the time domain can be expressed by an equivalent frequency-domain signal $V(f)$ consisting of a sum of elementary sinusoidal functions. To make a 'spectral analysis' of the waveforms $v(t)$ with frequency f_o, as shown in Figure 5.1, means to write the mathematical expression which describes this waveform as a function of time in the form:

$$v(t) = V_{dc} + V_1 \cos(\omega_0 t + \phi_1) + V_2 \cos(2\omega_0 t + \phi_2) + V_3 \cos(3\omega_0 t + \phi_3) + \cdots + V_i \cos(n\omega_0 t + \phi_i) \quad (5.6)$$

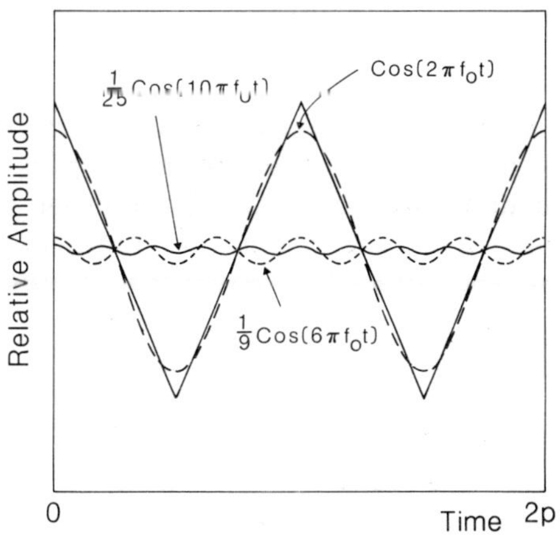

Fig. 5.1 Two periods of a triangular waveform, compared with the first three terms in the Fourier series

Fig. 5.2 Rectangular waveforms with period P and frequency $f_0 = 1/P$

where $\omega_0 = 2\pi/P$. Some of the coefficients V_i could be zero and (5.6) is in the *frequency domain* because it is a function of integer multiples of the fundamental frequency $f_o = \omega_0/2\pi$. Note that the choice of the symbol V as a 'dummy variable' for the amplitudes does not mean that this formula only applies to voltages; it can equally well apply to current pulses. The significance of spectral analysis is that it plays a pivotal role in EMC work for the following reasons.

(1) It can identify the limits of lumped circuit theory for a given circuit structure, by identifying the highest frequency component present.
(2) It is used extensively in filter design.
(3) It has applications for basic transmission line and antenna theory.
(4) It is the basis for EMC measurements.

As an example where spectral analysis is used, consider a typical electronic arrangement found in many devices, both domestic and industrial. The control module in Figure 5.3 is controlling an actuator, with current pulses $i(t)$ flowing along the wires. Spectral analysis of this current may reveal high frequency components, for which the wires will act like an antenna (the efficiency of radiation increases with frequency, see Section 5.3) and present a potential interference source for nearby radio receivers. Resolving problems like this requires spectral analysis of the signals.

5.2.1 The Fourier series

Spectral analysis is based on the Fourier theorem which states that periodic waveforms $f(t)$ can be expressed as the sum of a series of sinusoidal signals. There is an extensive literature based on the Fourier theorem [1] and this section will only summarise some results of immediate interest. If P is the

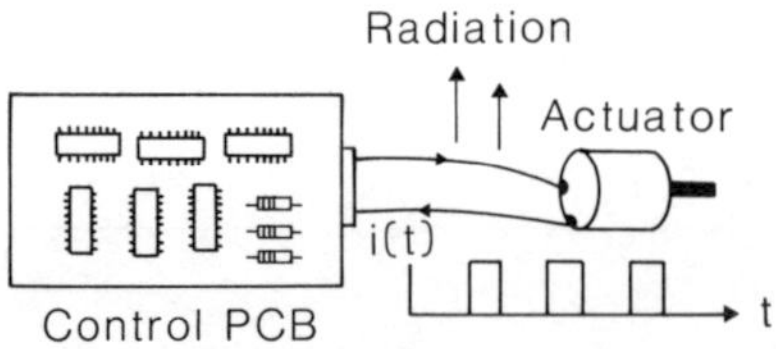

Fig. 5.3 Schematic diagram of a typical situation where an actuator is controlled by digital signals generated by control electronics mounted on a printed circuit board (PCB)

period of a signal $f(t)$ which, it can be expressed as

$$f(t) = M_0 + \sum_{h=1}^{\infty} M_n \sin(2\pi n f_0 t + \phi_n) \tag{5.7}$$

where $f_0 = 1/P$ is the fundamental frequency and the summation is from $n = 1$ to $n = \infty$. This series has a constant (dc) term M_0 and sinusoids of frequency f_0, $2f_0$, $3f_0$ etc., i.e. harmonics. It is often more convenient to re-express this equation as:

$$f(t) = A_0 + \sum_{n=1}^{\infty} A_n \cos(2\pi n f_0 t) + \sum_{n=1}^{\infty} B_n \sin(2\pi n f_0 t) \tag{5.8}$$

where $A_0 = M_0$, $A_n = M_n \sin \phi_n$, $B_n = M_n \cos \phi_n$, also $M_n^2 = A_n^2 + B_2^2$ and $\tan \phi_n = A_n/B_n$. The coefficients A_0, A_n, B_n can be found by evaluating the following integrals:

$$A_0 = \frac{1}{P} \int_0^P f(t)\, dt \tag{5.9}$$

$$A_n = \frac{2}{P} \int_0^P f(t)\cos(2\pi n f_0 t)\, dt \tag{5.10}$$

$$B_n = \frac{2}{P} \int_0^P f(t)\sin(2\pi n f_0 t)\, dt. \tag{5.11}$$

It is often assumed that $f(t)$ is periodic from $-\infty$ to $+\infty$, in which case the integration limits can be from $-P/2$ to $P/2$ instead of from 0 to P. Notice that A_o simply corresponds to the average value of the function over the period P. Fourier analysis breaks down the signal $f(t)$ into frequency components which are said to constitute the frequency spectrum of $f(t)$. The range of frequencies present in the spectrum, i.e. those with non-zero coefficients A_n or B_n, is the *bandwidth* of $f(t)$. All electronic systems, particularly communication systems, operate only on restricted bandwidths so that frequencies outside of a given frequency range are severely attenuated or eliminated. Further, the phases of the remaining components may be altered. The frequency spectrum present in any situation needs to be known in order to make predictions the types of distortion which may occur.

5.2.2 An example of a Fourier series

A triangular waveform is represented by the Fourier series

$$f(t) = 8/\pi^2(\cos(2\pi f_0 t) + 1/9 \cos(6\pi f_0 t) + 1/25 \cos(10\pi f_0 t) + \cdots). \tag{5.12}$$

This is shown in Figure 5.1 over a time interval corresponding to two periods, where the relative amplitude of the first three terms in this expansion are compared to the triangular waveform obtained by summing the whole series. Equation (5.12) can also be written in the form of (5.7) by noting that since

$$\tan \phi_n = A_n/B_n = \infty$$

(because $B_n = 0$ for all n), then $\phi_n = \pi/2$. Hence

$$f(t) = 8/\pi^2(\sin(2\pi f_0 t + \pi/2) + 1/9 \sin(6\pi f_0 t + \pi/2) + 1/25 \sin(10\pi f_0 t + \pi/2) + \cdots).$$

This spectrum contains an infinite number of frequency components, so the bandwidth is infinite. A practical system would remove all the frequencies above some upper limit. This would distort the shape of the waveform somewhat, but as the lowest frequencies are the most important (because the coefficients M_n are biggest) this distortion need not be too bad.

5.2.3 Fourier series of trains of pulses

The Fourier series described by (5.13) can be written succinctly in complex notation

$$f(t) = \sum_{n=-\infty}^{\infty} V_n e^{jn\omega t} \tag{5.13}$$

where the summation is from $-\infty$ to $+\infty$ and the coefficients V_n are given by

$$V_n = \left(\frac{1}{P}\right) \int_{-P/2}^{P/2} f(t) e^{-jn\omega t}\, dt \tag{5.14}$$

In general these coefficients are complex, and are related to the A_n and B_n coefficients (see Equation (5.8)) by

$$V_n = (A_n - jB_n)/2 \quad \text{when } n > 0$$
$$V_n = (A_n + jB_n)/2 \quad \text{when } n < 0$$

with a modulus such that

$$2\,|V_n| = \sqrt{(A_n^2 + B_n^2)}.$$

In Figure 5.4 is shown a train of rectangular pulses of amplitude A, width T_p and period P. Trains of pulses similar to these will be present in many types of equipment. Over the interval $-T_p/2$ to $T_p/2$ the function $f(t) = A$, a constant. For such pulses the Fourier coefficients are found by evaluating the integrals ((5.10) and (5.11) (which is zero in this case), i.e.

$$V_n = \frac{2}{P} \int_{-T_p/2}^{T_p/2} A \cos(2\pi nt/P)\, dt$$

i.e.

$$V_n = (2A/P)\{\sin(n\pi f_0 T_p)/n\pi f_0\}$$

which can be written more conveniently as

$$V_n = (2AT_p/P)\{\sin(n\pi f_0 T_p)/n\pi f_0 T_p\} \tag{5.15}$$

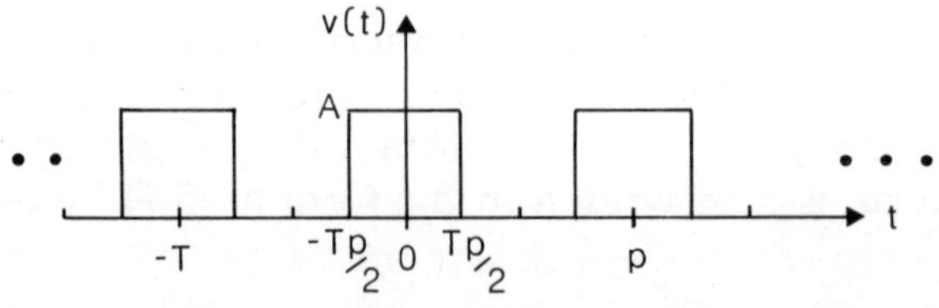

Fig. 5.4 A train of rectangular pulses, with period P, pulse width T_p and amplitude A

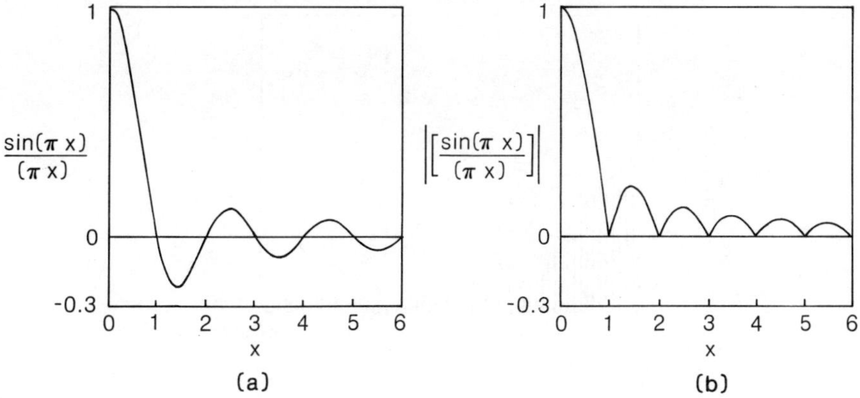

Fig. 5.5 The function $\sin(\pi x)/\pi x$ and its modulus

where $f_0 = 1/P$. This is a function of the form $\sin(\pi x)/\pi x$, plotted in Figure 5.5(a). A negative value for an amplitude means that the phase of the term in the Fourier series is changed by π. In most cases it is the magnitude of the coefficient which is of most interest, so in general the modulus $|V_n|$ will be plotted, as shown in Figure 5.5(b). The amplitude spectrum for the rectangular pulse train are zero when nf_0T_p equals an integer so that the product $(nf_0T_p\pi)$ is an integral multiple of π. In situations where the product $f_0T_p = T_p/P$ is fixed, the only variable in (5.15) (apart from the pulse amplitude A) is the integer n, which is the harmonic number. For a fixed ratio $T_p/P = 1/10$ say, the amplitudes of the different harmonics are distributed as shown in Figure 5.6, which shows that for this choice of T_p/P, when $n = 10$, 20, 30 etc. the amplitudes are zero. When drawing Figure 5.6 the amplitude A is chosen so that $(2AT_p/P) = 1$.

If the ratio T_p/P is allowed to reduce by keeping the pulse width T_p constant but allowing the period P to become very large, this train of pulses tends

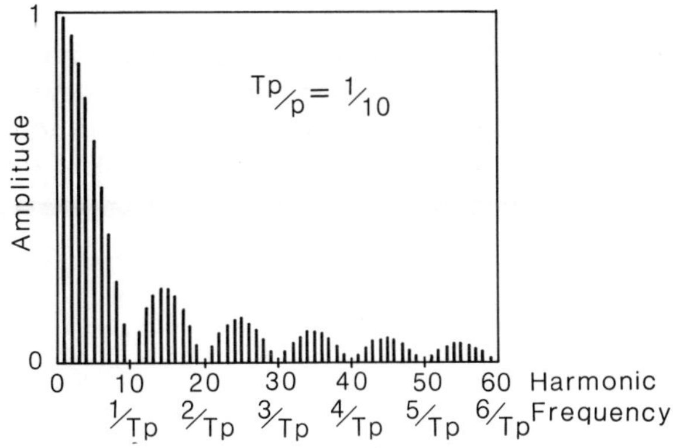

Fig. 5.6 The relative amplitudes of the harmonics in the fourier series of the pulse train shown in Fig. 5.4 where the ratio of pulse width to period, $T_p/P = 1/10$

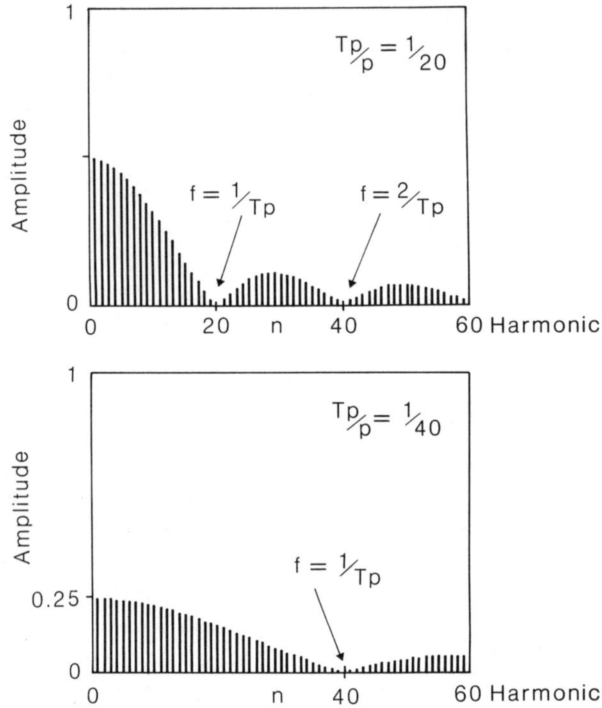

Fig. 5.7 Effect on the amplitude spectrum of allowing the ratio of pulse width to period, T_p/P to tend to zero

towards an isolated pulse. The amplitude spectrum changes, as shown in Figure 5.7, where the amplitude A has been kept the same as used to make Figure 5.5. Notice that the nulls are at the same frequencies of $1/T_p$, $2/T_p$ etc., but the number of harmonics between the nulls increases. In the limit of a single pulse as shown in Figure 5.8, the following things happen to the Fourier series.

(1) The amplitude of the individual components approaches zero. In Equation (5.15), as $f_0 \to 0$ (i.e. $P \to \infty$), $\sin(x)/x \to 1$, and $V_n \to (AT_p/P) \to 0$.
(2) The number of harmonics between nulls increase without bound.
(3) The shape of the curve is unchanged, remaining like $|\sin(x)/x|$.

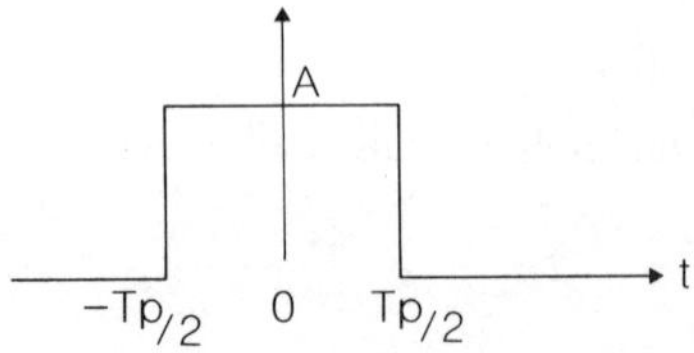

Fig. 5.8 An isolated pulse of width T_p seconds and amplitude A

The spectrum of an isolated pulse becomes continuous and cannot be represented by a summation such as (5.13). This is the vital difference between a pulse train and a single pulse, in that the pulse train contains harmonics at discrete frequencies whereas a single pulse contains all frequencies. Because the amplitudes have tended to zero it is necessary to talk about the spectral density rather than the amplitudes of individual harmonics. To deal with this problem mathematically, Fourier transforms are used.

5.2.4 Fourier transforms

For a function $f(t)$ the following pair of equations are called a Fourier transform pair:

$$V(f) = \int_{-\infty}^{\infty} f(t)e^{-j2\pi ft}\,dt = \int_{-\infty}^{\infty} f(t)e^{-j\omega t}\,dt \tag{5.16}$$

$$f(t) = \int_{-\infty}^{\infty} V(f)e^{j2\pi ft}\,df. \tag{5.17}$$

A common notation in engineering is to use lower case letters for a function and upper case for its transform, so the $V(f)$ is the Fourier transform of the function $f(t)$, and Equation (5.17) gives the inverse Fourier transform of the function $V(f)$. It should be realised that throughout the literature there are different conventions for the term which has the conjugate $(-j)$ and the initial factor in the Fourier transform. For example the inverse Fourier transform is commonly written in technical literature as an integral with respect to ω, such that

$$f(t) = 1/2\pi \int_{-\infty}^{\infty} V(\omega)e^{j\omega t}\,d\omega \tag{5.17'}$$

where

$$V(\omega) = \int_{-\infty}^{\infty} f(t)e^{-j\omega t}\,dt. \tag{5.16'}$$

There are certain advantages to this, such as ease of converting from Fourier to Laplace transforms, but for the purposes of this book it is convenient to use the form (5.16) and (5.17).

These definitions can be used to find the Fourier transform for an aperiodic signal, e.g. a single rectangular pulse of width T_p s as shown in Figure 5.8. The Fourier transform of this pulse is found by replacing the infinite limits of integration by $-T_p/2$ and $T_p/2$ because the function $f(t)$ is zero outside this range, and constant inside it. Thus

$$\begin{aligned} V(f) &= 2/P \int_{-T_p/2}^{T_p/2} Ae^{-j2\pi ft}\,dt \\ &= A\{\sin(\pi f T_p)/\pi f\} \end{aligned} \tag{5.18}$$

which can be written as

$$AT_p\{\sin(\pi f T_p)/\pi f T_p\}. \tag{5.19}$$

If Equation (5.18) is compared with (5.15) it will be seen that the Fourier transform $V(f)$ of an isolated rectangular pulse is a continuous function of frequency rather than having discrete harmonics, but it also has the form of $\sin(\pi x)/\pi x$.

Equation (5.19) can be used to illustrate an important point. Referring to Figure 5.6 it can be seen that the dominant parts of the spectrum is largest around the smallest frequencies, with the first null occurring when $f = 1/T_p$ where T_p is the pulse width. As this width is deceased this null point moves to a higher frequency and vice versa. This is a general property of all signals; the shorter the duration of the signal in the time domain the wider the spread in the frequency domain of the dominant terms in amplitude spectrum.

5.2.5 Spectrum analysers

The analytic techniques discussed in the discussed in the previous section have been used to compute algebraic expressions for the Fourier series and transforms of a number of waveforms; for examples see reference [2]. These are useful in design work for estimating potential EMC problems, such as radiation from circuits. However such calculations can usually be simplified by using an approximation for the maximum Fourier amplitudes as a function of frequency, as will be shown in Sections 5.2.7 and 5.4.4.2.

Signals encountered in real systems will usually have a more complex substructure than simple analytic functions, and such a substructure will give rise to additional Fourier components which are of interest. In these situations two techniques are used to obtain the amplitude spectrum. The first is to use analogue spectrum analyser to measure the spectrum actually produced by the waveforms, and the second to apply numerical techniques, such as the fast Fourier transform (FFT), to a set of digitised samples of the waveform. The FFT is discussed in the next section.

A spectrum analyser is a device which is sensitive to the different frequencies present in an arbitrary signal $f(t)$ and is able to measure their relative amplitudes. One method of doing this is to couple the signal simultaneously to a large number of different circuits (filters) tuned to selected frequencies and measure the effects in each. The output from each filter depends on its passband and the amplitude of a Fourier component at a particular frequency. However to obtain a wide bandwidth and good resolution a large number of such filters would be required, and this is costly.

A common and cheaper method, known as the superheterodyne or sweeping signal system, is to mix the signal with a local signal generated in a local oscillator. The frequency of the resultant output is the algebraic sum or difference of the two input frequencies (see Section 5.1.2), and this can be fed to a filter with fixed parameters. By varying the frequency of the local oscillator the different frequency components present in the original signal can give rise to signals in the passband of the filter and so be measured.

Spectrum analysers can be purchased which work right up to microwave frequencies, but they are very expensive. Some spectrum analysers make use of the FFT, which greatly reduces their cost but limits their frequency range. A feature of them is that often they pass the input signal through a bandwidth limited filter, called an anti-aliasing filter (see the next section), which removes high frequencies *before* producing the amplitude spectrum. This may cause difficulties for the unwary.

When choosing a spectrum analyser, the three most important parameters to consider are the frequency, resolution and dynamic ranges [23]. As has been noted, spectrum analysers can be obtained for all frequencies from the kHz range right up to millimetre wavelengths (above 18 GHz), so the appropriate one for the problem at hand has to be chosen. The *resolution* of a spectrum analyser is its ability to identify clearly closely spaced signals in the frequency domain. This can be particularly important at communication frequencies where the frequency bands are very crowded. The *dynamic range* is the range of input power which the instrument can detect a weak signal above the instrument noise, up to the input power it can tolerate without damage yet still give satisfactory measurements.

Finally, it should be remembered that even a first class instrument requires competent handling by the operator to extract the full precision available and avoid errors, so attending a training course provided by the manufacturer would be a worthwhile investment.

5.2.5.1 *The fast Fourier transform*

Suppose that a waveform is digitised, for example by using a digital storage oscilloscope to capture a real waveform or by modelling a waveform during design work, giving a large number of samples of the wave over some time period, and these data are then transferred into a computer. The integrals (5.9)–(5.11) which are used to evaluate the Fourier coefficients could then be done numerically. These techniques, such as Simpson's rule, do not require an analytic form for the function being integrated but only its value at certain points, so they can make use of the digitised waveform. However it is obvious that to be successful the digitisation must be a very good representation of the waveform, otherwise the method would not give reliable results. In principle such a method could be used to find the Fourier amplitude for any harmonic, but step by step integration methods on a computer are slow.

In 1965 J. W. Tukey and J. W. Cooley published a very important numerical algorithm, commonly called the FFT algorithm, that under some circumstances can speed up these calculations dramatically (many orders of magnitude for large data samples) [3]. A full discussion of the FFT algorithm is not appropriate in this book, as excellent references exist [4, 5] but a few remarks should be made.

The first is that the number of digitised points should satisfy $N = 2^m$ when m is an integer, and that care must be taken to ensure that the samples taken are close enough in time to give sufficient samples of the highest frequency which is likely to be present. Sampling theorems indicate that there should be at least

10 samples per highest frequency present, otherwise the algorithm will not detect reliably its presence [4].

A potential problem with the FFT concerns the phenomenon of aliasing. If a signal is sampled at a rate D samples per second then there is a frequency f_c given by

$$f_c = 1/2D$$

which is known as the Nyquist critical frequency [5]. If the signal being sampled has frequency components $|f| > f_c$, so that it is not limited to the sampling bandwidth, then they will give rise to false components within the range. This is why spectrum analysers which use the FFT can have anti-aliasing filters which remove from the signal those high frequency components which would be aliased in the instrument, a procedure which may be unsatisfactory for EMC work.

For an excellent and comprehensive discussion of all aspects of the FFT, such as power spectrum estimation and windowing, together with many examples of carefully debugged computer codes for these tasks, the reader is referred to reference [5].

5.2.6 Effect of finite rise times

Real pulses in circuits will have finite rise times t_r, and often the ratio of pulse width T_p to the rise time is fixed, see Figure 5.9. Waveforms like this are called trapezoidal waveforms. The Fourier coefficients for a pulse train of pulses of frequency f_0 (period P) and with finite rise times t_r are [2]

$$V_n = 2A\partial \frac{\{\sin(\pi n f_0 T_p)\sin(\pi n f_0 t_r)\}}{(\pi n f_0 T_p)(\pi n f_0 t_r)} \tag{5.20}$$

where $\partial = (T_p + t_r)/P$ is the duty cycle. This Equation is similar to (5.15) but has an extra term $\sin(\pi n f_0 t_r)/(\pi n f_0 t_r)$ which has the effect of reducing the amplitude of the higher harmonics and introducing extra nulls whenever $\sin(\pi n f_0 t_r) = 0$. These effects can be demonstrated by recalculating Figure 5.6 using the same

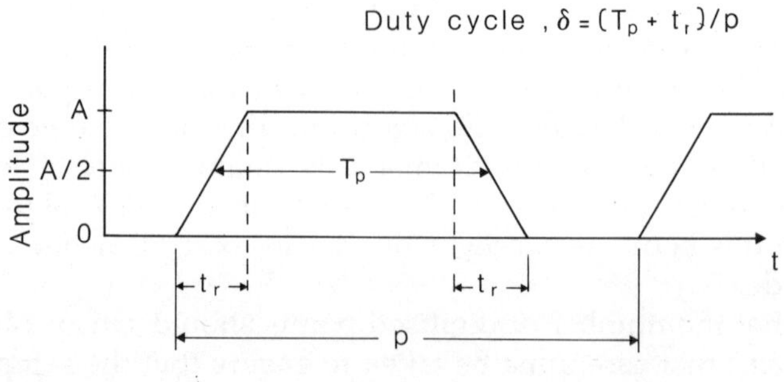

Fig. 5.9 A trapezoidal waveform with period P, width T_p, rise time t_r and amplitude A

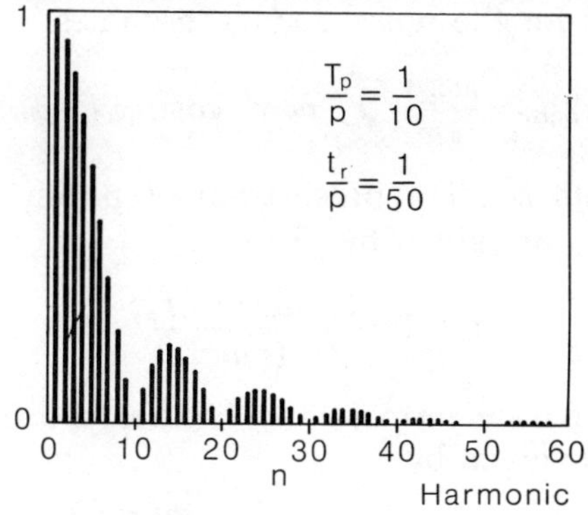

Fig. 5.10 Relative amplitudes of the Fourier components of a trapezoidal waveform with period P, width T_p, rise time t_r and amplitude A. The ratios T_p/P and t_r/P have been chosen to be 1/10 and 1/50 respectively

value of $T_p/P = 1/10$ but assuming that there is a finite rise time of order 1/5 of the pulse width, so that $t_r/P = 1/50$. In this case (5.20) becomes

$$V_n = 2A\partial \frac{\{\sin(n\pi/10)\sin(n\pi/50)\}}{(n\pi/10)(n\pi/50)}. \tag{5.21}$$

The relative amplitude spectrum of the first 60 harmonics given by (5.21) is shown in Figure 5.10, and should be compared with Figure 5.6. Notice that

(*a*) the term in $\sin(n\pi/10)$ which arise from the ratio of the pulse period to the pulse width is zero when $n = 10$, 20, 30 etc. and is unchanged by the presence of the term due to the finite rise time;
(*b*) the term in $\sin(n\pi/50)/(n\pi/50)$ which arises because the pulses have a finite rise time dampens the amplitude of these high harmonics.

5.2.7 Noise voltage in a coil

It is instructive to re-examine the example of noise picked up in a small coil a long way from a pair of wires carrying an alternating current which was introduced in Section 2.7.1. Equation (2.94) for the noise voltage generated by the inductive coupling between the current and the coil was derived assuming that there was a sinusoidal current flowing in the wires. If, however, there are current pulses in the wire, the noise voltage in the coil becomes the sum of a series with different harmonics. In such a situation the single current $i = i_0 \sin(\omega t)$ which used in the derivation of (2.70) is replaced by a Fourier series, which is convenient to write in the form of (5.7), i.e.

$$i(t) = M_0 + \sum_{n=1}^{\infty} i_n \sin(n\omega_0 t + \phi_n)$$

so that the expression for the noise voltage becomes

$$V_{\text{noise}} = \frac{\mu_0 a A}{\pi r^2} \sum_{n=1}^{\infty} n\omega_0 i_n \cos(n\omega_0 t + \phi_n). \tag{5.23}$$

The Fourier coefficients i_n for a pulse train of purely rectangular pulses of period P and width T_p are given by (5.15):

$$i_n = 2i_0 \frac{T_p}{P} \frac{\sin(\pi n f_0 T_p)}{(\pi n f_0 T_p)}. \tag{5.24}$$

Thus the coefficient of each term in the Fourier spectrum of noise voltages generated in the coil is given by

$$\begin{aligned} n\omega_0 i_n &= 2i_0(2\pi n f_0) \frac{T_p}{P} \frac{\sin(\pi n f_0 T_p)}{(\pi n f_0 T_\mathrm{p})} \\ &= \frac{4i_0}{P} \sin(\pi n f_0 T_p) \end{aligned} \tag{5.25}$$

i.e. the amplitude of each Fourier component $n\omega_0 i_n$ is proportional only to $\sin(\pi n f_0 T_p)$ with all the other terms being constant. Thus the fact that the coupling between the coil and the loop is linearly dependent on frequency cancels the $1/\omega$ dependence which usually damps the amplitudes of the higher harmonics of the rectangular pulses, so the amplifying their importance compared with what might be expected. Real pulses will have finite rise time t_r, so using Equation (5.21) shows that (5.24) should be written

$$i_n = \frac{2i_0\, \partial \sin(\pi n f_0 T_p)\sin(\pi n f_0 t_\mathrm{r})}{(\pi n f_0 T_p)(\pi n f_0 t_\mathrm{r})} \tag{5.26}$$

so that (5.25) becomes

$$\begin{aligned} n\omega_0 i &= \frac{4i_0\, \partial\, \sin(\pi n f_0 T_p)\sin(\pi n f_0 t_\mathrm{r})}{T_p(\pi n f_0 t_\mathrm{r})} \\ &= \frac{4i_0(1 + t_r/T_p)\sin(\pi n f_0 T_p)\sin(\pi n f_0 t_\mathrm{r})}{P(\pi n f_0 t_\mathrm{r})} \end{aligned} \tag{5.27}$$

because the duty cycle $\partial = (T_p + t_\mathrm{r})/P$. There is some modulation of the harmonic spectrum due to the term $\sin(\pi n f_0 t_\mathrm{r})$ and damping of the high harmonics due to the term $1/(\pi n f_0 t_\mathrm{r})$, both of which arise from the finite rise time of the pulses. In Figure 5.11 is shown the *envelope* of the current amplitudes $n\omega_0 i_n$ generated by the coil, given by (5.27), where the coil is inductively coupled to typical high frequency clock pulses with period 50 ns, width 10 ns and rise time of 1 ns; this should be compared with Figure 5.10 which shows the relative amplitude of the Fourier components of the current which is coupled to the coil. The difference arises because the coupling between the coil and the pulses generating the noise is frequency dependent. The lesson to be learnt from this example is that a Fourier spectrum obtained by using a spectrum analyser to examine a noise voltage may be difficult to interpret in situations where the coupling is frequency dependent, even when the current waveforms causing the noise are well understood.

5.2.8 An approximation for the Fourier spectrum

There is a very useful approximation which enables the envelope of the Fourier amplitudes to be calculated easily which is very useful in design work as it avoids the necessity of finding individual Fourier amplitudes. In the function $\sin(x)/x$ which occurs in formulae for the Fourier amplitudes the value of $\sin(x)$ cannot exceed 1, so at large x values the graph of the function is dominated by the $1/x$ term, which is small as x becomes large. The maxima of the function $\sin(x)/x$ lie on the curve $1/x$ so that the amplitude of any Fourier component will be bound by this curve. Thus this much simpler equation can be used to calculate the maximum possible amplitude of each Fourier component.

When $x = \pi f T_p$ (as in the expression of the Fourier transform of the pulses considered so far), then

$$1/x = 1/(\pi f T_p)$$

is equal to 1 when

$$f = (1/\pi)/T_p \approx 0.318/T_p. \qquad (5.28)$$

Thus for a pure rectangular pulse the envelope of the Fourier transform as a function of frequency is shown as the dotted line in Figure 5.12. It has a flat part up to frequency $0.318/T_p$, followed by the curve $1/\pi f T_p$ up the highest frequencies. However this is still not in the most convenient form. As will be seen in Section 5.4, when making predictions of radiated field strengths from circuits it is very useful to have the different Fourier amplitudes expressed in dB, and it is in this form that the useful approximation appears.

The function $\log_{10}(1/\pi f T_p)$ is zero when $f \approx 0.318/T_p$, is -0.497 when $f = 1/T_p$, and is -1.497 when $f = 10/T_p$. Recalling the way in which dB was defined, the maximum points on the amplitude spectrum in dB can be approximated by the function

$$V(f) = 20\ \log_{10}(1/\pi f T_p)\ \text{dB} \qquad (5.29)$$

which has a negative gradient of -20 dB for each factor of $10//T_p$ (decade) increase in frequency f above a lower limit of $1/\pi T_p$ *Hz.*; the graph of $V(f)$ dB as a function of frequency is shown in Figure 5.13.

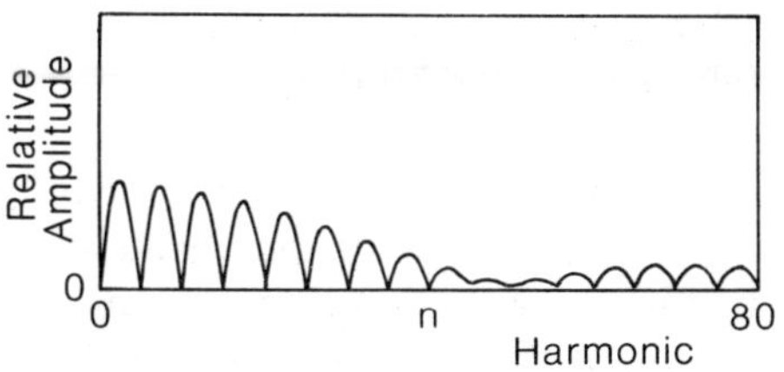

Fig. 5.11 Envelope of the relative sizes of the Fourier components as a function of harmonic number of the noise voltage produced in a small coil with frequency dependent inductive coupling to a pair of wires carrying a trapezoidal waveform

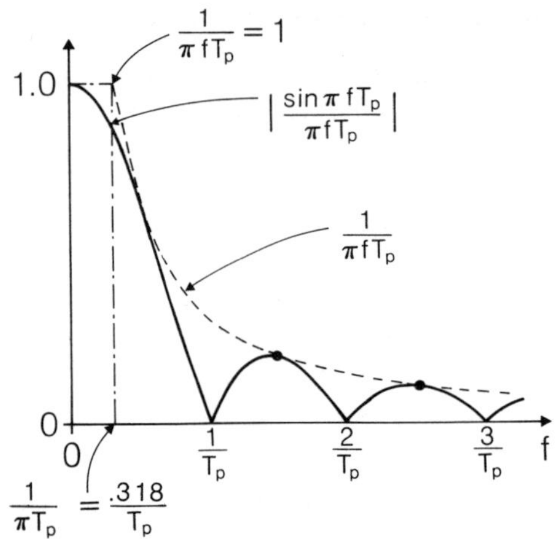

Fig. 5.12 Envelope of the maxima of the amplitudes of the Fourier components as a function of frequency

Extending this concept to pulses with finite rise times enables the effect of a finite rise times on the amplitude of the Fourier components to be seen easily. For such pulses the maxima of the amplitude spectrum are obtained from Equation (5.21) (ignore the sine terms)

$$V(f) = 2A\partial(1/\pi f T_p)(1/\pi f t_r).$$

Expressing this in dB gives

$$V(f) = 20\ \log_{10}(2A\partial) + 20\ \log_{10}(1/\pi f T_p) + 20\ \log_{10}(1/\pi f t_r)\ \text{dB} \qquad (5.30)$$

which is valid for frequencies above $1/(\pi t_r)$ Hz. Above this frequency the combined graph decreases at -40 dB per decade, see Figure 5.14. There are two things to notice about this figure.

(1) The amplitude in dB is relative to unit amplitude. To find the absolute value in dB at any frequency add $20\ \log_{10}(2A\partial)$, where $\partial = (T_p + t_r)/P$ is the duty cycle and A is the amplitude of the pulse.
(2) The first break point in the curve occurs at a frequency of $1/\pi T_p$, from which the amplitudes decrease by -20 dB/decade and the second at $1/\pi t_r$ above which the fall of is -40 dB/decade.

Given the pulse parameters of a trapezoidal waveform (amplitude A, period P, pulse width T_p and rise time t_r) the upper limit on the amplitude of any Fourier component in dB at any frequency can be calculated using the appropriate terms in Equation (5.30). These days it is very easy to set up this calculation on a personal computer, using a spreadsheet. Such a tabular form is very convenient to use for the prediction of radiated field strengths from circuits at different frequencies in order to compare them with emission limits,

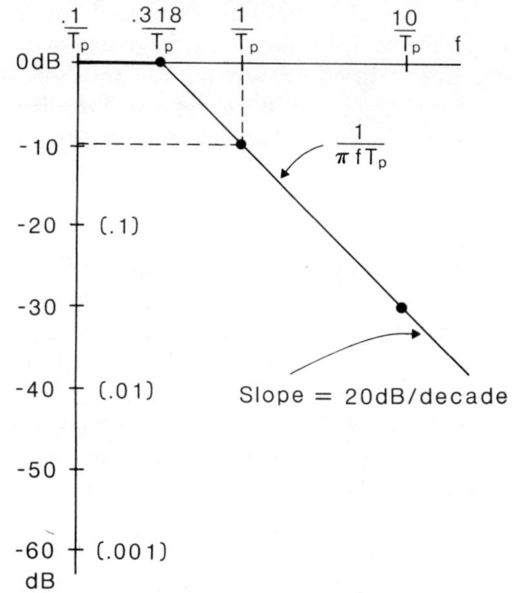

Fig. 5.13 The approximation 20 $\log_{10}(1/\pi f T_p)$ dB for the Fourier amplitudes as a function of frequency f for a rectangular waveform with pulse width T_p and zero rise time. The approximation is valid for $f > 1/\pi T_p$ Hz

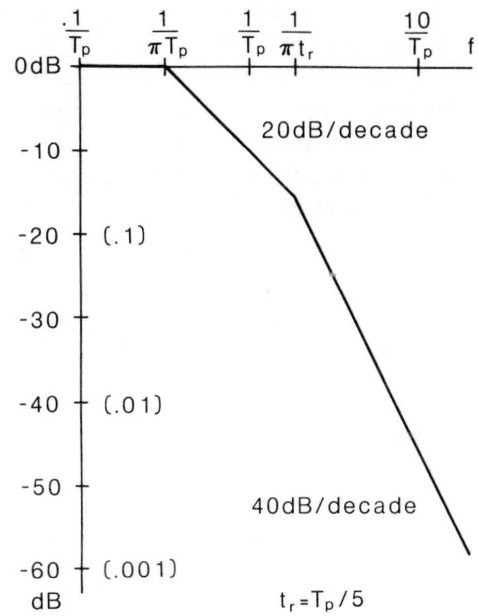

Fig. 5.14 The approximation 20 $\log_{10}(1/\pi f t_r)$ dB for the Fourier amplitudes as a function of frequency f for a trapezoidal waveform with pulse width T_p and rise time t_r. The approximation is valid for $f > 1/\pi t_r$ Hz

as will be seen in Section 5.4. Alternatively one may construct graphs such as Figure 5.14 using nomograms (template graphs with axis labelled in dB and frequency, containing guide lines drawn with slopes of −20 dB/decade and −40 dB/decade) and from these read off the relative amplitude at any frequency and tabulate by hand. For an example of this see Appendix A in reference [6].

5.2.9 Interference bandwidth

Detectors of electromagnetic radiation have a range of frequencies, called the bandwidth of the detector, over which they are sensitive to a signal; outside this range the detection efficiency often drops dramatically. The precise IEEE definition of 'bandwidth' of a device is 'the width of the frequency band over which a given characteristic of an equipment or transmission channel does not differ from its reference value by more than a specified amount or ratio' [26].

When using a detector to make quantitative measurements it is important to known how its bandwidth relates to the spectral distribution of the signal. Figure 5.15 shows clearly that, if the detector bandwidth is limited with respect to the range of frequencies present in the signal, then the output from the detector cannot give an accurate representation of the signal.

Rather than discuss the bandwidth of a particular detector it is usual to classify the bandwidth of any interference with respect to that particular detector. Thus 'narrowband' noise contains a range of frequencies which fall within the bandwidth of a given detector, so that a detector tuned to receive the signal will pick up all of it. In this case a measurement of the signal can be made without reference to the bandwidth of the detector. On the other hand a 'broadband' signal contains a range of frequencies which is considerably larger than the bandwidth of a given detector, so that the signal is detected across the entire tuning range of the detector. In this case the signal is measured on a per-unit-bandwidth basis, such as 'per MHz', depending on the bandwidth of the detector. For example, man-made background radiation, shown in Figure 1.10, is broadband interference, so the measured field strengths are given on a 'per MHz' basis. See Section 5.4.2.1 for a full discussion of the units used in Figure 1.10.

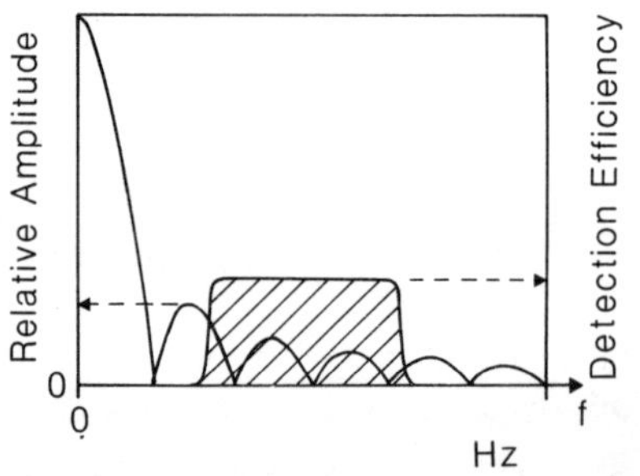

Fig. 5.15 Schematic diagram showing the range of Fourier components which would be detected by a receiver with narrow bandwidth

5.3 ANTENNAS AND RADIATION

There are four principal reasons why special attention must be given to these topics.

(1) Radiation is one of the coupling paths that can introduce interference into a system, so an understanding of how it interacts with antennas is helpful in minimising the coupling. For example it is useful to appreciate the not-so-obvious fact that slots in conductors can behave as very efficient antennas.
(2) Antennas of various sorts are used as sources of electromagnetic radiation for EMC testing and it is important to appreciate that their properties so that the correct type can be used in different circumstances. However such tests may in themselves act as a disturbance to others, so that methods of confining the radiation within a test region without substantially changing the properties of the radiation field or generating unwanted resonances in the enclosure are also needed.
(3) Antennas are used to detect radiation emitted from any equipment under test (EUT), and their properties must be fully understood in order than an accurate quantitative measurement of the magnitude of the radiated fields can be made and compared to the appropriate allowable limits.
(4) Any wires or loops carrying an alternating current can act unintentionally as an antenna, so that an understanding of the general principles of antennas and radiation enables simple but very useful calculations of radiated field strengths to be made at the design stage and give rise to good EMC design practises.

Each one of these topics has an extensive literature and contains much of topical research interest. In particular the details of EMC testing and measurement in general are complex and under constant development. In Chapter 1 a general introduction to the principles of testing has been given, and this chapter will discuss simple calculations for prediction the likely results of test measurements of radiated field strengths from simple circuits. Also methods of containing radiation within cavities, in particular the Crawford and GTEM cells, will be discussed. There are many textbooks covering electromagnetic radiation, the one by Jordan and Balmain being a popular and comprehensive work [22].

5.3.1 Differential-mode and common-mode radiation

Radiation can be classed as differential-mode radiation and common-mode radiation. Differential-mode radiation is produced by currents flowing around loops which then act as loop antennas, and is the easiest type to predict and control. In Section 5.4.3 formulae for predicting differential mode radiation from simple loops will be developed and discussed.

Common-mode radiation is caused by *unintentional* currents flowing in the same direction on conductors between different parts of a system which should ideally be at the same potential, and as such they are difficult to predict. Unfortunately they are usually the greatest source of radiated emission because the unwanted current paths are often along power leads, cable shields or other cables which can be long enough to act as large, efficient, antennas. In Section 5.4.5 one particular approach to modelling common mode radiation from small table-top systems with attached cables will be discussed, and general methods of controlling common-mode radiation will be considered in Chapter 6.

5.3.2 General properties of antennas

In Section 2.4 the equations of describing the components of the fields radiated by two of the simplest possible antennas, a short wire and a small loop, were discussed. Even for these simple structures the mathematics needed for a full description of the radiated field distributions are complicated. However some general points can be made using these simple examples which are true in many situations.

(1) Antennas have near and far field regions with very different properties.
(2) Radiation from antennas is often polarised.
(3) Antennas do not radiate uniformly in all directions, but each type has a characteristic radiation pattern, so that there may be very little radiation in particular directions.
(4) The actual performance of antennas depends on their local environment.

These properties of antennas, and others, are usually described by a number of greatly simplified scalar quantities which will now be briefly summarised.

5.3.2.1 *Power pattern, directivity, and gain*

The manner in which the power density varies with angle and distance from a given antenna can be calculated from using the expression for the time averaged Poynting vector (Equation (2.26)) and the expression for the field components. For example for the short wire antenna the far field components are given by Equations (2.22), and is easily shown [7] that power flowing in a particular direction is given by

$$P_d = (Z_0/2)(I_0 L\beta/4\pi r)^2 \sin^2\theta \text{ W m}^{-2} \tag{5.31}$$

where θ is the azimuthal angle measured from the antenna axis, as shown in Figure 2.1(a). By integrating (5.31) over all angles the total power flowing outwards is

$$P_T = (Z_0/12\pi)(I_0 L\beta)^2 \text{ W}. \tag{5.32}$$

A graph of the power density as a function of angle at a fixed distance is called the power pattern of the antenna. As can be seen from (5.31) there is no power flow parallel to the axis of the short wire antenna along the $\theta = 0$ direction. The directivity d of the antenna, is defined as

$$d = \frac{\text{Maximum power density}}{\text{Average power density}}. \tag{5.33}$$

This dimensionless quantity is a convenient way of quantifying the way in which an antenna radiates power in certain directions. For the short wire antenna the maximum power density is in the $\theta = 90^\circ$ direction, and the average power density $\langle P_d \rangle$ is found by averaging (5.32) over a sphere of surface area $4\pi r^2$, i.e.

$$\langle P_d \rangle = P_T / 4\pi r^2.$$

Thus for the short wire antenna the directivity is given by

$$d = \frac{(Z_0/2)(I_0 L\beta/4\pi r)^2}{(Z_0/12\pi)(I_0 L\beta)^2/4\pi r^2} = 3/2.$$

This is often expressed in dB, such that

$$D = 10 \log_{10}(d) \text{ dB} = 10 \log_{10}(1.5) = 1.76 \text{ dB}. \tag{5.34}$$

The gain g of an antenna is related to the directivity d by

$$g = \eta d \tag{5.35}$$

where $0 < \eta \leqslant 1$ is the efficiency of the antenna, which will be discussed in the next section. The directivity can be calculated from the geometrical arrangement of the elements of the antenna, whereas the gain is the quantity which is used to calculate the actual power flow because it takes into account losses due to resisitive effects in the antenna.

5.3.2.2 *Radiation resistance*

The total radiated power is delivered by the current flowing in the antenna. If there are no losses in the antenna then it can be considered to be behaving as if it had a resistance R_{rad} dissipating $P_T = I^2 R_{\text{rad}}$ W, where I is the rms current $I_0/\sqrt{2}$. Thus, using (5.32), R_{rad} for a short wire antenna is

$$R_{\text{rad}} = 80\pi^2 (L/\lambda)^2 \ \Omega \tag{5.36}$$

where the approximation $Z_0 \approx 120\pi$ has been used and $\beta = 2\pi/\lambda$. The radiation resistance varies as L^2, which means that for a given wavelength the longer the antenna the more power it can radiate.

It must be stressed that the short wire antenna does not correspond to a real antenna as it has been assumed that the current density is uniform along it, which is not possible with real antennas. A very common antenna is the dipole of length L, and for a sinusoidal current distribution along such an antenna Liao [7] shows that

$$R_{rad} = 60 \int_0^{\pi} \{\cos(\beta L \cos(\theta)/2) - \cos(\beta L/2)\}^2/\sin\theta \, d\theta. \tag{5.37}$$

Definite integrals like this are easily done using MATHCAD. For a half-wave antenna where $L = \lambda/2$, this gives $R_{rad} = 73\ \Omega$. In a short dipole antenna where the $L \ll \lambda$, the current distribution is approximately triangular, varying in a linear manner from a maximum at the centre to zero at the ends of the antenna. In this case the radiation resistance is given by [7]

$$R_{rad} = 20\pi^2 (L/\lambda)^2\ \Omega. \tag{5.38}$$

The total resistance of an antenna is the sum of the radiation resistance and the ohmic resistance R_{ohm} of all the conductors including the ground system. The efficiency of the antenna is defined as the ratio of the power transferred to the load by the antenna to the power collected by the antenna. If an antenna is matched to its terminating impedance then this can be written [7] as

$$\eta = R_{rad}/(R_{rad} + R_{ohm}). \tag{5.39}$$

For dipole antennas this is typically of order 0.75, whereas microwave horn antennas have $\eta = 1$.

5.3.2.3 *Effective aperture*

The effective aperture of a receiving antenna, A_e, is defined as the ratio of the power P delivered to the load to the power density P_d in the incident wave, i.e.

$$A_e = P/P_d.$$

As the units of P are W and those of P_d are W m^{-2}, the effective aperture has units of m^2 as befits an area. This quantity is a measure of the way in which energy is transferred from the incident wave into the receiving system, and must not be confused with the physical size of the antenna. When the antenna has no losses and there is perfect impedance match with the terminating impedance the aperture is called the *maximum effective aperture* A_{em}. Due to reciprocity, the shape of the radiation pattern for a transmitting antenna is the same as its pattern when used as a receiving antenna, so it can be shown that A_{em} is related to antenna directivity d by [7]

$$d = 4\pi A_{em}/\lambda^2. \tag{5.40}$$

As an example, for a short-dipole antenna $A_{em} = 3\lambda^2/2\pi\ m^2$. Tables of formulae for different antennas can be found in many books.

5.3.3 Slot antennas and apertures

Apertures in shields are one of the main routes by which unwanted radiation can enter or escape from systems, and slots in particular can act as very efficient antennas, radiating as effectively as a dipole antenna. When the length of the slot is about $\lambda/2$, or more generally, when the perimeter of the orifice is at least one wavelength, the slot may transmit considerable amounts of energy. For EMC design work is useful to have expressions for the antenna characteristics of apertures, such as their gain. An example of this was seen in Section 4.6.1 where the effect of small holes on the shielding effectiveness of a screened room was calculated.

For finding the radiation properties of slot antennas the guiding principle is due to Booker [8], and states that the radiation pattern of a slot antenna is the same as a complementary dipole antenna but with the magnetic and electric field directions reversed. Thus the radiation from a slot is polarised with the magnetic vector parallel to the long axis of the slot and the electric vector parallel to the short axis of the slot. The expressions for the directivity and effective aperture of the long narrow slot with width $\ll \lambda$ are the same as for the corresponding dipole, but the expressions for the radiation resistance of the slot are different [7].

For rectangular or circular apertures the calculation of radiation patterns, aperture gain etc. is greatly simplified by the fact that the radiation pattern is the Fourier transform of the field distribution in the aperture [9]. This powerful result enables the results of Fourier theory to be used for detailed studies of the effects of apertures, but such requirements are beyond the scope of this book. However such methods can be used to justify the following result. If an aperture of *geometric area A* is uniformly illuminated such that the field is zero everywhere outside the aperture (i.e. there are no currents flowing in the material around the aperture) then the far field axial gain is

$$g = 4\pi A/\lambda^2. \tag{5.41}$$

Thus for a circular aperture of area πa^2 the gain is

$$g = 4\pi^2 a^2/\lambda^2 = (2\pi a/\lambda)^2. \tag{5.42}$$

This result was used in Section 4.6.1.

5.4 PREDICTION AND MEASUREMENT OF RADIATED FIELDS

5.4.1 Introduction

Antennas are used to measure the field strengths of radiation by connecting them to a load via a transmission line (typically a coaxial cable) and the making a measurement of the voltage generated across the load, as is shown schematically in Figure 5.16. In this figure P_d W m^{-2} is the power density of the incident radiation at the antenna, g_a is the antenna gain, g_t is the gain of the transmission line connecting the antenna to the load Z in which the received power P_r is dissipated and across which a voltage V is generated. These quantities are related by [7]

$$P_d = \frac{4\pi P_r}{\lambda^2 g_a g_t} \text{ W m}^{-2} \tag{5.43}$$

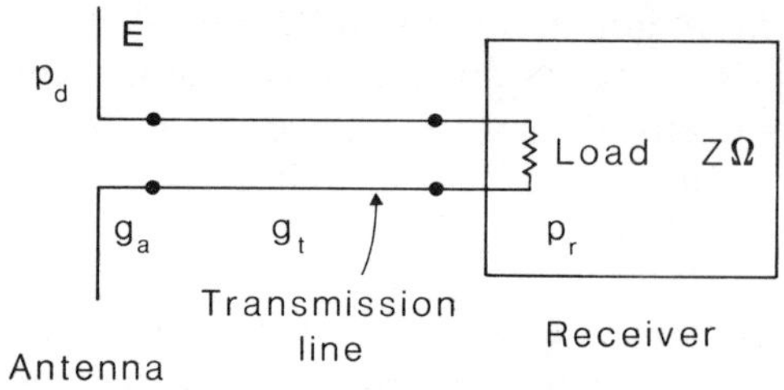

Fig. 5.16 Schematic diagram of a receiver connected to an antenna with gain g_a by a transmission line with gain g_t, used to measure the field strength E V m^{-1} in an incident electromagnetic wave carrying a power density P_d W m^{-2}

where λ is the wavelength of the incident radiation. The quantity $\lambda^2/4\pi$ is the isotropic antenna aperture [7].

The factor which is used to convert the voltage V across the load to the electric field strength E V m^{-1} of the incident radiation is known as the antenna factor, AF, which has units of m^{-1}. This quantity is often determined experimentally for each antenna system at each frequency and orientation by calibrating the antenna in a known field. The manufacturer of such calibrated antennas supplies a graph of antenna factor in dB as a function of frequency. This is discussed further in Section 5.4.2.1.

It is important to realise that this field strength measurement procedure is not as simple as it sounds. Antenna factors are often calculated or measured under far field and free space conditions, whereas they may be used to make measurements in situations which may differ radically from this. Effects such as reflection from the ground or nearby structures, resonances, and mutual coupling of the antenna and the radiation source when in the near field could cause errors in radiated susceptibility measurements. Also the presence of a ground plane below an antenna can change its free space AF. Nevertheless it is useful to summarise the mathematics for this topic so as to become familiar with the important factors, to define the units used to quantify field strengths, and develop some formulae which are useful for making simple predictions of possible radiated field strengths at the design stage. For more details, a review of EMC measurement methodologies can be found in reference [10].

5.4.2 Mathematical basis

The power density carried by the radiation is given by the Poynting vector

$$\mathbf{S} = \mathrm{Re}(\mathbf{E} \times \mathbf{H}^*) \ \mathrm{W\,m^{-2}}.$$

In the far field the time averaged power density $\langle P_d \rangle$ carried by plane wave radiation is

$$\langle P_d \rangle = E^2/Z_0 \ \mathrm{W\,m^{-2}} \tag{5.44}$$

where $Z_0 \approx 120\pi\ \Omega$ is the wave impedance in the far field. If an antenna radiates a total time averaged power density $\langle P_t \rangle$ then at a distance r m

$$\langle P_d \rangle = \langle P_t \rangle / 4\pi r^2 \ \mathrm{W\,m^{-2}}.$$

At a distance r (in the far field) from the transmitter the electric field strength will be

$$E^2 = Z_0 \langle P_t \rangle / 4\pi r^2 \ \mathrm{V\,m^{-1}}$$

i.e.

$$E \approx (1/r)\sqrt{(30\langle P_t \rangle)} \ \mathrm{V\,m^{-1}}. \tag{5.45}$$

Note that for a radiating antenna

$$\langle P_t \rangle = p_t g_a g_t$$

where p_t is the power input to the transmission line feeding the antenna, g_t is the gain in this line and g_a is the antenna gain.

It is useful to realise that a field of order 1 V m^{-1} can be thought of as a large field since FCC emission limits for domestic digital equipment are specified in the 100–200 μV m^{-1} range (see Table 5.2), and typical TV and radio receivers can easily detect field strengths of a few mV m^{-1}. Also small amounts of power radiated from a system can give rise to quite large fields in the vicinity. For example a system radiating a total of only 1 mW could generate fields of order 58 mV m^{-1} at a (far field) distance of 3 m, enough to affect seriously nearby radios tuned to the same frequency range. On the other hand, fields due to lightning are in the range 1 to 10 kV m^{-1} and EMP pulses up to 50 kV m^{-1}.

5.4.2.1 *Units*

Power and field strengths are often measured in decibels (dB) which express the ratio of two power or voltage levels (see Section 3.2.3). For power levels P_1 and P_2 then

$$\mathrm{dB} \equiv 10 \log_{10}(P_2/P_1).$$

If a power level P_2 is measured then the second power level P_1 with which to compare it can be defined arbitrarily. For example units of dBW are decibels above 1 watt, i.e. P_1 is chosen to be 1 watt, whereas units of dBmW have P_1 chosen to be 1 mW. One chooses the most convenient unit for the power levels being considered.

If power P is dissipated across a resistance R then since $P = V^2/R$, two voltage levels V_1 and V_2 *across the same impedance* can be compared by using the definition

$$\mathrm{dB} \equiv 20 \log_{10}(V_2/V_1).$$

If V_2 is measured then V_1 can be defined as an arbitrary reference level, e.g. when $V_1 = 1\ \mu$V then V_2 is measured in units of dBμV. Units of field strength measured in μV m^{-1} can be expressed in dBμV m^{-1} by choosing the reference field strength to be 1 μV m^{-1}.

The conversion to decibels is often applied to equations to cast them into a form in which dB units can be used directly. For example the electric field strength E V m^{-1} at the antenna (see Fig. 5.16) is related to power density P_d W m^{-2} (Equation (5.45)) by

$$E = \sqrt{(Z_0 P_d)} \approx 19.4\sqrt{P_d}\ \mathrm{V\,m^{-1}}.$$

Thus using Equation (5.43)

$$E \approx 19.4\sqrt{(4\pi P_r/\lambda^2 g_a g_t)}\ \mathrm{V\,m^{-1}}.$$

Converted to dBV m^{-1} this becomes

$$\begin{aligned} E &\approx 20\ \log_{10}\{19.4\sqrt{(4\pi P_r/\lambda^2 g_a g_t)}\}\ \mathrm{dBV\ m^{-1}} \\ &\approx 36.75 - 20\ \log_{10}(\lambda) + P_r(\mathrm{dBW}) - G_a - G_t\ \mathrm{dBVm^{-1}} \end{aligned} \tag{5.46}$$

where G_a and G_t are the gains of the antenna and transmission line measured in dB and P_r is measured in dBW (since $20\ \log_{10}(\sqrt{P_r/1}) = 10\ \log_{10}(P_r/1)$ dBW

The received power is measured by dissipating it in a meter with input impedance Z matched to the transmission line connecting it to the antenna, generating a voltage V. Since $P_r = V^2/Z$,

$$E \approx 19.4\sqrt{(4\pi V^2/Z\lambda^2 g_a g_t)}\ \mathrm{V\,m^{-1}}.$$

If $Z = 50\ \Omega$ then the field strength in dBV m^{-1} is

$$E \approx 19.75 - 20\ \log_{10}(\lambda) + V(\mathrm{dBV}) - G_a - G_t\ \mathrm{dBV\ m^{-1}}. \tag{5.47}$$

Thus the antenna factor for a 50 Ω load is defined as

$$\mathrm{AF} = 19.75 - 20\ \log_{10}(\lambda) - G_a\ \mathrm{dB}. \tag{5.48}$$

As pointed out in Section 5.4.1 it is usual for antenna manufacturers to calibrate them as a function of frequency f and provide a graph of AF versus f. Thus the field strength present at the antenna is obtained from the voltage V across the load, measured in dBV at each frequency, by

$$E = V(\mathrm{dBV}) + \mathrm{AF} - G_t\ \mathrm{dBV\ m^{-1}}. \tag{5.49}$$

The gain of a lossy transmission line, G_t dB, will be a negative quantity which can be found from the manufacturers' specifications.

Finally, as was discussed in Section 5.2.9, the measurement of the field strength of broadband noise has to take account of the bandwidth of the detector used to measure it, on a 'per-unit-bandwidth' basis. Thus the man-made background field strengths displayed in Figure 1.10 are measured in units of dBμV m^{-1} MHz^{-1}.

5.4.3 Radiation from a loop

5.4.3.1 Loops with impedance $Z < Z_0$

From Equation (2.21) the time averaged expression for the far field (i.e. where terms in $1/r$ dominate over those in $1/r^2$) electric component E_ϕ of a small loop antenna of area $A = \pi a^2$ carrying a sinusoidal current of amplitude I_0 can be written

$$E_\phi = \frac{\omega\mu_0 A I_0 \sin(\theta)\beta}{4\pi r} \text{ V m}^{-1}$$

where $\beta = 2\pi/\lambda$ and the loop dimensions $\ll\lambda$. This formula was derived assuming a uniform ring of current without any voltage drop due to finite impedance. It has a maximum with $\theta = 90^\circ$ (i.e. in the plane of the loop), and using $\omega\mu_0 = 2\pi Z_0/\lambda$ this can be recast to give

$$E_\phi = \frac{\pi Z_0 A I_0}{r\lambda^2} \text{ V m}^{-1} \tag{5.50}$$

or

$$E_\phi \approx \frac{120\pi^2 A I_0}{r\lambda^2} \text{ V m}^{-1}.$$

This electric field is polarised parallel to the plane of the loop. Mardiguian [6] points out that by using λ m $= 300/F$, where F is the frequency in MHz, Equation (5.50) can be written in the useful form

$$E_\phi \approx \frac{1.3 A I_0 F^2}{r} \ \mu\text{V m}^{-1} \tag{5.51}$$

where A is the area of the loop in cm^2. He also stresses that designers tend to work with loops driven by a voltage V and terminating with a known impedance Z, so that the expression for the maximum radiated electric component in the far field as a function of the driving voltage of a small loop becomes

$$E_\phi \approx \frac{1.3 A V F^2}{rZ} \ \mu\text{V m}^{-1}. \tag{5.52}$$

As this formula applies strictly to an isolated loop, it is usual [6, 11] to allow for in-phase reflection from a ground plane near a real loop by multiplying this expression by a factor two, giving

$$E_\phi \approx \frac{2.6 A V F^2}{rZ} \ \mu\text{V m}^{-1}. \tag{5.53}$$

This factor of two increase corresponds to a possible increase of 20 $\log_{10}(2)$ dB = 6 dB in the radiated field strength (see Section 5.4.4.1). This equation is valid when the loop impedance $Z < Z_0$, the impedance of free space. Loops with impedance $> Z_0$ will be discussed in the next section.

Equation (5.53) was derived assuming the loop dimensions $\ll\lambda$. For high frequencies where these dimensions are *greater* than $\lambda/4$ the validity of the formula can be extended to obtain better agreement with measurements by substituting $\lambda/4$ for the longest loop dimension, reducing the effective area [6]. Also, at high frequencies the current is not in phase all around the loop, which will reduce its efficiency as a radiator, and there will be resonances (see

Section 6.2). Since the area should always be made as small as possible to satisfy radiated susceptibility limits, Equation (5.53) is normally satisfactory for making useful predictions of the maximum likely radiation from an *isolated* loop. However, if the loop is in the vicinity of a cable such as a power lead then common mode currents induced in this cable by the current in the loop may radiated substantially more than the loop itself. This possibility is discussed further in Section 5.4.5.

5.4.3.2 *Loops with impedance* $Z > Z_0$

The set of Equations (2.22) for the fields radiated from a small loop were derived assuming a uniform loop of current. If there is a large impedance in the loop then a voltage drop will be present across it which will generate a dipole-like field in the space around it. Thus it is plausible that the radiation from a high impedance loop will have some of the characteristics of a high impedance field generated by a short wire antenna! Figure 5.17 shows that the total current flowing in a high impedance loop is the sum of a normal loop current I_1 which flows around the whole loop, and a current I_2 which flows up and down the wires in the vicinity of the voltage source, making that section of the loop behave as a short wire antenna of length L, see reference [16]. With such a model the loop forms a short, open ended, transmission line of length d, with an input impedance $|Z_{in}| = Z_c/\beta d$ given by Equation (3.39), where Z_c is the

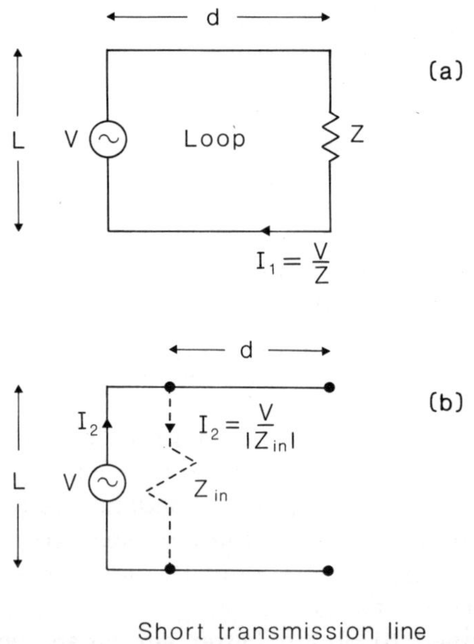

Fig. 5.17 High frequency alternating current in a high impedance loop consists of two components, a current I_1 which flows around the loop and I_2 which flows in the wires in the vicinity of the driving voltage V

characteristic impedance of the line. Thus the current flowing in the short wire acting as the antenna is

$$I_2 = V/|Z_{\text{in}}| = V\beta d/Z_c. \tag{5.54}$$

From (2.22*a*) the maximum magnitude of the far field electric component of a short wire antenna is

$$E_\theta = \frac{\beta L I_0}{4\pi\varepsilon_0 rc}$$

$$= \frac{\beta L Z_0 I_0}{4\pi r} \tag{5.55}$$

where I_0 is the current flowing in the wire. Combining (5.54) and (5.55) gives

$$E_\theta = \beta^2 AVZ_0/Z_c 4\pi r \text{ V m}^{-1}$$

$$= \frac{\pi AV(Z_0/Z_c)}{r\lambda^2} \text{ V m}^{-1}$$

where A is the area of the loop.

As can be seen from Figure 9.4 in reference [20] the characteristic impedance of infinitely long, lossless, parallel-wire transmission lines in free space is within a factor two of Z_0 over a large range of ratios of wire-separation to wire-radius, so a reasonable approximation in this calculation for a *short* transmission line is to put $Z_c \approx Z_0$, which gives

$$E_\theta \approx \frac{\pi AV}{r\lambda^2} \text{ V m}^{-1}. \tag{5.56}$$

This can be compared with (5.50*a*) for the maximum far field electric component due to magnetic dipole radiation, which can be written

$$E_\phi = \frac{\pi AV(Z_0/Z)}{r\lambda^2} \text{ V m}^{-1}. \tag{5.57}$$

Taken at face value (5.57) implies that the radiated electric field from a loop decreases as the loop impedance increases. However (5.56) shows that the radiated electric field from loops with high impedance is independent of the loop impedance. Thus (5.57), or its equivalent (5.52), can always be used, except that the loop impedance Z is set at 377 Ω for *all* high impedance loops [6].

Confirmation of these ideas can be found in reference [25], where Upton and Marvin have measured the transverse wave impedance (the ratio of the electric and magnetic field components perpendicular to the radial vector from the source) as a function of frequency in the vicinity of low and high impedance loops. These data have then been compared with calculations made using the Numerical Electromagnetics Code (see Section 5.4.6). The source-sensor distance was 50 cm over the frequency range 10–400 MHz, corresponding to a separation in wavelengths from λ/60 to 2λ/3, and the loops were loaded with impedances of 0, 40, 98 and 468 Ω. They found that a loop loaded with 468 Ω has a similar response to an electric dipole.

5.4.4 Prediction of radiated fields

5.4.4.1 *The basic calculation*

The maximum radiated electric field strengths from a loop as a function of frequency for a typical PCB loop carrying a trapezoidal waveform, such as a clock pulse, can now be calculated. The basic method is to make a Fourier analysis of the waveform to determine the amplitude of the different harmonics, then for each harmonic in turn use Equation (5.53) to predict the maximum field strength at some distance r m. These predicted strengths can then be examined to see if they are within appropriate limits. A very convenient way to do this is to work in decibels. Equation (5.53) can be written

$$E_\phi = 20 \log_{10}(2.6F^2/rZ) + 20 \log_{10}(A) + V(f)(\text{dBV}) \text{ dB}\mu\text{V m}^{-1} \quad (5.58a)$$

or separating out the 6 dB possible effect of reflections from the ground plane (see Section 5.4.3.1)

$$E_\phi = 20 \log_{10}(1.3/rZ) + 20 \log_{10}(A) + V(f) \text{ (dBV)} + 40 \log_{10}(F) + 6 \text{ dB}\mu\text{V m}^{-1} \quad (5.58b)$$

where $V(f)$ (dBV) is the amplitude of each harmonic in dBV, F is the frequency in MHz, Z is the loop impedance and A cm^2 is the loop area. As has been shown in Section 5.2.8 (Equation (5.30)) $V(f)$ can be approximated by

for frequencies

$$f > 1/\pi T_p \quad \text{and} \quad f < 1/\pi t_r$$

$$V(f) = 20 \log_{10}(2V_0\, \partial) + 20 \log_{10}(1/\pi f T_p) \text{ dBV}; \quad (5.59a)$$

for all frequencies

$$f > 1/\pi t_r$$

$$V(f) = 20 \log_{10}(2V_0\, \partial) + 20 \log_{10}(1/\pi f T_p) + 20 \log_{10}(1/\pi f t_r) \text{ dBV} \quad (5.59b)$$

where T_p is the pulse width, t_r is the rise time and the pulse amplitude has been written as V_0 rather than A to avoid confusion with the loop area A cm^2. If the frequency is F MHz, and T_p and t_r are measured in nanoseconds, (5.59*b*) can be written

$$\begin{aligned} V(F) &= 20 \log_{10}(2V_0\partial) + 20 \log_{10}(1000/\pi F T_p) + 20 \log_{10}(1000/\pi F t_r) \text{ dBV} \\ &= 20 \log_{10}(2V_0\, \partial) + 20 \log_{10}(1/\pi F T_p) + 20 \log_{10}(1/\pi F t_r) + 120 \text{ dBV}. \quad (5.59c) \end{aligned}$$

The radiated field strength dBμV m^{-1} can be found for a given circuit by summing up each of these terms. It is useful to make a table showing each term so that the contribution made to the overall sum can be clearly seen.

5.4.4.2 *A spreadsheet to calculate radiated field strengths*

A very convenient and quick way to perform this calculation is to make use of a general purpose spreadsheet program which are now widely available on computers. In Figure 5.18 is shown an example of a spreadsheet which has been

Calculation of radiation from a small loop as function of frequency
to compare with FCC class B limits at 3 m range.

INCLUDES 6 dB TO ALLOW FOR POSSIBLE IN PHASE GROUND REFLECTIONS

Tp	30.0	ns	Amp	3.5	V
tr	1.5	ns	r	3.0	metres
Period	100.0	ns	Z	200.0	Ohms
length	12.0	cm	Delta	0.315	
width	1.0	cm	..		
Area	12.0	square cm			
Fix length at lambda/4 above F3			=	625	MHz
F1 =	10.6	MHz			
F2 =	212.2	MHz			

1	2	3	4	5	6	7	8	9	10
n	F MHz	dB	dBV	dBV	Total	dB	(Area)	Field	FCC limit dBuV/m
0	10	6.9	0.0	0.0	6.9	-7.3	21.6	21.2	none
3	30	6.9	-9.0	0.0	-2.2	11.8	21.6	31.2	40.0
5	50	6.9	-13.5	0.0	-6.6	20.7	21.6	35.7	40.0
7	70	6.9	-16.4	0.0	-9.5	26.5	21.6	38.6	40.0
8	80	6.9	-17.5	0.0	-10.7	28.9	21.6	39.8	40.0
10	100	6.9	-19.5	0.0	-12.6	32.7	21.6	41.7	43.0
15	150	6.9	-23.0	0.0	-16.1	39.8	21.6	45.2	43.0
20	200	6.9	-25.5	0.0	-18.6	44.8	21.6	47.7	43.0
25	250	6.9	-27.4	-1.4	-22.0	48.7	21.6	48.2	46.0
30	300	6.9	-29.0	-3.0	-25.2	51.8	21.6	48.2	46.0
35	350	6.9	-30.4	-4.3	-27.8	54.5	21.6	48.2	46.0
40	400	6.9	-31.5	-5.5	-30.2	56.8	21.6	48.2	46.0
45	450	6.9	-32.5	-6.5	-32.2	58.9	21.6	48.2	46.0
50	500	6.9	-33.5	-7.4	-34.0	60.7	21.6	48.2	46.0
60	600	6.9	-35.0	-9.0	-37.2	63.9	21.6	48.2	46.0
70	700	6.9	-36.4	-10.4	-39.9	66.5	20.6	47.3	46.0
80	800	6.9	-37.5	-11.5	-42.2	68.9	19.4	46.1	46.0
90	900	6.9	-38.6	-12.5	-44.3	70.9	18.4	45.1	46.0
95	950	6.9	-39.0	-13.0	-45.2	71.8	17.9	44.6	46.0

Fig. 5.18 A spreadsheet calculation on a computer to calculate the radiation expected from a small loop at different frequencies for comparison with the FCC Class B limits

set out to calculate the expected radiation from a loop carrying a trapezoidal waveform and compare it with some chosen limits, in this case the FCC class B limits for domestic digital electronics at 3 m range [11].

At the top of the spreadsheet one finds the pulse parameters, such as pulse width '*Tp*' (subscripts cannot be used on this particular spreadsheet), rise time '*tr*" and period *P* in nanoseconds, the pulse amplitude 'Amp' in volts as well as the length and width of the loop in centimetres. The loop impedance Z Ω and the test distance *r* m are also displayed. As discussed in Section 5.4.3.2 if the loop impedance $Z > Z_0$ it should be fixed to equal Z_0. With a properly constructed spreadsheet the effects of changing any of these parameters will propagate automatically throughout the entire calculation.

For use later in the calculation, the duty cycle 'Delta', the area of the loop, the frequencies $F1 = 1/\pi T_p$, $F2 = 1/\pi t_r$ (in units of MHz) and $F3$ (for which the longest dimension of the loop, 12 cm in this example, is equal to $\lambda/4$) are all calculated separately.

The calculation proper takes place in the region below the dotted line. The columns are labelled 1 to 10 from left to right and contain the following:

Column 1: Harmonic number (integer). All the results appropriate to each harmonic are displayed across the page on the same line.
Column 2: Harmonic frequency F MHz, with the fundamental frequency at the top.
Column 3: The term $20 \log_{10}(2V_0\partial)$ where V_0 is pulse amplitude and ∂ is the duty cycle.
Column 4: The term $20 \log_{10}(1000/\pi F T_p)$ when $F > F1$, see (5.59*c*).
Column 5: The term $20 \log_{10}(1000/\pi F t_r)$ when $F > F2$.
Column 6: The sum of columns 3, 4 and 5. This is the amplitude of each harmonic in dBV.
Column 7: The term $20 \log_{10}(2.6F^2/rZ)$ where r is the test distance and Z the loop impedance, see (5.58*a*).
Column 8: The term $20 \log_{10}(A)$ where A is the loop area in cm^2. If $F > F3$ the longest loop dimension should be replaced by $\lambda/4$ and the area recomputed. This has the effect of reducing the effective loop area, see reference [6].
Column 9: The sum of columns 6, 7 and 8. This is the predicted field strength $dB\mu V\ m^{-1}$.
Column 10: The FCC class B limits $dB\mu V\ m^{-1}$.

The FCC Class B limits at 3 metres distance are shown in Table 5.2. The highest allowable limit of 46 $dB\mu V\ m^{-1}$ corresponds to only 200 $\mu V\ m^{-1}$. As Figure 5.18 shows, these limits are violated by most frequencies above 150 MHz with the design considered, so some action would have to be taken to rectify this. However, even if the calculation showed that the limits were not violated, the prediction for the radiation from the whole system, of which the loop is a part, may not be reliable if the loop is near a cable, as will be discussed in Section 5.4.5.

Table 5.2. FCC Part 15, Subpart J Class B limits for radiation from digital electronics between 30 and 1000 MHz at 3 m distance from the circuit

$dB\mu V\ m^{-1}$	$\mu V\ m^{-1}$	Frequency range (MHz)
40	100	30–88
43.5	150	88–216
46	200	216–1000

5.4.5 Common mode radiation from cables

Common mode currents (see Section 6.5.1) are often produced unintentionally within a system, due to effects such as finite impedance of ground planes and coupling between different circuits, and they are difficult to predict and model.

Hubing and Kaufman [16] have proposed a simplified technique called by them *end-driven wire* modelling which is useful for electrically small desk-top products with an attached external cable, such as a power lead. The technique cannot be used for products which are electrically large (dimensions > λ), without cables, floor standing or well shielded. The original application was to video display terminals when it was discovered that common mode currents produced on the external cables by coupling to the video signal were the major source of radiation.

They approached the problem by using popular computer codes (so-called moment-method programs, some of which will be briefly discussed in Section 5.4.6) to model a test circuit and predict the radiation that would be emitted from it. These predictions were then compared with careful measurements of the actual radiation emitted from a real circuit. The test circuit modelled by the computer programs was a loop of 10 cm by 1 cm with impedance Z Ω driven by a 30 MHz 100 μV voltage and connected to ground by a long wire, see Figure 5.19(*a*). They found that the radiation was not very sensitive to the impedance of the loop, and concluded that circuit could be simplified to a single wire driven at one end, as shown in Figure 5.19(*b*). There was very satisfactory agreement between the predicted radiation from this circuit and the radiation from the real test circuit where the power cable was connected *indirectly* to the loop. The important point is that the radiation from the total system was much bigger than that from the loop itself.

They suggested the following procedure to analyse small desk-top systems equipped with a cable.

(a) Reduce the product to its end-driven wire equivalent, complete with lumped-element components and shields.
(b) Analyse the end-driven wire model using a moment-method technique.

Once the circuit driving the wire has been identified, all components that do not have an effect on the common-mode current can be neglected.

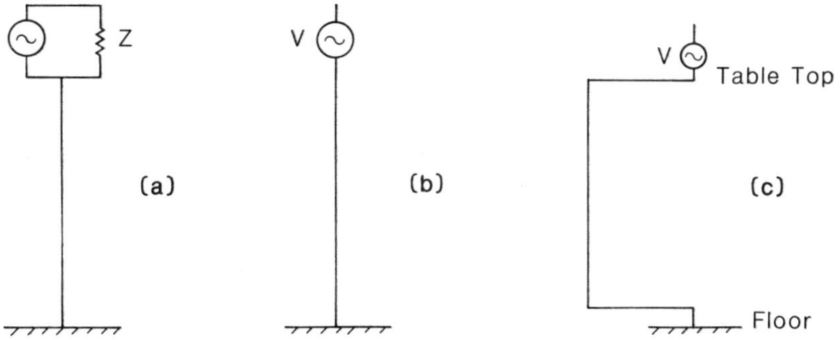

Fig. 5.19 Some of the circuits studied by Hubing and Kaufman to develop the *end-driven wire* model for predicting radiation from table-top products with an attached cable; see reference [16]. (Reproduced by permission of IEEE)

For a cable run with a geometry shown in Figure 5.19(*c*) the field strength at 3 m calculated with this method was 61.9 dBμV m^{-1} at 30 MHz, well above the FCC Class B limit of 40 dBμV m^{-1}, but within 3 dB of the actual measurement. They conclude that the method does not eliminate the necessity of making measurements, but has proven to be very useful for comparing different designs or investigating the effects of changes to a product.

5.4.6 Computer codes to predict radiation

Realistic predictions of the radiation produced from arrays of wires can only be made by using powerful general purpose computer codes, originally developed for antenna design. There are a number of these available for various computers ranging from the desk-top personal computer (PC) to large supercomputers. One of the most famous is the Numerical Electromagnetics Code (NEC) written in FORTRAN by the Lawrence Livermore National Laboratory, USA [17]. There is a version of this called MININEC which runs on a PC [18]. These programs accept as input the wire configurations and thickness, voltage sources and loads. These data are then used to calculate such things as the radiation pattern and field strengths, using a numerical technique known as the Method of Moments (MOM), or moment-method. It is probably difficult to model any reasonably complex circuit boards with these packages.

Other commercially available packages such as MSC/EMAS™ [19] are designed to run on computers with full graphics facilities and can deal with DC, AC and transient currents including eddy currents and resonant models. The frequency range is from DC to the frequency of visible light, and the program uses a new formulation of Maxwell's equations based on the vector potential. It is claimed that this offers significant advantages over other methods such MOM.

Access to, and familiarity with, packages like these is mandatory if accurate and useful design work is to be achieved. There is an informative tutorial on computational electromagnetics, with an extensive bibliography, in reference [21].

5.4.7 Broadband antennas

The bandwidth of an antenna is the frequency range over which its properties remain reasonably constant. (See Section 5.2.9 for a precise definition.) As can be seen from Section 5.3 the parameters of simple antennas, such as radiation resistance, vary with wavelength. It is desirable for antennas to have a large gain in order to radiate or receive as much power as possible in a given direction. The requirements of large bandwidth and high gain can only be achieved by careful design. Even then, no single antenna can be used over frequency ranges from kHz to many GHz and different types must be used at different frequencies. For example, to generate test fields inside a reverberation chamber (see Section 5.5.4), the workers in reference [10] used the following

Table 5.3. Antennas and associated equipment shown in Figure 5.20, supplied by Electro-Metrics Inc., New York

Item	Name	Antenna type	Frequency range
2	MTA-60	Parabolic	1–10 GHz
3	RVR-30M	Manual vertical	9 kHz–30 MHz
4	MFA/B/C	Magnetic field probes	9 kHz–230 MHz†
5	BDA-30	Broadband dipole	20–220 MHz
6	EFP-30	Electro-static probes	9 kHz–1 GHz
9	BIA-30C	Collapsible bi-cone	30–200 MHz
10	ALP-51H	Loop	100 kHz–30 MHz
11	RGA-60	Double ridged guide	1–18 GHz
12	LPA-30	Log periodic	200 MHz-1 GHz
13	TDA-30/1-4	Tunable dipoles	28 MHz–1 GHz
16	ALR-30M	Manual loop	9 kHz–30 MHz
17	BDA-30S	Broadband dipole	20 MHz–200 MHz
18	ALP-51L	Loop	30 Hz–30 MHz
19	ELS-11	Sensor (magnetic field)	20 Hz–50 kHz
20	ALP-70	Loop	10 kHz–30 MHz
21	LCA-30	Conical log spiral	200 MHz–1 GHz
		Associated equipment	
1	TRP-163	Support tripod	
7	BPA-1000	Broadband preamp	10 kHz–1 GHz
8	CIG-25	Impulse generator	
14	PA-1000	Power amplifier	10 MHz–1 GHz
15	FE-60	Frequency extender	

† 3 probes cover this frequency range.

antennas: log-periodic (0.2–1.0 GHz), corrugated horn (1.0–4.0 GHz), and double ridged circular horn (4.0–18 GHz).

There are a large number of different types of antennas available, some of which are shown in Figure 5.20. They are identified in Table 5.3. In Figure 5.20, the log-periodic antenna (item 12) gets is name from the fact that its properties repeat themselves periodically with increasing frequency, so that on a log graph the frequencies at which the performance is identical are equally spaced. Reference [7] contains a detailed discussion of the design of log-periodic, phased-array and Yagi–Uda antennas.

In Figure 5.21 a bi-cone antenna is shown being used on an open field test site (protected by a dielectric tent) to make EMC tests on a desk-top VDU. The antenna can be moved in the vertical plane and rotated, and the EUT is also situated on a turntable in order to change its orientation with respect to the antenna during the tests. In Figure 5.22 are shown some test data of emission levels taken on an open field site. Figure 5.22(a) shows the amplitudes of the Fourier components detected with the EUT unmodified, and Figure 5.22(b) the effect of making modifications to reduce emissions.

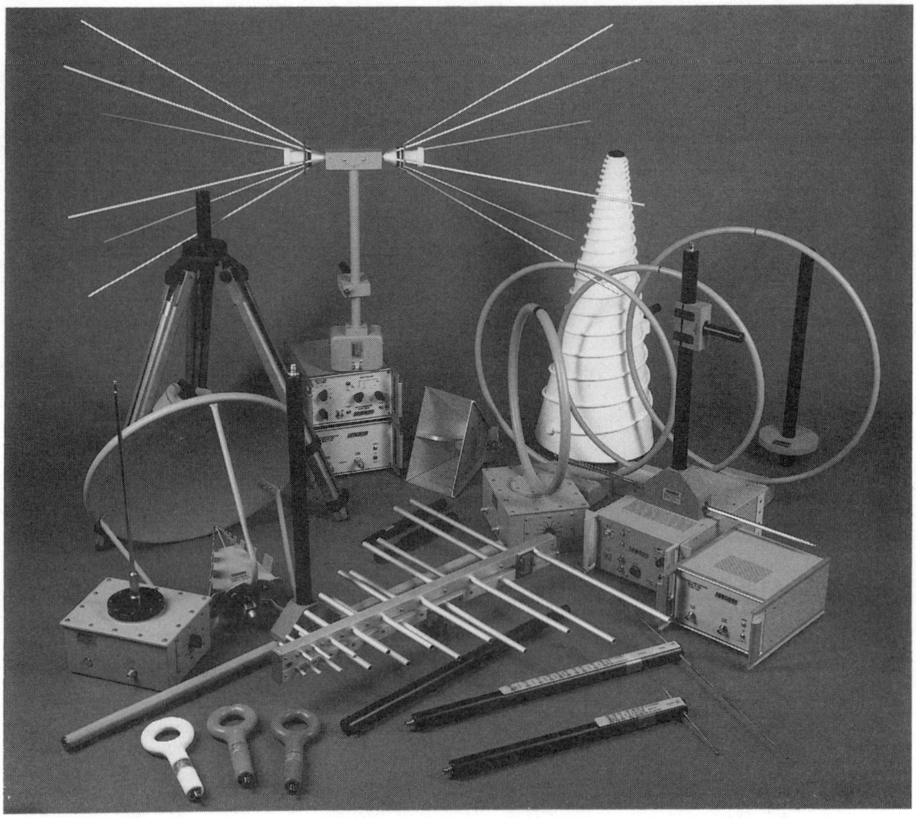

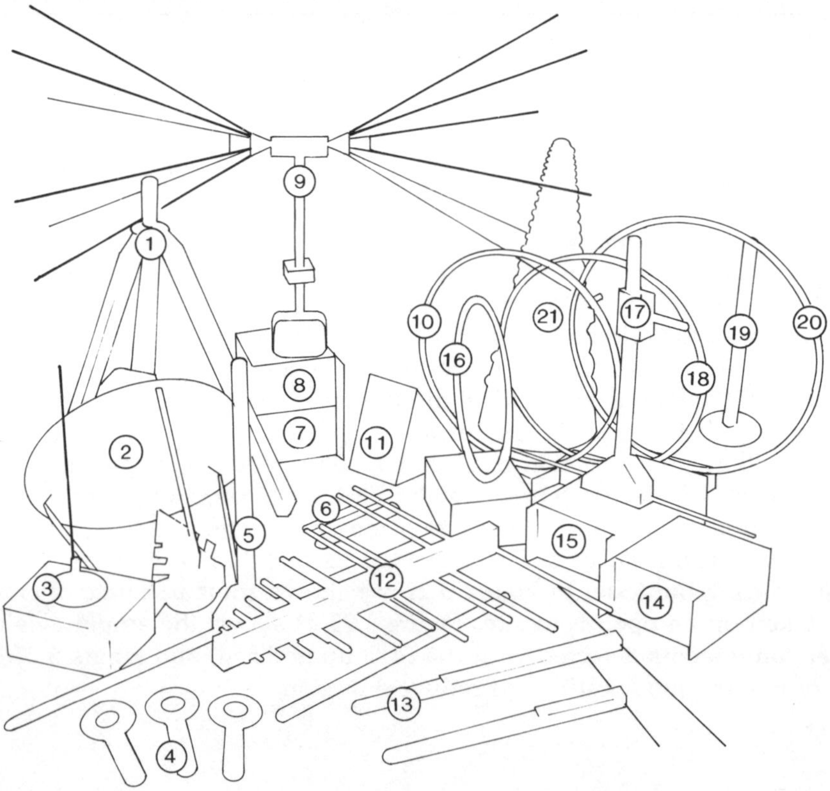
1
2
3
4
5
6
7
8
9
10
11
12
13
14
15
16
17
18
19
20
21

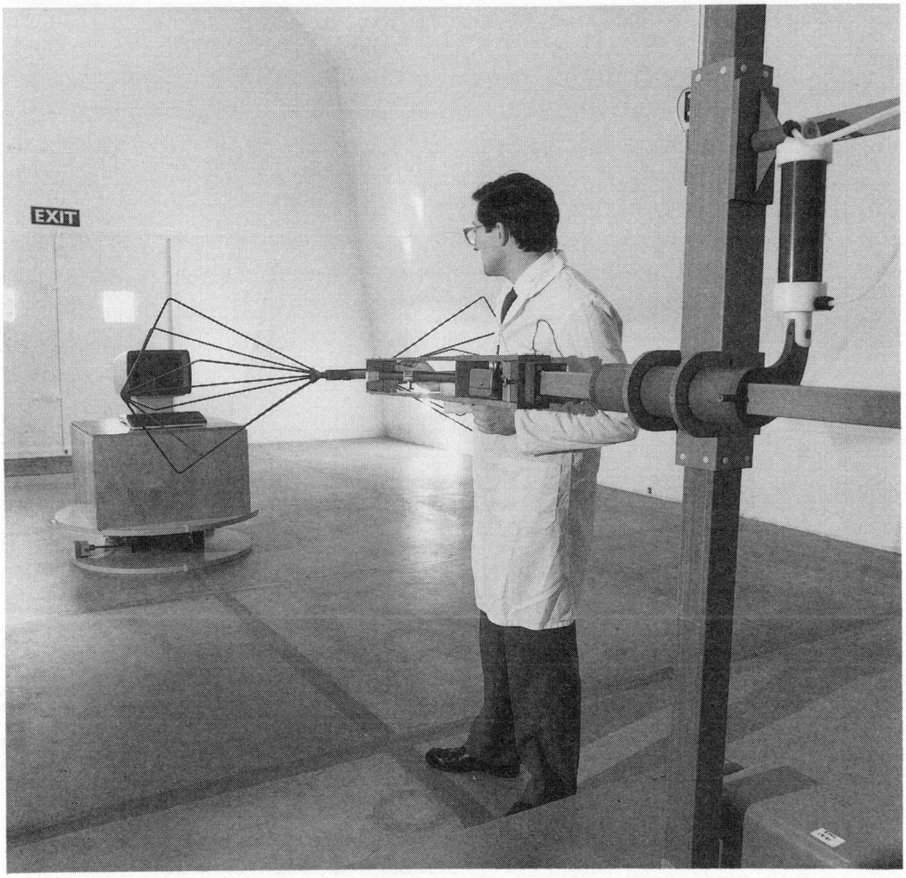

Fig. 5.21 Bi-cone broadband antenna being used on an open field test site. (Reproduced by permission of Siemens Plessey Assessment Services Ltd.)

5.5 GENERATION OF ELECTROMAGNETIC FIELDS FOR EMC TESTING

5.5.1 Introduction

Susceptibility tests require that equipment under test (EUT) be placed in a specified electromagnetic environment (often a plane wave) and its performance monitored. Such requirements mean that techniques to make fields with accurately known properties are of the greatest importance. The first methods used to do this utilised radiation from standard gain horn antennas for frequencies above a few hundred MHz or radiation travelling in parallel plate lines below a few MHz. However, major problems with these methods are the

---

Fig. 5.20 Some examples of broadband antennas. (Reproduced by permission of Electro-Metrics Inc., New York)

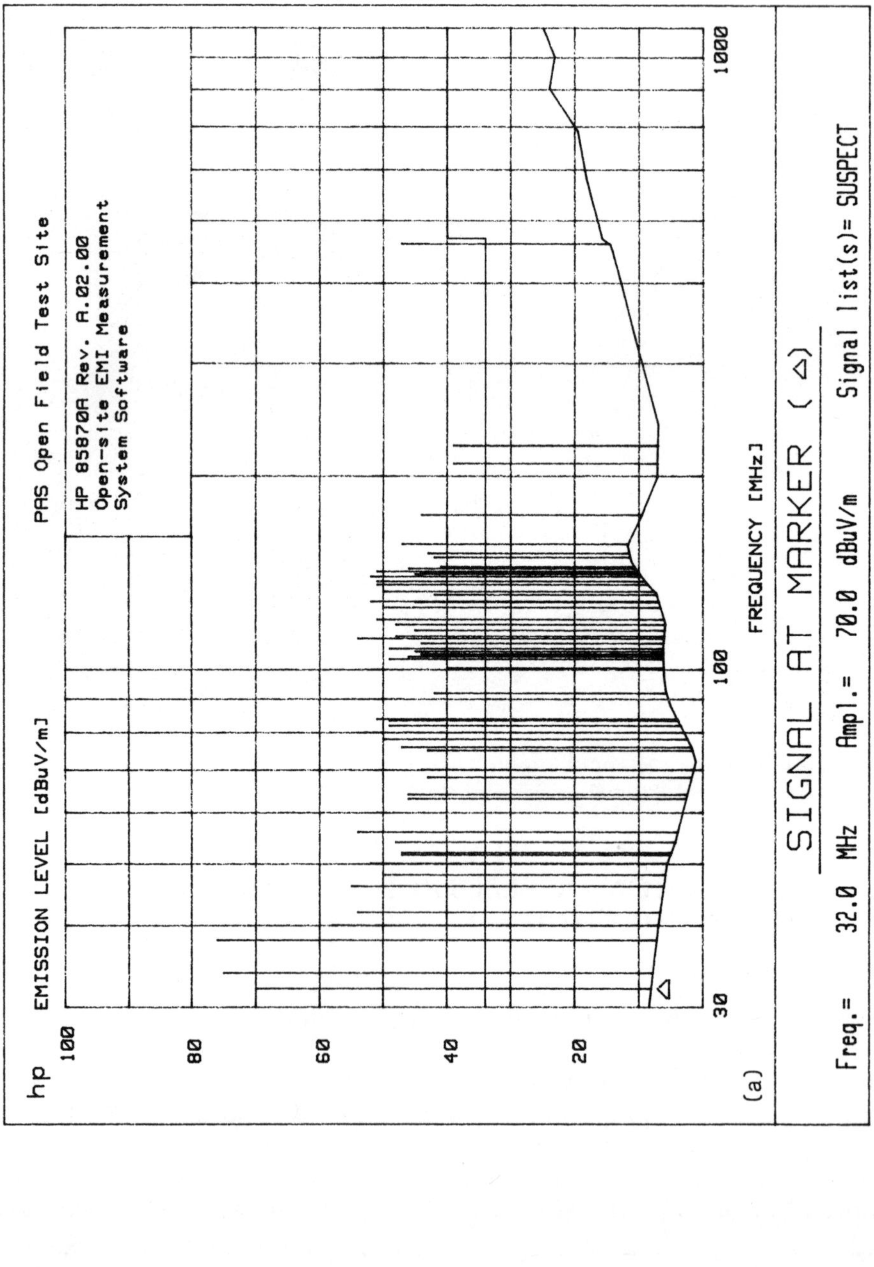
hp
EMISSION LEVEL [dBuV/m]
PAS Open Field Test Site
HP 85870A Rev. A.02.00
Open-site EMI Measurement
System Software
100
80
60
40
20
30
100
1000
FREQUENCY [MHz]
(a)
SIGNAL AT MARKER (△)
Freq.= 32.0 MHz Ampl.= 70.0 dBuV/m Signal list(s)= SUSPECT

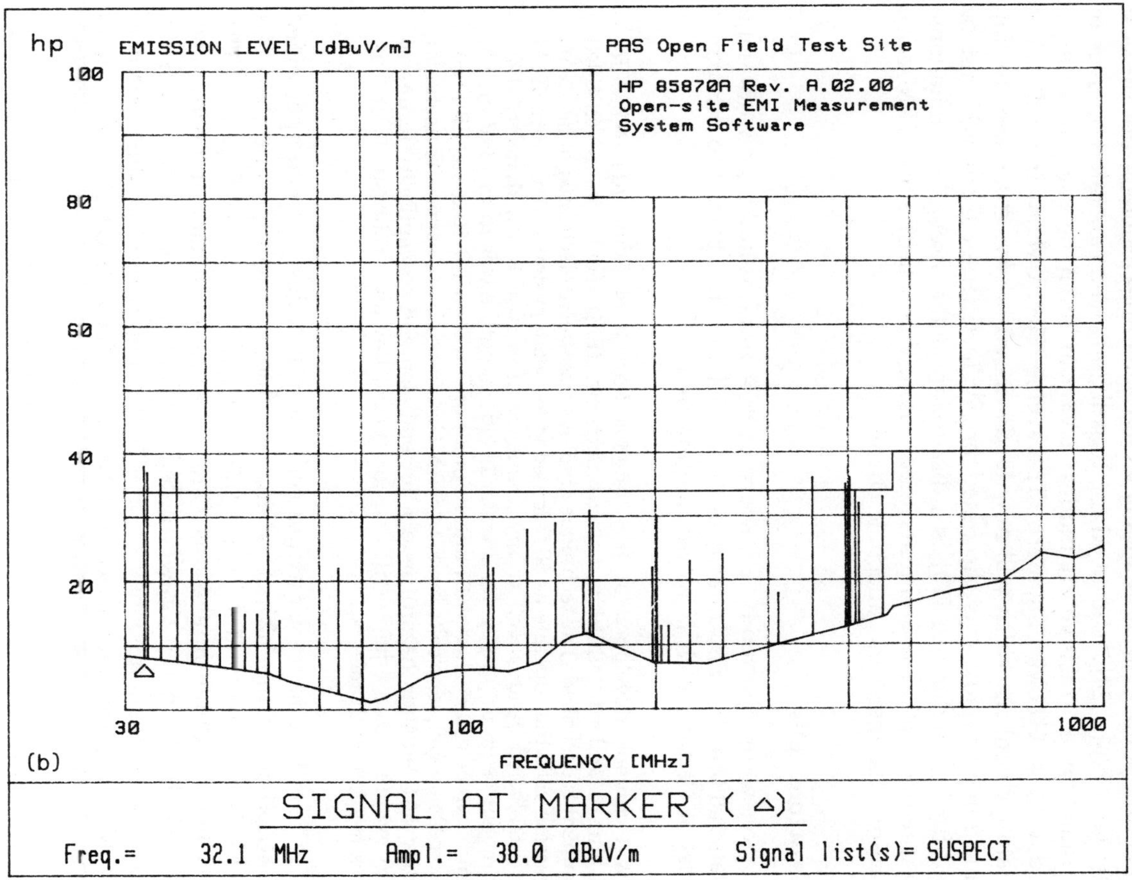

Fig. 5.22 Radiated susceptibility measurements on an open field site; (*a*) before modification of EUT, (*b*) after modifications to reduce emissions. (Reproduced by permission of Siemens Plessey Assessment Services Ltd.)

limited bandwidth of the radiation and the fact that they radiated into the surroundings and could themselves be a major disturbance to others. Screening these sources by putting them in a screened room with conducting walls was problematical as, among other things, resonances within the room could change the field at any point. Anechoic rooms would solve this problem, but the cost would be very high [24]. The need for broadband, well calibrated sources of *confined* radiation at (relatively!) modest cost can often be met by using a Crawford cell, a GTEM (Gigahertz Transverse ElectroMagnetic) cell, or a reverberating chamber, which will now be discussed briefly. Open area testing in an unconfined region was discussed briefly in Chapter 1 but will not be discussed further. Reference [10] is a discussion of the strengths, limitations, and interpretation of measurements made using open field sites, TEM cells, reverberating chambers and anechoic chambers.

5.5.2 Crawford cells

These devices were reported by Crawford in 1974 and resulted from work done at the National Bureau of Standards in the USA [12]. The abstract to this paper is a model of clarity which is worth quoting here.

> 'A new technique developed at the National Bureau of Standards (NBS) for establishing standard, uniform, electromagnetic (EM) fields in a shielded environment is described. The technique employs transverse electromagnetic (TEM) transmission cells that operate as 50 Ω impedance-match systems. A uniform TEM field is established inside a cell at any frequency of interest below that for which higher modes begin to propagate. Standard field strength levels from $10 \mu V\ m^{-1}$ to $500\ V\ m^{-1}$ can be established with uncertainties of less than 1.0 dB to 2.0 dB inside the NBS cells for frequencies from dc to 500 MHz. The cells are especially useful for calibrating EM radiation hazard meters, for emission and susceptibility testing of small to medium sized equipment, and for special low level calibrations of very sensitive field strength meters'.

The basic idea behind the Crawford cell is to put 'sides' onto an open sided shielded strip line, Figure 5.23(*a*), where a conductor of width w and thickness t is placed between two conductors separated by a distance b, see Figure 5.23(*b*). After choosing the outer dimensions b and thickness of metal t, the width w of the centre electrode in the Crawford cell is trimmed carefully to obtain the desired characteristic impedance to match the impedance of the source of radiation. Matching the impedances will prevent reflections which could lead to

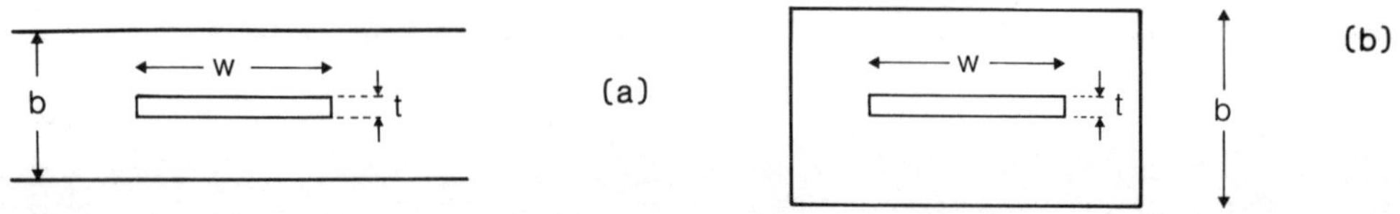

Fig. 5.23 (*a*) Cross section of an open sided shielded strip line, (*b*) cross section of a Crawford cell. Adapted from reference [12]. (Reproduced by permission of IEEE)

resonances which would spoil the uniformity of the TEM wavefront. The completed cell is a section of rectangular or square cross section coaxial transmission line with tapers at each end to connect to standard coaxial connectors one of which feeds in the rf power, the other acts as a 50 Ω load, see Figure 5.24. The square cross section provides maximum test area but with somewhat reduced electric field uniformity. The rectangular cross section gives greater field uniformity but less test space for the EUT. Table 5.4 adapted from Crawford's paper [12] give the dimensions of square or rectangular cross section cells (as defined in Figure 5.21(*b*)) and their maximum frequencies.

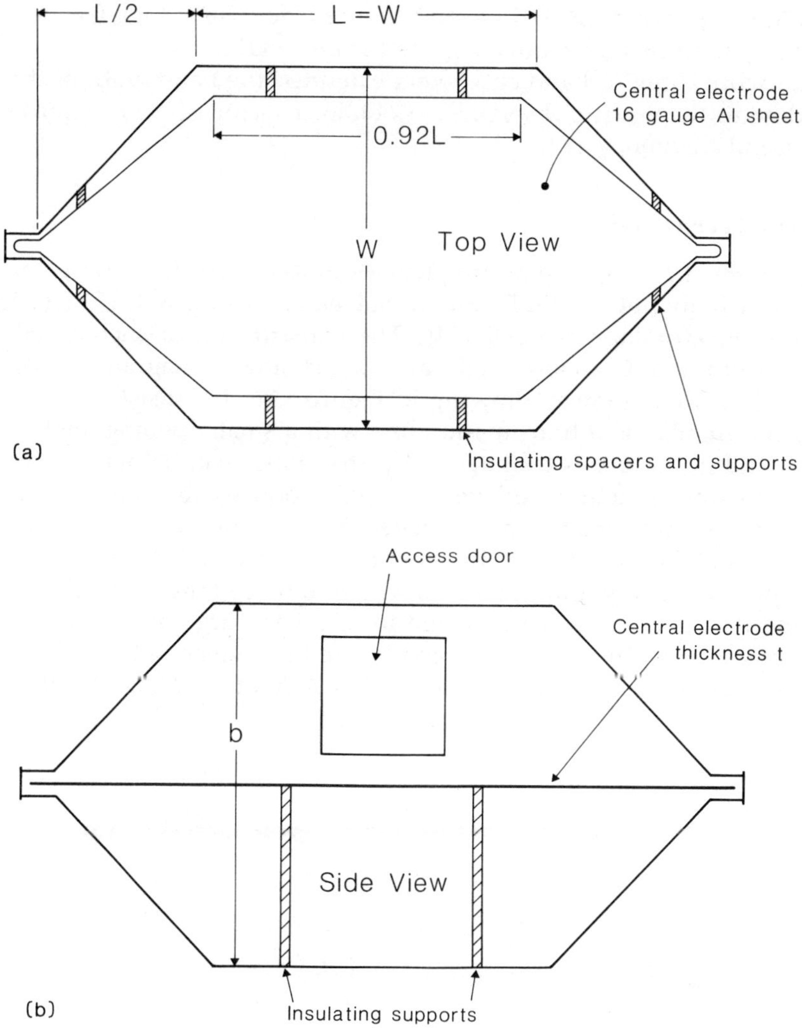

Fig. 5.24 Schematic layout of a Crawford cell, adapted from reference [12]. (Reproduced by permission of IEEE)

Table 5.4. Dimensions of Crawford cells, adapted from reference [12]

	Square cell			Rectangular cell		
Max frequency (MHz)	b (cm)	w (cm)	t (cm)	b (cm)	w (cm)	t (cm)
100	150	123.83	0.157	90	108.15	0.157
300	50	41.28	0.157	30	36.05	0.157
500	30	24.77	0.157	18	21.83	0.157

Notice that for the metal thickness t of 0.157 cm (16 gauge Al sheet) the width w is given to a precision of 0.01 cm and that cells for use at high frequencies are quite small, with greatly reduced space for any EUT.

In later work Crawford and coworkers extended the bandwidth of these cells by modifying them with absorbers to obtain a factor of two increase in the upper useful frequency [13].

5.5.3 The GTEM cell

The requirement of having high frequencies upto the gigahertz range and large size in which to put an EUT can be solved by using a GTEM (Gigahertz Transverse ElectroMagnetic) cell [14]. The construction of such a cell is very different from the Crawford cell, and a schematic diagram is shown in Figure 5.25, with a cutaway drawing in Figure 5.26. It consists of a section of tapered rectangular 50 Ω transmission line with a small opening angle of about 15°. The end of the line is terminated by absorbing material which can absorb the incident wave and any currents. The cell produces an undistorted, slightly spherical wavefront which can be excited by a continuous wave source or a pulse generator for EMP tests. The most notable feature is the fact that cells of all sizes produce a very uniform field over a frequency range from DC to several GHz, and that cells can be constructed to take very large objects such as racks of electronics. Over the region occupied by the EUT which is 1/3 of the available test space the field uniformity can be ±1 dB, with a ±3 dB variation with frequency [14].

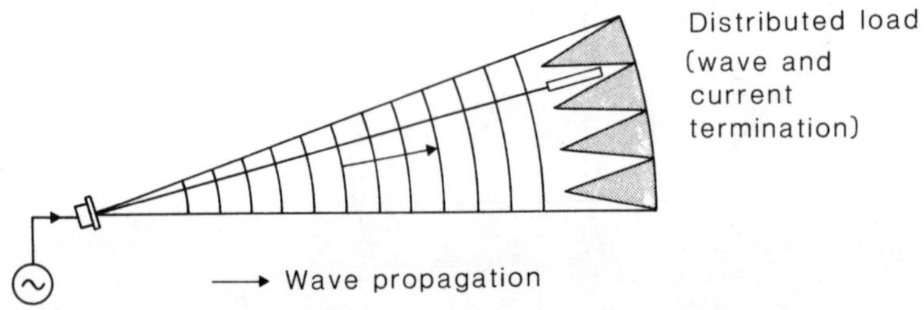

Fig. 5.25 Schematic diagram of GTEM cell

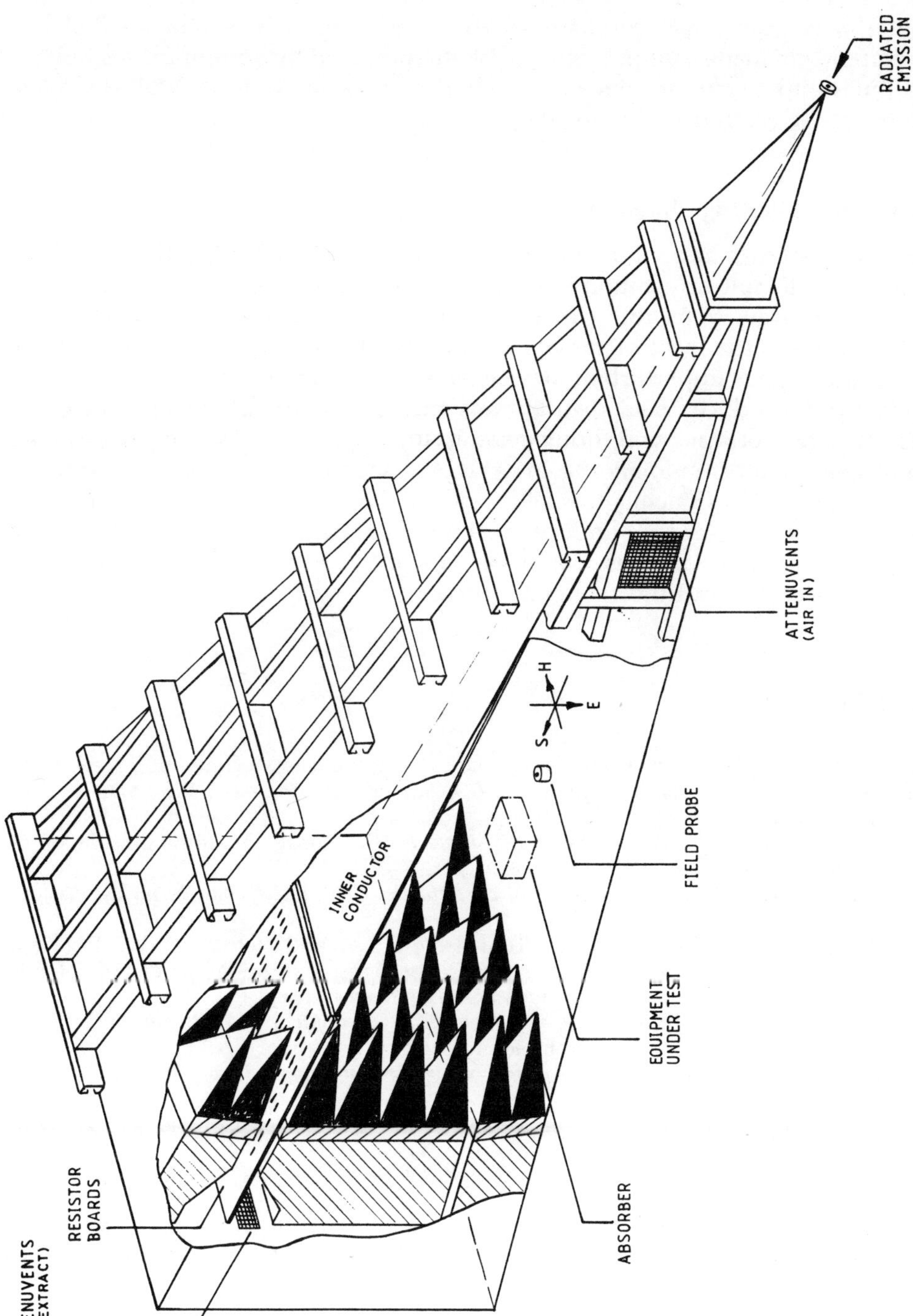

Fig. 5.26 Cutaway drawing of a GTEM cell. (Reproduced by permission of Ray Proof and ABB Research Centre, Switzerland)

The GTEM cells are also very suitable for radiated emissions testing, and for objects which are small compared to the wavelength, it is claimed that the measurements made within a cell can be manipulated by computer to simulate accurately and quickly results which would be obtained in FCC, VDE etc. open site tests; see reference [15] for details.

5.5.4 Reverberating chambers

A feature of EMC measurement procedures is that usually the EUT must be rotated to different orientations during testing to take account of polarisation effects. For large items this can be difficult. A reverberating chamber is an enclosure in which the test radiation supplied by antennas is 'stirred' by an electrically large field perturbing device, such as a rotating fan with large metal blades. This produces field distributions throughout the chamber which, on average, are isotropic and homogeneous. Thus the EUT is bathed in an unpolarised, uniform radiation field, so that it need not be rotated with respect to the antennas producing the fields. A similar technique is used in some microwave ovens to ensure uniform heating of the item being cooked. For a full discussion of the advantages and limitations of this technique for EMC testing see reference [10] and the references therein.

REFERENCES

[1] Papoulis, A. (1962) *The Fourier Integral and its Applications* McGraw-Hill, New York.

[2] *Reference Data for Radio Engineers* (1977) 6th edition, Howard W. Sams & Co., Indianapolis.

[3] Cooley, J. W. and Tukey, J. W. (1965) An algorithm for the Machine Calculation of Complex Fourier Series *Math. Comput* **19** 297–301.

[4] Weaver, H. J. (1983) *Applications of Discrete and Continuous Fourier Analysis* Wiley, New York.

[5] Press, W. H., Flannery, B. P., Teukolsky, S. A. and Vetterling, W. T. (1988) *Numerical Recipes in C* Cambridge University Press, Cambridge. (Versions of this book are available in the languages FORTRAN and Pascal.)

[6] Mardiguian, M. (1984) *Interference Control in Computers and Microprocessor-Based Equipment*, Interference Control Technologies, Gainesville, VA.

[7] Liao, S. Y. (1988) *Engineering Applications of Electromagnetic Theory* West Publishing Co., St Paul.

[8] Booker, H. G. (1946) Slot Aerials and Their Relation to Complementary Wire Aerials, *J. IEEE (London)* **93** part IIIA, no. 4.

[9] Jull, E. V. (1981) *Aperture Antennas and Diffraction Theory* Peter Peregrinus Ltd., Stevenage.

[10] Ma, M. T., Kanda, M., Crawford, M. L. and Larsen, E. B. (1985) A Review of Electromagnetic Compatibility/Interference Measurement Methodologies, *Proc. IEEE* **73** 388–411.

[11] Ott, H. W. (1988) *Noise Reduction Techniques in Electronic Systems* 2nd edition, Wiley, New York.

[12] Crawford, M. L. (1974) Generation of Standard EM fields Using TEM Transmission Cells *IEEE Trans. on EMC* **EMC-16** 189–195.

[13] Crawford, M. L., Workman, J. L. and Thomas, C. L. (1978) Expanding the Bandwidth of TEM Cells for EMC Measurements, *IEEE Trans. on EMC* **EMC-20** 368–375.

[14] Hansen, D., Wilson, P., Koenigstein, D. and Schaer, H. (1989) A Broadband Alternative EMC Test Chamber Based on a TEM-Cell Anechoic-Chamber Hybrid Concept *Int. Symp. EMC, Nagoya, Japan*.

[15] Simulating open area test site emission measurements based on data obtained in a novel broadband TEM cell, 1989 *Proc. IEEE* (*Nat. Symp. on EMC, Denver, CO*) pp 171–177.

[16] Hubing, T. H. and Kaufman, J. F. (1989) Modelling the Electromagnetic Radiation from Electrically Small Table-Top Products *IEEE Trans. on EMC* **EMC-31**.

[17] Burke, G. J. and Poggio, A. J. (1987) Numerical Electromagnetics Code (NEC)—Method of Moments *NOSC Tech. Document 116* Naval Ocean Systems Center, San Diego. (Purchase address; Naval Postgraduate School, Code 62-AB, Monterey, CA93943, USA.)

[18] Logan, J. C. and Rockway, J. W. (1986) The New MININEC (Version 3): A Mini-Numerical Electromagnetic Code, *Technical Document 938*, Naval Ocean Systems Center, San Diego, CA 92152-5000, U.S.A.

[19] MSC/EMAS™ is marketed by the MacNeal-Schwendler Corporation, 815 Colorado Blvd., Los Angeles, CA 90041-1777, USA.

[20] Johnk, C. T. A. (1988) *Engineering Electromagnetic Fields and Waves*, 2nd edition, Wiley, New York.

[21] Miller, E. K. (1988) A Selective Survey of Computational Electromagnetics *IEEE Trans. on Antennas and Propagation* **AP-36** 1281–1305.

[22] Jordan, E. C. and Balmain, K. G. (1968) *Electromagnetic Waves and Radiating Systems* 2nd edition, Prentice-Hall, Englewood Cliffs, NJ.

[23] Leikus, V. C. (1987) The three most important properties of spectrum analysers *News from Rhode & Schwarz, Munich* **116** 25–27.

[24] Kunz, K. S., Hudson, H. G., Breakall, J. K., King, R. J., Pennock, S. T. and Ludwigsen, A. P. (1987) Lawrence Livermore National Laboratory Electromagnetic Measurement Facility *IEEE Trans. on EMC* **EMC-29** 93–103.

[25] Upton, M. E. G. and Marvin, A. C. (1988) Measurement of transverse wave impedance in the vicinity of electrically small radiating structures *6th Int. Conf. on EMC* University of York, IERE publication 81, pp 163–166.

[26] IEEE Editorial Policy Statement (1987) More on EMC Terminology *IEEE Trans. on EMC* **EMC-29** 202–205.

6 Important Factors in Practical EMC Design

6.1 INTRODUCTION

In earlier chapters some of the basic electromagnetic theory relevant to EMC has been covered and now it is necessary to see how it can be applied to real situations. In this chapter the problems that have to be considered in various types of system are individually highlighted (e.g. the design of circuit board layouts) though several topics of general EMC concern are also introduced. For example the concepts involved in the proper use of coupling estimates, ground layouts (including the concept of zoning), filters and balanced systems are all of general applicability to a variety of systems of differing complexity.

As discussed in Chapter 1, any problem involving the electrical interference of one system by another has to have a source, a coupling path and a receptor. It goes without saying that any particular problem may involve more than one of each of these. It is also important to remember that EMC problems are of two general types:

(1) the search for, and negation of, a particular source of interference during system development or operation;
(2) the design of the system to meet specific EMC regulations. These regulations might cover emission or susceptibility or both.

In either case it is important to be able to understand and control the coupling paths that might be involved. For example a particular system might be radiating excessively due to the high clocking speeds of the particular microprocessor used. However, the radiation may be escaping from common mode currents on the power lines rather than directly from the circuit board containing the microprocessor. Thus to modify the emission one would have either to lower the clock speed (rather than lower the areas of the high current loops on the circuit board) or alter in some way the magnitude of the common mode currents. Here then is a typical problem where the coupling path from source to receptor involves conduction and radiation. Thus as a first step it is useful to revise what has been discussed earlier with regard to possible coupling paths. In every case the electrical interference caused by one system on another could in principle be calculated from Maxwell's equations by using the correct boundary conditions and some numerical technique. This is usually impossible in practise so that use has to be made of a number of approximations. Whilst in every case the electromagnetic fields of the source are responsible for the interference it is convenient (if somewhat arbitary) to divide the coupling paths into three types.

(*a*) *Conductive.* This is the process by which interference propagates along any conductor (cable system, ground plane etc.) and which can be common or differential mode.
(*b*) *Reactive.* This refers to a direct field coupling process which occurs dominantly in the near field. In this situation the contribution of the electric and magnetic fields to the coupling can be calculated separately. In the case of the electric field the coupling is called *capacitive* and for the magnetic field *inductive.*
(*c*) *Radiative.* This is the process that describes the direct field coupling experienced by systems in the far field of the disturbing source. In this regime energy is carried by a plane EM wave and both E and H must be considered together.

In the literature (*b*) and (*c*) are often called radiative.

6.2 SOME SIMPLE COUPLING ESTIMATES

In order to control and understand the various forms of coupling it is useful to be able to estimate their magnitudes in a number of simple systems. Such estimates are necessarily approximate and it is important to remember how easily the choice of an incorrect model can lead to erroneous results in electromagnetic problems. Consider first of all the coupling between a plane wave and a small loop. The loop of height h and length l is orientated as shown in Figure 6.1 with respect to the impinging EM wave. The integral forms of Maxwell's equations can be used to calculate the voltage V_i which appears around the loop. From (2.87) and (2.88),

$$\oint \mathbf{E} \cdot d\mathbf{l} = -\frac{\partial}{\partial t} \oiint \mathbf{B} \cdot d\mathbf{S} = V_i$$

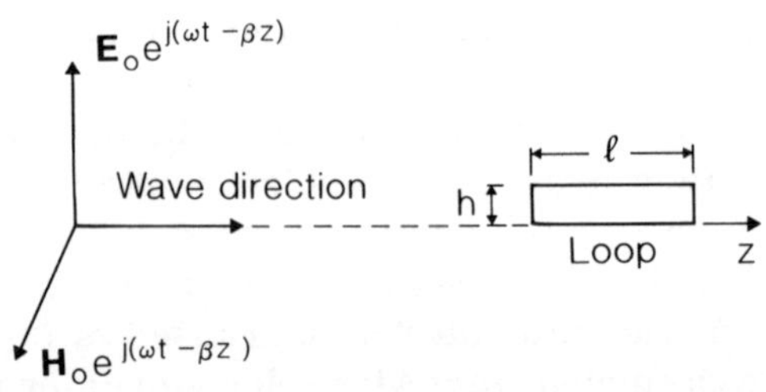

Fig. 6.1 A plane wave incident onto a rectangular loop. The **H** component is perpendicular to the plane of the loop

So that using

$$E_0/H_0 = Z_0 = \surd(\mu_0/\varepsilon_0) = c\mu_0$$

with

$$c = \frac{1}{\sqrt{\mu_0\varepsilon_0}} \quad \text{and} \quad \beta = \frac{2\pi f}{c}$$

leads to

$$V_i = -\mathrm{j}\omega\mu_0 H_0 h \mathrm{e}^{\mathrm{j}\omega t}\left[\frac{\mathrm{e}^{-\mathrm{j}\beta l} - 1}{\mathrm{j}\beta}\right]$$

where $H = H_0\mathrm{e}^{\mathrm{j}(\omega t - \beta z)}$ represents the plane wave H component as shown in Figure 6.1.

Then the magnitude of the voltage V_i is given by:

$$V_i = hH_0c\mu_0\surd[2(1 - \cos\,\beta l)]\,. \tag{6.1}$$

This is the same result that would have been obtained if instead the electric field has been used to perform the integration. The values of V_i calculated from Equation (6.1) for particular conditions are given in Figure 6.2. It is clear that the coupling increases with frequency and that at high frequency resonances occur (when $\beta l = 2n\pi$, $n = 1, 2, 3$ etc.) where the coupling goes to zero. Unfortunately it is seldom possible to use these minima in EMC design since in practice the actual loops involved and the waves that excite them never correspond to the ideal case considered above. However results obtained from (6.1) can be used to

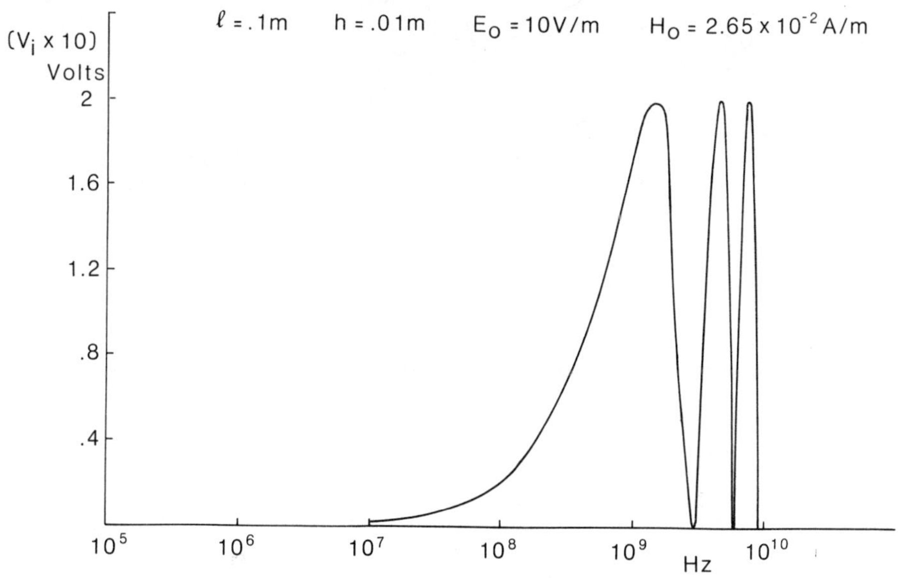

Fig. 6.2 Voltage induced on a rectangular coil by a plane wave with the geometry as in Fig. 6.1

estimate upper limits of radiative coupling in some cases [1]. Calculation of the far field radiation strengths emitted from loops or arrays of loops can be carried out using the techniques given in Chapter 5. The difficulty here being to identify the relevant loops involved.

In the near field, coupling between various circuits can be modelled in terms of capacitive or inductive effects. The system must be smaller than a fraction of a wavelength for such a circuit theory approach to work. In the first instance a problem involving capacitive coupling will be investigated. Consider the general case of a partially screened cable as shown in Figure 6.3 located close to an unscreened wire which is taken to be the source of the electrical disturbance in the system. In this case it is assumed that the screen where it exists is perfect (i.e. no holes such as would occur in a braided screen) and that the coupling does not load the noise source. The simplified equivalent circuit representing this situation is shown in Figure 6.3(*b*).

Then simple circuit analysis shows:

$$V_0/V_N = \frac{\omega R_N C_2}{\sqrt{[1+\omega^2 R_N^2 (C_N + C_2)^2]}} \tag{6.2}$$

where V_N = the voltage applied to the source wire A and V_0 = the voltage appearing across the load C_L, R_N of the victim cable B, and $C_N = C_L + C_3$.

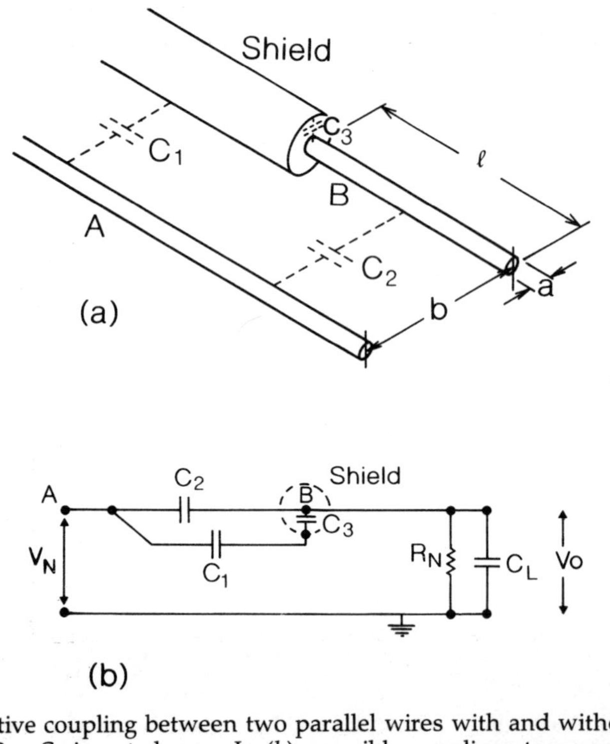

Fig. 6.3 Capacitive coupling between two parallel wires with and without shielding. In (*a*) the load R_N, C_L is not shown. In (*b*), possible couplings to ground are ignored

The coupling capacitance between A and B which for the system shown is given by [2]

$$C_2 = \pi \varepsilon \ln(a/b) l \tag{6.3}$$

where a = the radius of the conductors A and B, b = the distance between the unscreened conductors, l = the unshielded length of B.

For $\omega R_N(C_N + C_2) \ll 1$

$$V_0/V_N = \omega R_N C_2 \tag{6.4}$$

so that the coupling increases with frequency whereas for

$$\omega R_N(C_N + C_2) \gg 1$$

$$V_0/V_N = C_2/(C_N + C_2) \tag{6.5}$$

the coupling is now fixed by capacitive division. The break point angular frequency between these two situations is given by:

$$\omega = 1/R_N(C_N + C_2). \tag{6.6}$$

When $l = 0$ ($C_2 = 0$) the coupling is zero but in practice this is difficult to achieve due to the need to make connections at the ends of cable runs and because real shields may contain occasional holes. Figure 6.4 gives calculated values of V_0/V_N using (6.2) for two different load resistances. This shows that by keeping impedances low in the threatened circuit capacitive coupling effects at low

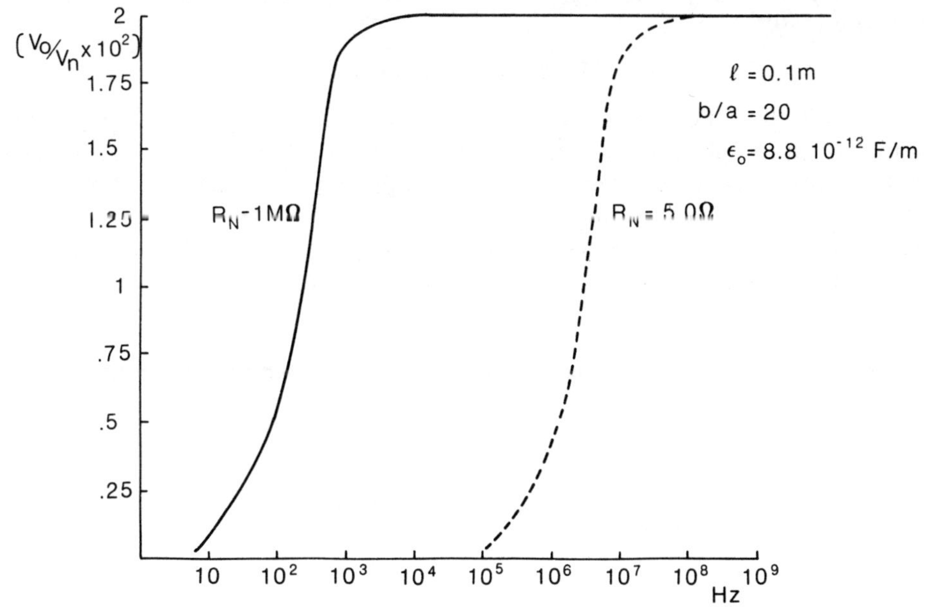

Fig. 6.4 Capacitive crosstalk for a 10 cm length of unshielded line

frequency can be minimised. It is clear that by varying the values of the parameters in (6.2) a whole range of practical systems can be modelled. A further discussion on the behaviour of screens which are dominated by the distribution of capacitance (the so-called electrostatic screening) is to be found in Section 4.2.3.

Inductive coupling can be estimated by calculating the flux ϕ cutting the victim circuit and then using the relation;

$$V_i = -\mathrm{d}\phi/\mathrm{d}t \qquad \text{with } V_i = \text{the induced voltage.}$$

However $\phi = MI$, with M = mutual inductance between the culprit and victim circuits, and I = current in the culprit circuit.

Then

$$V_i = -M\,\mathrm{d}I/\mathrm{d}t$$
$$= -\mathrm{j}\omega MI \qquad \text{with } I = I_0 \mathrm{e}^{\mathrm{j}\omega t}.$$

Now $\phi = \oiint B \cdot \mathrm{d}S$ over the victim effective circuit area so that $M = \oiint B \cdot \mathrm{d}S/I$ = the mutual inductance between the victim and culprit circuits. For the example of the two loops discussed in Section 2.7.1 (see Figure 6.5(*a*) for geometry definitions)

$$V_i = \mathrm{j}\omega_0 aAI\omega/\pi r^2$$

or

$$M = \mu_0 aA/\pi r^2 \text{ H.} \tag{6.7}$$

For two coplanar loops (see Figure 6.5(*b*) for geometry definitions) Ott [3] has given the result

$$M = 4 \times 10^{-7} \ln(b/a) \text{ H.} \tag{6.8}$$

Using results like these it is possible to calculate the inductive coupling for a number of practical systems. One general result is that in order to minimise inductive coupling it is necessary for the culprit circuit to carry a low current and so high culprit circuit impedances are to be preferred.

The general problem of screening out inductive coupling has already been discussed in Chapter 4 for the case where the screen surrounds the whole

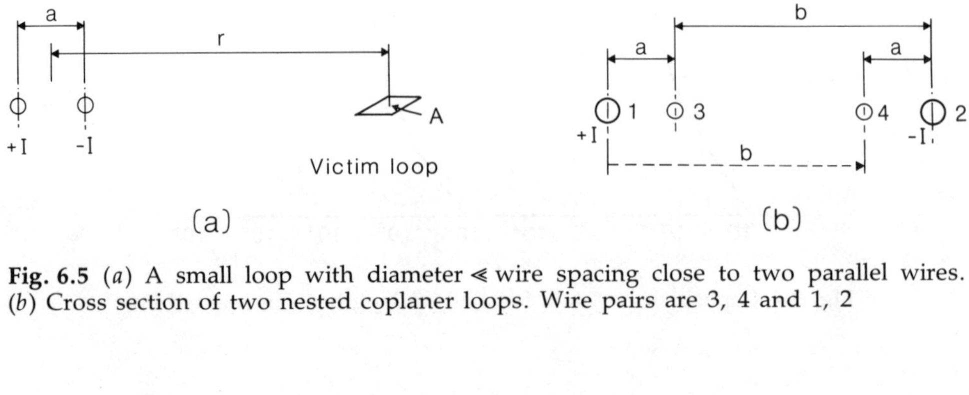

Fig. 6.5 (*a*) A small loop with diameter ≪ wire spacing close to two parallel wires. (*b*) Cross section of two nested coplaner loops. Wire pairs are 3, 4 and 1, 2

victim or culprit circuit. For example consider the case of a coaxial cable (with a non-magnetic shield) immersed in a time changing uniform magnetic field. In a coaxial cable the go and return currents flow on the central conductor and on the inside of the shield (at high frequencies). Any magnetic field at low frequencies will penetrate the shield without attenuation but no induced voltages will appear because for a symmetrically situated centre conductor the induced voltages in the effective loops on each side of the centre conductor will cancel each other out. The same process gives an unshielded twisted pair cable its reduced response to inductive effects. In this case the induced voltage in one loop is cancelled out by that generated in the next of the series which will be twisted through 180°. Such cables perform better if the number of loops/unit length is increased and the loop area is made as small as possible. Twisted pair systems can be shielded as well to give extra protection. Unfortunately all types have a large interconductor capacitance so that at high frequency coaxial cables are to be preferred. The behaviour of the shield on a coaxial (and a shielded twisted pair system) when exposed to higher frequency magnetic fields is dominated by the occurrence of eddy currents and the skin effect. As discussed in Chapter 4 these phenomena reduce the magnetic field within the screened volume so that any residual induced voltages continue to decrease as the frequency is raised.

In some circumstances shielding against inductive coupling can also be obtained when only part of the circuit is screened. Consider the circuit and shield shown in Figure 6.6(*a*). The shield can be considered to be only slightly larger in diameter than the centre conductor. It is assumed that at the frequencies considered the shield is transparent to any magnetic fields involved. The circuit is coupled to some culprit circuit by a mutual inductance M. The shield and the stray capacitances form another circuit which can be considered to have this same mutual inductance to the culprit circuit. Almost identical voltages will be induced around both loops but because of the high impedance of the stray capacitances at low frequency no appreciable current will flow in the shield. Thus the shield has no effect on the induced voltage appearing on the inner circuit. If on the other hand the shield is now grounded as shown in Figure 6.6(*b*) a current I_S will flow in the shield. This current can inductively couple to the inner circuit so changing the overall induced voltage on the inner circuit. Ott [3] has shown that the mutual inductance M_S between the inner conductor and the shield (and vice versa) is equal to the self-inductance L_S of the shield so that it is fairly simple to calculate the overall effect using the equivalent circuit shown in Figure 6.6(*c*). For a culprit circuit current of I A the induced voltage in both the shield and the main circuit is $V_S = j\omega MI$ (ignoring the effect of the shield current I_S).

But $I_S = V_S/(R_S + j\omega L_S)$ and this produces a voltage V_N on the inner conductor given by

$$\begin{aligned} V_N &= j\omega M_S I_S = j\omega L_S I_S \\ &= j\omega MI(j\omega L_S/(R_S + j\omega L_S)). \end{aligned} \tag{6.9}$$

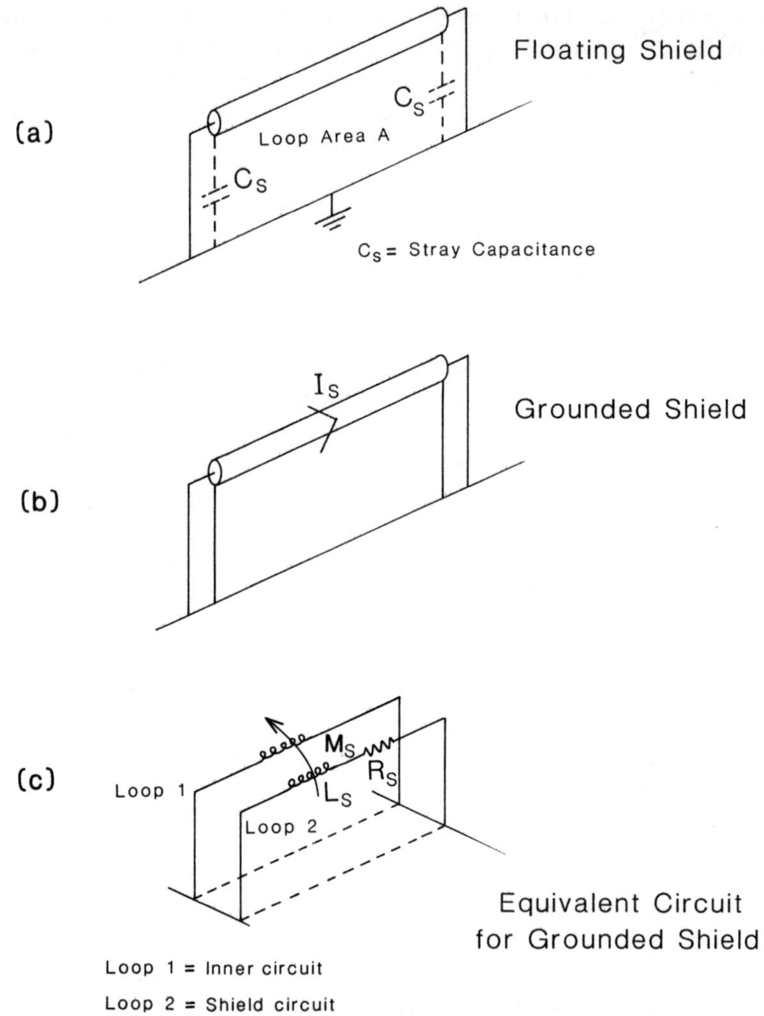

Fig. 6.6 Shielding effect of a partial shield for inductive coupling. The circuits are inductively coupled with mutual inductance M to another circuit carrying a current I, not shown in the figure

Thus the resultant noise voltage on the inner conductor is given by

$$\begin{aligned} V_T &= V_S - V_N \\ &= j\omega MI(R_S/L_S)(L_S/(R_S + j\omega L_S)) \end{aligned} \tag{6.10}$$

giving the magnitude which can be written as

$$V_T = MI\omega(R_S/L_S)(\omega^2 + R_S^2/L_S^2)^{-1/2} \tag{6.11}$$

and for ω large

$$V_T = MIR_S/L_S \tag{6.12}$$

which is independent of frequency whereas the low frequency limit (and that for the unshielded case) is given by

$$V_{\mathrm{T}} = MI\omega \tag{6.13}$$

Hence above the break point frequency f_{C} where

$$f_{\mathrm{C}} = R_{\mathrm{S}}/2\pi L_{\mathrm{S}} \tag{6.14}$$

the shield provides a reduction in the induced noise which could be characterised by a shielding effectiveness SE given by

$$\begin{aligned} \mathrm{SE} &= 20\ \log_{10}(MIR_{\mathrm{S}}/MIL_{\mathrm{S}}\omega)\ \mathrm{dB} \\ &= 20\ \log_{10}(\omega_{\mathrm{C}}/\omega)\ \mathrm{dB} \end{aligned} \tag{6.15}$$

with

$$\omega_{\mathrm{C}} = 2\pi f_{\mathrm{C}}.$$

A normalised induced voltage can be defined which is independent of the primary coupling

$$V_{\mathrm{T}}/MI = \omega R_{\mathrm{S}}/(R_{\mathrm{S}}^2 + \omega^2 L_{\mathrm{S}}^2)^{1/2}. \tag{6.16}$$

For the case of a copper cylinder of radius a, thickness t and height h above a conducting plane one has Coekin [4] (c = speed of light)

$$\begin{aligned} R_{\mathrm{S}} &= \rho/2\pi at\ \Omega\ \mathrm{m}^{-1} \\ L_{\mathrm{S}} &= 60\varepsilon_r^{-1/2}c^{-1}\ \ln(2h/a)\ \mathrm{H\ m}^{-1}. \end{aligned} \tag{6.17}$$

Figure 6.7 gives calculated values of 6.16 for $a = 5\times10^{-3}$ m, $t = 2.5\times10^{-4}$ m, $h = 2\times10^{-2}$ m and $\rho = 1.8\times10^{-8}\ \Omega$ m. For this shield $f_{\mathrm{C}} = 0.874$ kHz which is similar to values reported for a number of coaxial cables [3].

The shielding effect discussed above is not due to any exclusion of the field by the shield but because of the cancellation effects caused by particular current return paths. Similarly it is possible to remove the magnetic fields generated by conductors by setting up equal and opposite fields generated by the return path. An example of this is furnished by the behaviour of a coaxial cable. Thus the current carried by the centre conductor sets up a magnetic field which is counteracted by the magnetic field set up by the return current provided it passes uniformly along the cable shield braid. If the shield is grounded at each end there is the possibility that some of the return current will go via the ground, thus negating this cancellation effect. Ott [3] has shown that, provided $f > f_{\mathrm{C}}$, the current returns via the shield. Thus above this cut off frequency the coaxial cable is a good container of any magnetic fields generated within its structure.

The general way to cut down inductive effects in any threatened circuit is to reduce the effective area of any pick-up loop. In the limit however the shield current often represents the dominant problem. Consider the coaxial cable illustrated in Figure 6.8(a). At high frequencies the inner circuit DCEF carries the signal and any shield current generated inductively (or by any other

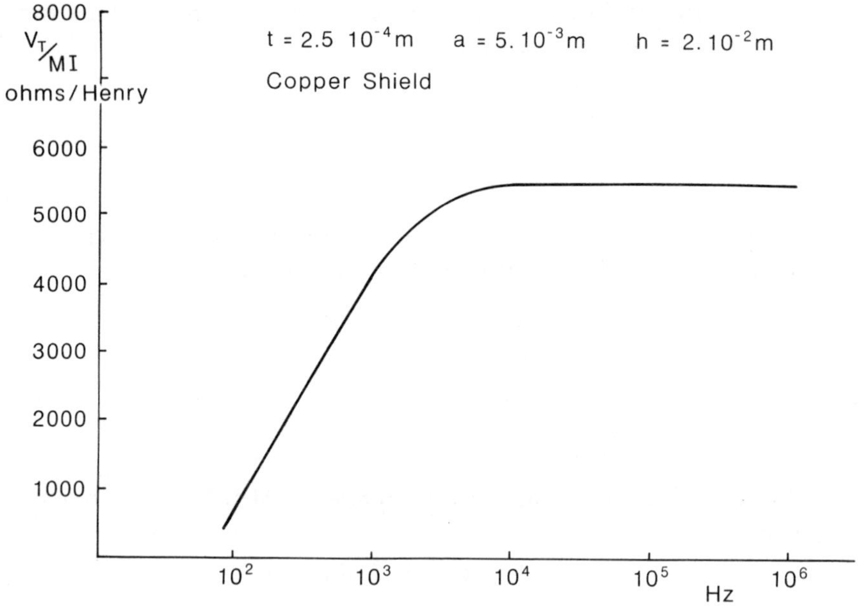

Fig. 6.7 Normalised induced voltage for a shield above a ground plane as a function of frequency

coupling process) flows in the circuit ABCD. The low frequency coupling model discussed above can be used to estimate the effect of the shield current. In the equivalent circuit shown in Figure 6.8(*b*) it is clear that the noise voltage V_2 appearing across R_2 is given by

$$V_2 = j\omega L_S I_S + I_S R_S - j\omega M I_S \qquad \text{for } R_2 \gg R_1$$

but $M = L_S$ so that

$$V_2 = I_S R_S.$$

Hence the shield resistive drop which occurs in the common impedance shared by the shield and the signal circuits is the source of the residual noise. Such common impedance coupling is frequent source of pick up in practical systems. The above model is not strictly correct at high frequency since the skin effect reduces the internal current along the shared impedance and the residual noise is now given by

$$V = I_S Z_T$$

where Z_T = the transfer impedance which is discussed more fully in Chapter 4.

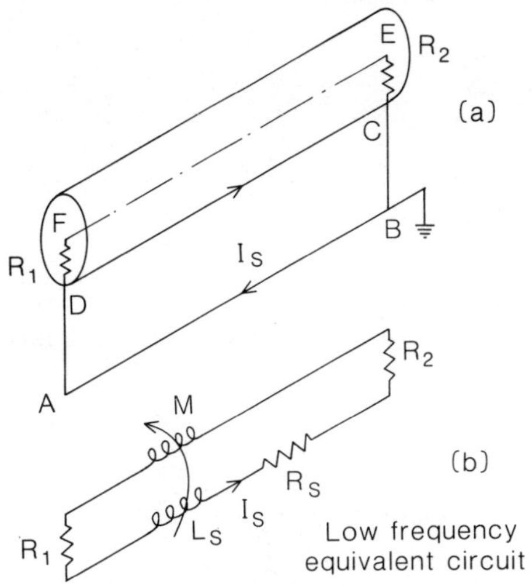

Fig. 6.8 Behaviour of a coaxial cable line excited by shield currents

6.3 GROUNDING

6.3.1 Introduction

In every circuit there has to be a path for the current leaving any supply voltage line to return to its source. This return path is usually at a low reference potential termed the ground potential. Ideally this ground return should be at a uniform potential but a moments reflection indicates that since it cannot have zero impedance the potential distribution will depend on the current flowing.

The ground return is often, but not always, connected to the earth potential, which is the potential the Earth as a whole has with respect to the rest of the universe. Figure 6.9(*a*) indicates a simple model of the Earth showing a central conducting core with a more resistive outer layer upon which the bulk of man's engineering activities are carried out. In a static situation the Earth's surface will be at V_E, the earth potential, but as soon as transients are involved this will no longer be true. The classic example of this occurs in the case of lightning strikes from a charged cloud to the Earth. As shown in Figure 6.9(*b*) a simplified model consists of representing the charged system as capacitor but remembering that it is really an extended region of space with charges distributed over a large area. The charging is caused by solar powered convection. When a lightning discharge occurs to a specific location as shown in Figure 6.9(*c*) currents must flow to neutralise the induced charges on the earth. Some simplified paths are shown in the Figure. The combination of the current flow and the unavoidable earth impedances means that transient voltages appear at various locations on the ground. The avoidance of the worst effects of these is discussed in Section 6.6.5.

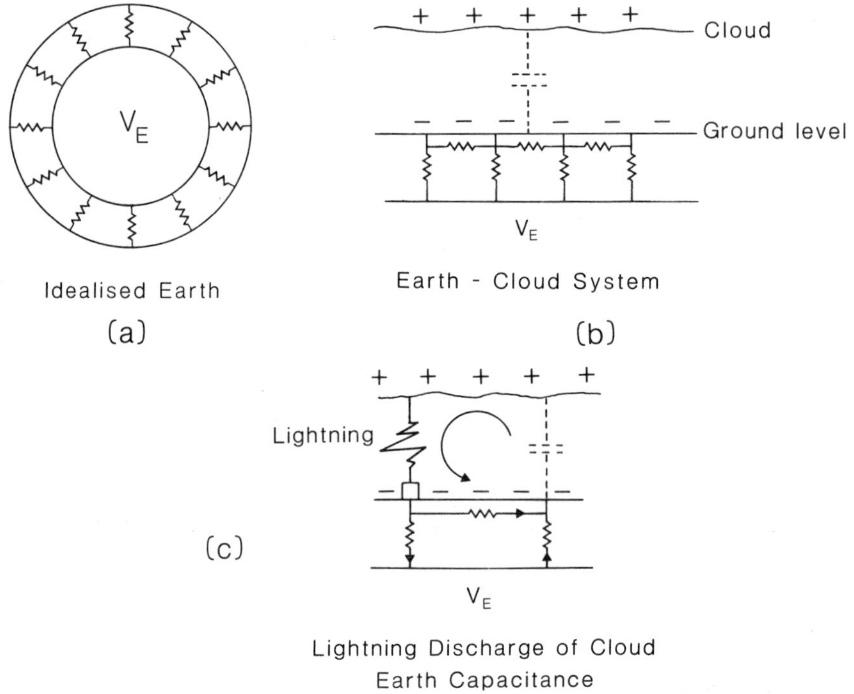

Fig. 6.9 (*a*) Idealised earth, (*b*) earth–cloud system, (*c*) lightning discharge of cloud–earth capacitance

It is clear then that a proper appreciation of the behaviour of grounding system rests (like all other aspects of EMC) in applying a reasonably correct model which is simple enough to understand but not so simple that it leads to large errors. Herein lies the major difficulty of all design problems. Figure 6.10(*a*) shows a very simple circuit of a voltage source V_S with a source resistance R_S, its load Z_L, and emphasising that the ground return should usually be modelled as having an inductance L and a resistance R. It is of course obvious that voltage line also has a complex impedance but this is neglected in this section. In this circuit it is clear that if any current flows then the voltage at the bottom of Z_L will be different from that at the source ground. In this circuit there is no connection to earth so that in principle the whole circuit can be at any potential relative to V_E which, without careful planning, can represent a safety hazard. Figure 6.10(*b*) shows a more typical circuit in which there is an earth connection and two voltage sources sharing the same ground return. Now a number of other problems can arise. Firstly when circuits share the same ground return, common impedance coupling can occur. This simply means that a current flowing in one circuit will produce a voltage difference across the ground impedance which then appears in the other circuit. The solution to this type of problem has to involve either reducing R and L or using a separate ground for the two (or more) circuits. As will be explained in the next

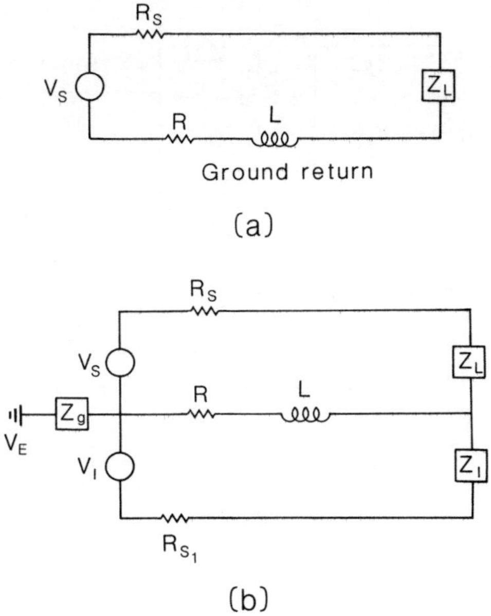

Fig. 6.10 (*a*) Simple equivalent circuit model of a grounding system, (*b*) two circuits with shared ground return

section most circuits have one or more connections to earth for safety reasons. When such connections exist one then has to consider the earth impedance Z_g and any currents that can flow to earth so generating voltage differences between the source ground and earth potential. If there is more than one ground connection in a particular circuit the existence of current flowing within the earth usually means that the two points are not at the same potential so injecting noise voltages into the circuit. Another consequence of two earthing points (these may be direct or capacitive connections) is the forming of earth or ground loops which will couple into any electromagnetic field present so generating a further current flow within the earthing system. These ground loop currents may create appreciable noise voltages. Thus grounding problems are most often concerned with common impedance coupling and where to make the earth connection. Whilst these last two factors appear simple, hidden within them is sufficient complexity to make it difficult to design good practical grounding systems which satisfy both safety and EMC requirements.

6.3.2 Safety and signal grounds

The need for safe grounding practices is clear if one considers the situation of an ungrounded enclosure using the low frequency equivalent circuit shown in Figure 6.11(*a*). The mains is depicted as having two leads, one live (L) and the other neutral (N). The enclosure is nominally floating with its potential V_C

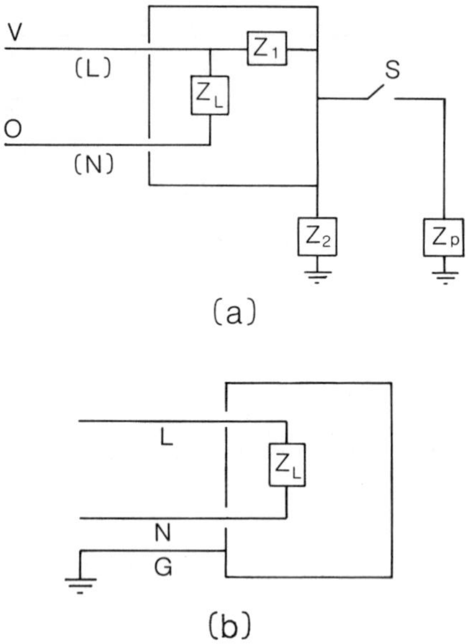

Fig. 6.11 Equivalent circuit demonstrating the operation of a safety ground. In part (*a*) Z_1, Z_2 are stray impedances from ground to the case and Z_p is the impedance of a person to ground

being determined by the two stray impedances Z_1 and Z_2 so that

$$V_C = VZ_2/(Z_1 + Z_2).$$

If persons having an impedance Z_p to earth then make contact with the enclosure (S is closed) they experience the voltage V_C and a current I flows through their bodies, given by

$$I = VZ_2/(Z_1(Z_p + Z_2) + Z_2Z_p). \tag{6.18}$$

If the electrical fault causes a short to the case then $Z_1 = 0$ and

$$I = V/Z_p. \tag{6.19}$$

This current will be a serious shock hazard both because of the exposure to the full mains voltage but also because the fault current may be insufficient to trigger any cut off devices (fuses etc.). To protect against these possibilities most electrical codes specify that an earthed ground wire must be attached to every enclosure as shown in Figure 6.11(*b*). Now the case is always at earth potential (except possibly during the initial transient of a sudden fault) and any fault connection to ground is much more likely to trigger the protective fuses or circuit breakers. There is thus almost no risk when a person touches such a system. The primary ground connection to earth in most systems will be at the local sub-station and/or at the point of entry of the mains into the building. However the exact details of the earthing arrangements differ from country to country so it is very important to check the requisite national codes before carrying out any work involving EMC countermeasures and mains earthing systems. In the UK the details of national requirements are to be found in the IEE Wiring Regulations [5]. More restrictive grounding/earthing rules are often laid down for medical apparatus and it is important that these restrictions are obeyed when considering EMC inspired changes to such equipment. These rules again differ from country to country but the current UK and IEC regulations are identical [6] and should be consulted in the first instance.

Signal grounds are simply the current returns that are needed in all circuits. They are assumed to be at a low uniform potential and are often the conducting chassis of a piece of equipment, the metallic enclosure or a specially constructed ground plane. In many cases, as will be seen below, a single point on the conducting ground may be taken as the low reference potential. However if the ground system is expected to exhibit this low potential over a large area then the important thing is that it should have as low an impedance as possible. It will probably have a connection to earth at some point but this is not necessarily essential for its operation and it is usually only done for safety reasons. For example aircraft electrical systems operate quite happily remote from earth. Of course even in this case an effective 'earth' potential has to be defined in terms that pertain to the interior of the aircraft so that safety grounding procedures can still be applied. The next question that arises is how to connect the various circuits in any system to the ground. First it is necessary to recognise that all grounds and their connections to any circuits represent quite complex networks. A simple example of a possible system which shows a number of poor design features is illustrated in Figure 6.12. These poor features are:

(1) The connection of circuit A to two different points on the ground via leads having impedances Z_{A1} and Z_{A2} so creating a ground loop.
(2) The connection of the ground on circuit A to that on B via the lead having an impedance Z_{AB} so creating another ground loop.
(3) The presence of the distributed impedances Z_g shows that the supposed ground plane has appreciable residual impedance which may cause common impedance coupling problems.

One simple way to overcome some of these problems especially at low frequencies is to use a single point ground to which the circuits can be

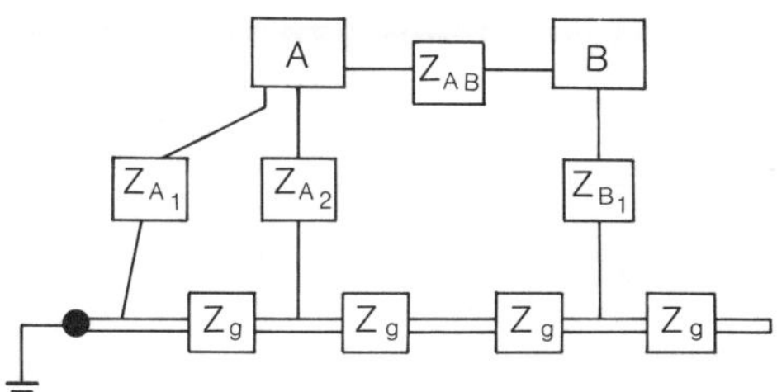

Fig. 6.12 Typical grounding system with poor design features

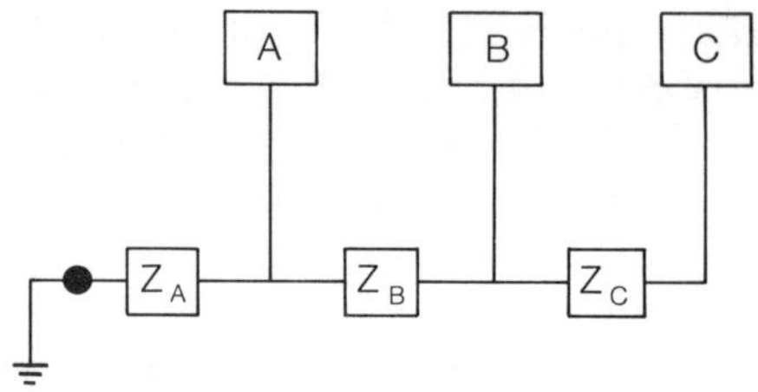

Fig. 6.13 A series connected single point grounding scheme

connected either in series or in parallel. The series connection usually uses less wire and so costs less. Such a system is shown in Figure 6.13. The impedances Z_A, Z_B, Z_C represent the total impedances (resistive/inductive) of each of the connecting leads. At low frequencies this system avoids ground loops but leaves the possibility that noisy circuits may common impedance couple into other circuits via the fact that all the return currents must flow through part or all of each of the connecting leads. Series connected single point grounds may be suitable for undemanding applications but more generally it is better to use the parallel connection system shown in Figure 6.14. Because the individual circuits *A*, *B*, *C* are separately connected to the ground point the return currents in one circuit cannot cause noise voltages to appear in another circuit. One might allocate the ground connections in a typical system so as to separate noisy or high power circuits from the low level sensitive circuits. Such a ground allocation is indicated in Figure 6.15.

For higher frequency systems it becomes increasingly difficult to have single point grounds because the stray capacitances eventually provide many low impedance connections to the ground. Such unintended links form ground loops as shown in Figure 6.16. Moreover one has increasingly to consider the impedance of the ground and any connecting lead since at high frequencies these may reach unacceptably high values. For example when any transmission line (real or accidental) is shorted and has a length close to a quarter wavelength then it has a very large input impedance, as was seen in Section 3.5.6. At high frequencies then it is necessary to ensure that the ground has a low impedance, usually by making it a ground plane. With such an arrangement $Z_g \to 0$ as

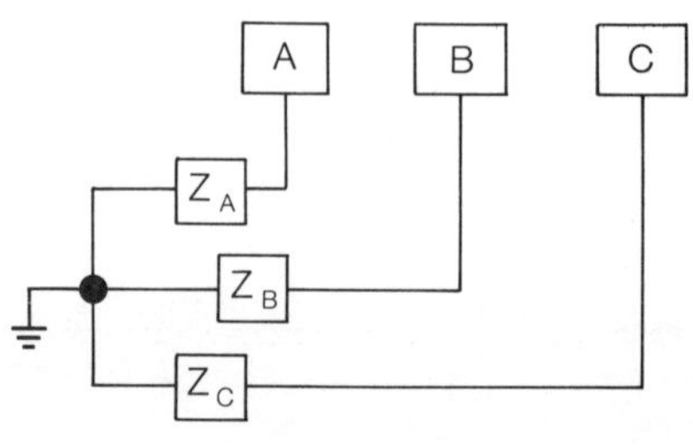

Fig. 6.14 A parallel-connected single point grounding system

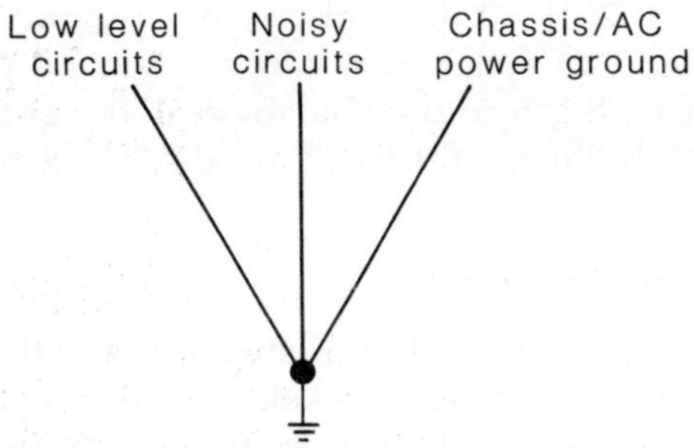

Fig. 6.15 Allocation of parallel connections to a typical practical system arranged so as to minimise common impedance coupling

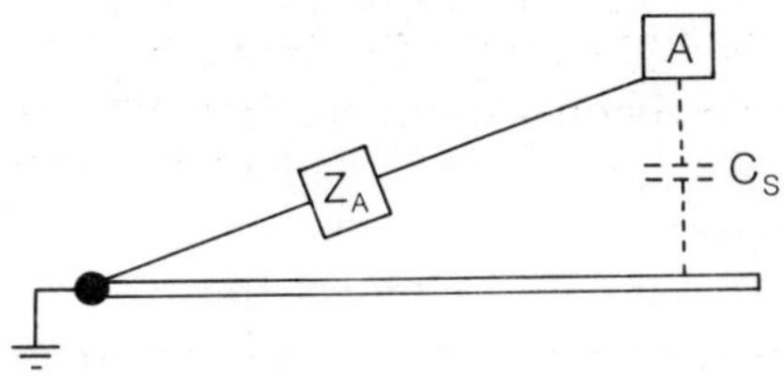

Fig. 6.16 Schematic diagram showing how stray capacitance turns a single point ground into a multipoint ground at high frequencies

shown in Figure 6.17. Individual connections from each system are then made by using as short a lead as possible. A typical rule of thumb [1, 2] is to keep all leads less than 1/20 of a wavelength long so as to prevent inadvertent high impedances from occurring. In all cases these wavelengths must correspond to the highest frequency likely to be involved in the operation of the system. It is clear that the choice of a particular grounding system depends on economics and system performance. A simple design guideline [1, 2] is to use single point grounds for systems involved with frequencies below 1 MHz and multipoint grounds for those involved with frequencies above 10 MHz. These two types dominate the grounding systems commonly used but a number of hybrid types are also possible. For example it is possible to have a system that behaves as a single point ground at low frequency but as a multipoint ground at high

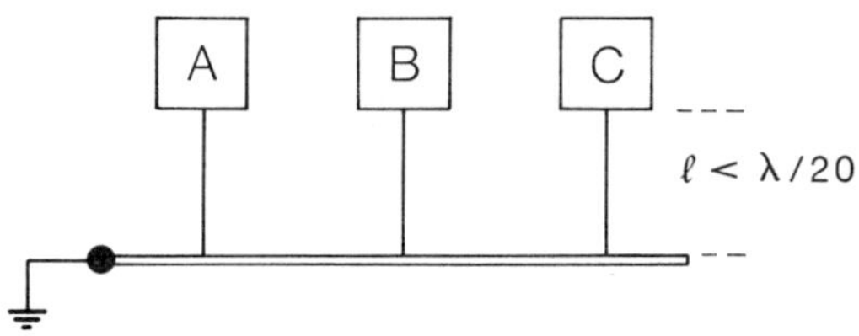

Fig. 6.17 Multipoint grounding system with short connecting leads

frequency or vice versa [2, 1]. Multipoint grounds produce ground loops which can be troublesome and in order to minimise their effects it is important to minimise the loop areas. It is also possible to break the ground loops by the use of various isolation techniques and these are discussed in Section 6.5.

6.3.3 Cable grounding and 'pigtails'

In the first instance the grounding of cable shields obeys the same rules outlined in the last section. Thus for frequencies below 1 MHz a single point ground is to be preferred whereas above 10 MHz multipoint ground systems are nearly always essential. In the single point case one has still to decide at which end the cable shield should be grounded. Consider the system illustrated in Figure 6.18(*a*) which consists of a source connected by a coaxial cable to a load which might be an amplifier. Here the amplifier is grounded to earth and Ott [3] has shown that it is best to connect the shield at the same end, as indicated. If the source had however been grounded then the shield would need to be connected at that end. For the case of a shielded twisted pair the same rules apply as is shown in Figure 6.18(*b*).

At this point is is necessary to recognise that the geometry of the ground connection can be crucial as well as its exact location. In many screened cable systems it has been the practice to terminate the normal 360° shield in a single wire or 'pigtail' so as to make multi cable/connector assemblies easier. This has

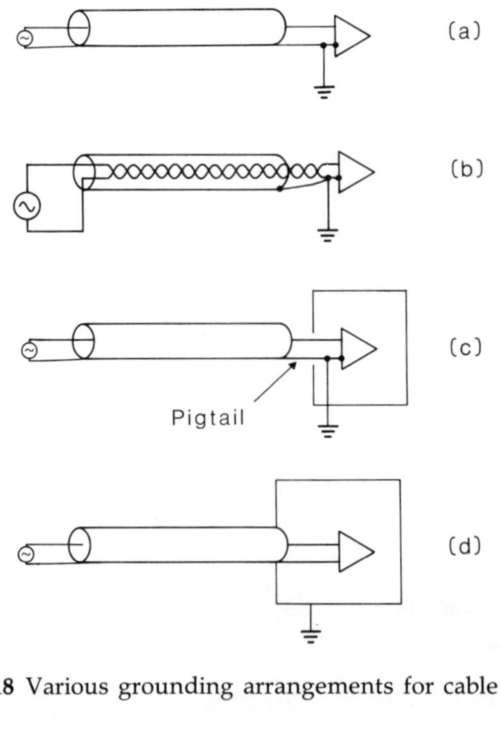

Fig. 6.18 Various grounding arrangements for cable shields

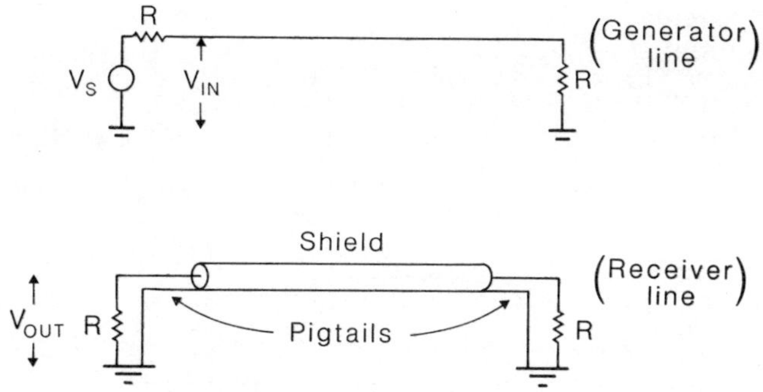

Fig. 6.19 Experimental arrangement for measurement of coupling due to the presence of pigtails on a shielded line, adapted from reference [7]. (Reproduced by permission of IEEE)

been shown to be source of EMC problems, both from the point of view of pick-up and of unwanted radiation [7, 8]. The main error is of course to imagine that an apparently small perturbation to a large shielded system would have no appreciable effect on its performance. Figure 6.18(*c*) shows a typical pigtail connection from a shield to a shielded container holding an amplifier. As a contrast the preferred method of connection is shown in Figure 6.18(*d*) and to understand why this is so it is necessary to consider some research [7, 8, 9, 10] in more detail. Paul [7] measured the coupling between various parallel cable systems having a variety of different pigtail lengths. In this work the results are given in terms of a voltage transfer ratio V defined as follows. Consider the system in Figure 6.19 consisting of a generator circuit and a receiver circuit. The generator has a source of impedance R connected via an unscreened line (length = 3.66 m) to a load of impedance of R. This source generates a voltage V_{IN} on the generator line. The receiver circuit of the same total length as the generator circuit has a shield around all its length except for the two equal length pigtails which connect the shield to ground at each end. The shielded receiver line is terminated at each end to ground by a resistance R. The voltage V_{OUT} developed across the terminating resistance at the sending end is a measure of the pick-up and the transfer ratio is given by

$$V = V_{OUT}/V_{IN}. \tag{6.20}$$

At high frequency this transfer ratio becomes large as shown in Figure 6.20 which gives experimental results for V as a function of frequency for $R = 50\ \Omega$. It is clear that the difference between a short pigtail (0.5 cm) and a much longer one (8 cm) can be as much as 30 dB. These conclusions are much the same if $R = 1000\ \Omega$ or if the circuits are moved further apart. It is thus clear that short pigtails can cause appreciable crosstalk and that they should normally be avoided. Simple low frequency circuit models of the above experimental system show [7, 3] that for low circuit impedance the dominant coupling is by the inductive effects of the pigtail. On the other hand, for high impedance circuits

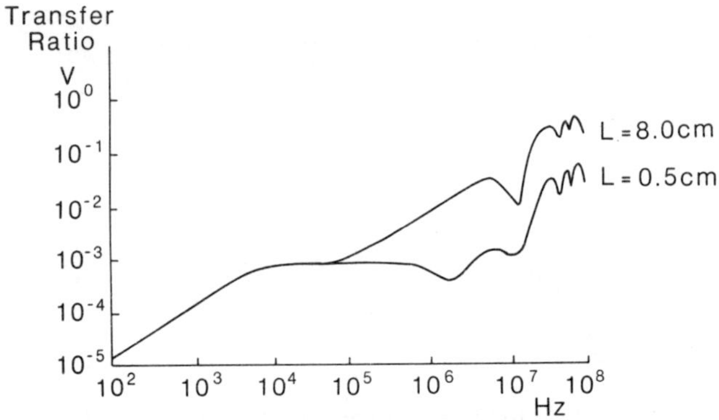

Fig. 6.20 Voltage transfer ratio for two different lengths L of pigtail with $R = 50\ \Omega$ in the system depicted in Fig. 6.19. The generator line was closely coupled to the receiver line, adapted from reference [7]. (Reproduced by permission of IEEE)

the coupling is mainly caused by the capacitive link between the unscreened portion of the receiver line and the generator.

Another slightly different case that needs consideration [8, 9] is the effect of pigtails on the stray radiation from an apparently well screened system. The situation depicted in Figure 6.18(*c*) is once again under consideration. In this case Hejase *et al.* [8] have calculated the radiated fields using the model shown in Figure 6.21. This model represents a coaxial transmission line of total length $L - L_1$ penetrating an infinite ground plane. At the end of the line is an air gap of length L_1 in which pigtails of different types and lengths can be inserted. The results are presented in terms of a shielding effectiveness SE defined in this case by:

$$\mathrm{SE} = 10 \log_{10}(P_0/P_1) \qquad (6.21)$$

where P_0 = the power radiated when no pigtail is present at all. The

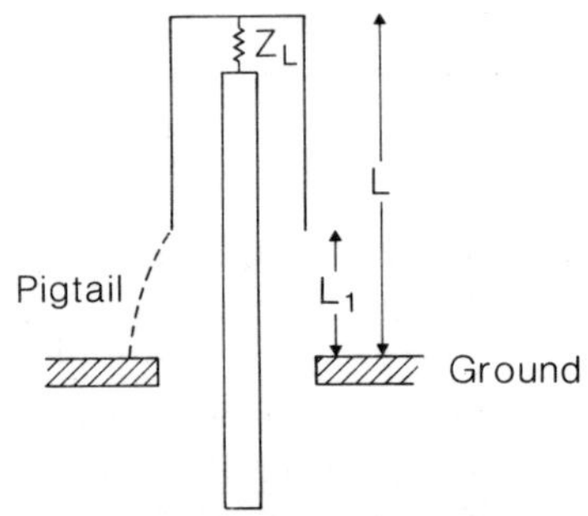

Fig. 6.21 Transmission line model for pigtail connected coaxial system, where L is the total length of the system, L_1 is the length of the pigtail and Z_L is the characteristic impedance of the line, adapted from reference [8]. (Reproduced by permission of IEEE)

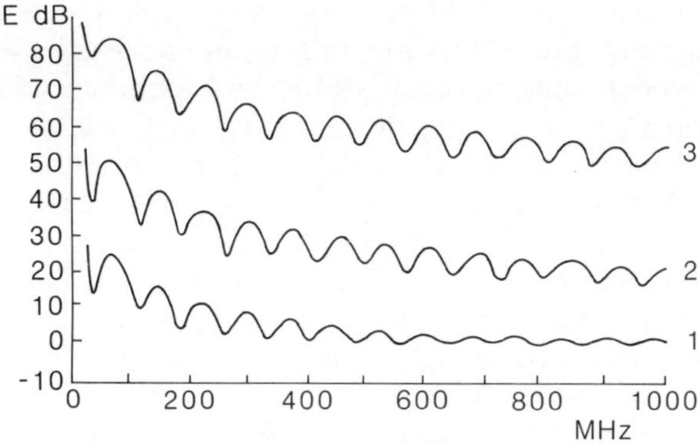

Fig. 6.22 Shielding effectiveness SE dB (defined by Equation (6.21)) as a function of frequency, for $Z_L = 50\ \Omega$ and a variety of pigtail lengths. Curve 1 is for a single 3 inch (7.6 cm) pigtail, curve 2 for a single 0.489 inch (1.24 cm) pigtail and curve 3 for four 0.489 inch pigtails arranged in parallel around the central signal wire, adapted from reference [8]. (Reproduced by permission of IEEE)

transmission line is then ungrounded and this represents the worst case. Also P_1 = the power radiated when a specific pigtail is inserted across the airgap.

Figure 6.22 shows the results of some of their numerical calculations for a 50 Ω line terminated in 50 Ω. At high frequencies there is a 60–70 dB difference between using one 3 inch pigtail as opposed to using 4 parallel 0.489 inch pigtails. Experiment and theory confirm that the effect of the pigtails is to allow more current to flow on the outside of the coaxial line where it is able to radiate efficiently. It was also shown that the region close to the pigtail was not itself the dominant source of the radiation.

Graf and Vance [10] considered the slightly different excitation of the same basic geometry (Figure 6.18(*c*)) that occurs when examining how to exclude cable shield currents from the interior of attached shielded enclosures. They examined both single and twisted pair conductor systems and showed that for both systems pigtails allowed currents to penetrate the shielded enclosure.

All this work confirms the need for 360° shield connections. This was discussed in Chapter 4 with especial emphasis on the need to keep skin currents on the outside (or on the inside as the situation demands) of shields. It has been shown that pigtails and cable braids tend to make it difficult to apply this principle. It has also been shown that the topology of any ground connection can be very important and another example of this is discussed in the next section.

6.3.4 Grounding for single and nested shielded enclosures

Screened enclosures in the first analysis should be treated using the same rules as those outlined in the previous section for the grounding of cable shields. They are after all simply extensions of each other. The only real difference being that screened enclosures usually have large stray capacitances to earth or other parts of the ground system. Thus isolation of an enclosure presents a problem at somewhat lower frequencies than would be experienced for a cable. It was shown in Chapter 4 that the performance of most shields is dominated by defects in their implementation. Unavoidable defects occur at cable entry points and joints. It is therefore important to consider any ground connection in this light [10, 11]. If the ground connection penetrates the shield then any current on that conductor can now flow on the interior surface of the shield as shown in Figure 6.23(a). The obvious solution to this is get the topology correct and lead any current onto the outer surface as shown in Figure 6.23(b). Experiments [12, 13] have shown that the first and incorrect type of shield penetration can be important in causing excess electrical noise in shielded enclosures.

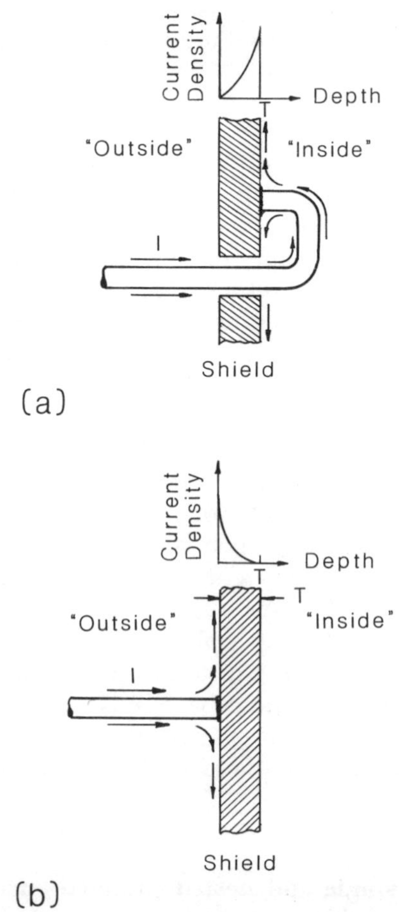

Fig. 6.23 (a) Topologically incorrect ground connections to a shield, (b) topologically correct ground connections to a shield. Adapted from reference [11]. (Reproduced by permission of IEEE)

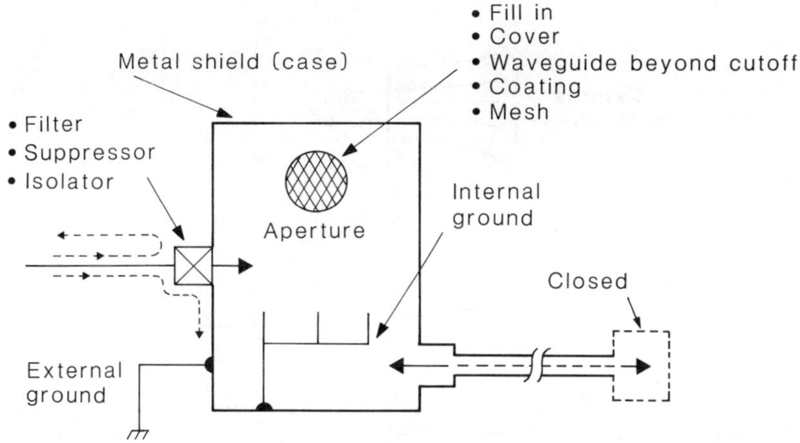

Fig. 6.24 List of optimum treatments of shield penetrations and ground connections to ensure the integrity of the shield. Adapted from reference [12]. (Reproduced by permission of Defense Nuclear Agency, USA)

It has been shown that in order to control the ingress of unwanted electromagnetic energy through the shield of any enclosure it is necessary to treat the cable shields and the ground connection (if one is made) so as to prevent current penetration from the outside surface onto the interior or vice versa. The most effective ground connection system to do this is illustrated in Figure 6.24. This diagram also indicates that any shield opening should be treated by using fill in covers, meshes, conducting windows or by a length of waveguide operating beyond its cut-off frequency so as to prevent electromagnetic leakage into the shielded area. The diagram further shows that cable or conduit attached to the shield should be considered as part of the shield volume and treated with the same rigour. Finally one further route for noise entry is considered in this last figure which has not yet been discussed. This is the signal conductor. It is of course evident that the use of the term signal conductor here implies any conductor mains, signal or otherwise that penetrates the shield. In order to combat this route it may be necessary to use a filter, a suppressor, an isolator or a combination of these. This is indicated schematically in the figure where one such cable entry treatment device is shown as a box with a cross. The operation and use of all the above devices will be discussed in Sections 6.4 and 6.5.

A typical enclosure may have many different cables entering its volume as well as the need for a properly attached ground connection. If the shields of these cables and the ground have large currents flowing in them (due to external fields) then the performance of the shield may be improved by preventing the current from flowing over the whole shield surface [11]. In situations where this is important this is encouraged by using single entry panels as shown in Figure 6.25.

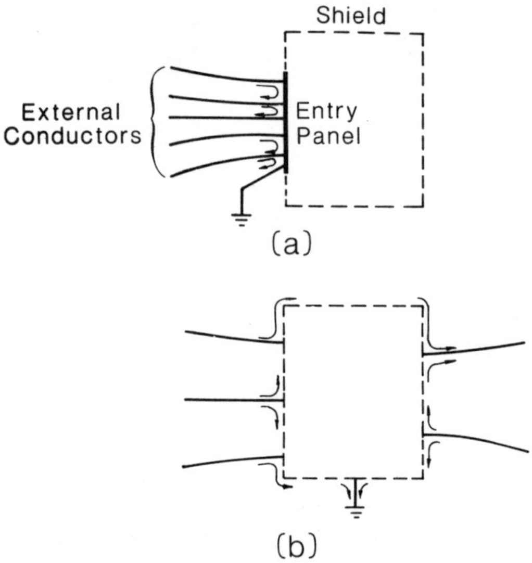

Fig. 6.25 (*a*) The use of single entry panel in an attempt to minimise the current flow over the surface of a shield, as shown in (*b*), adapted from reference [11]. (Reproduced by permission of IEEE)

When a complex system is being protected by a system of shields it is necessary to have a procedure which will help decide the degree of shielding required in each area. The need for this leads to the concepts of zoning and the use of nested shields. In one sense decisions involving these concepts will have to be made for all systems however small they are. Thus a designer always has to decide how much electromagnetic isolation one part of the system should have from another. The design might not involve complete shields but the principle of assigning zones may still be helpful. One particular problem in the use of nested shields is the correct method for making the ground connections and this will be emphasised here. First of all, however, it is necessary to consider a rationale by which nested shields can be designed. Vance *et al.* [12] have suggested that the degree of shielding required within any given region (or zone) between two barriers should be designed using two simple conditions:

(1) the minimum shielding effectiveness of the outer barrier should at least prevent damage to apparatus or circuits within the protected zone;
(2) the maximum shielding effectiveness of the outer barrier should not exceed a value sufficient to reduce the electrical noise below that normally present within that zone. In practise this might mean reducing the noise inputs to not much more than 20 dB below the ambient.

The use of these conditions is illustrated in Figure 6.26 for the case of an external telephone connected by overhead lines to some building.

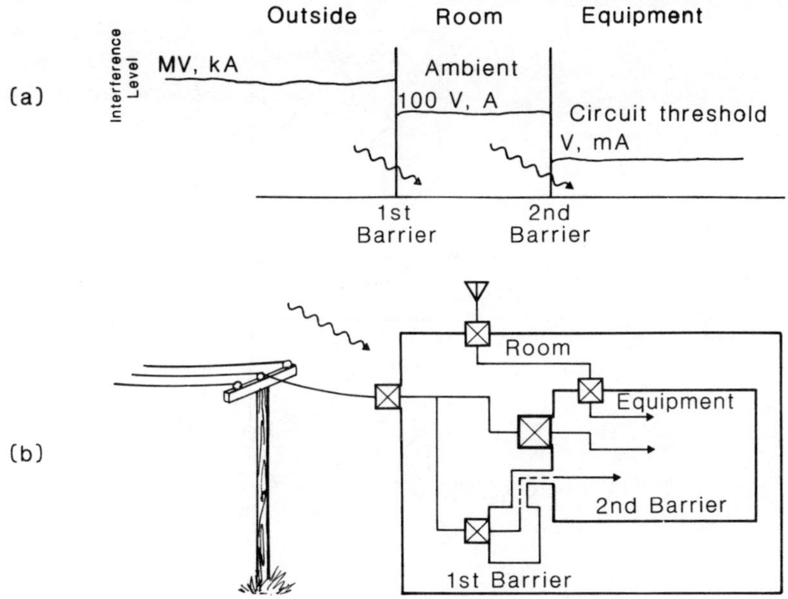

Fig. 6.26 (*a*) Interference levels through a two barrier shielded system, (*b*) schematic diagram of a two barrier system. Adapted from reference [12]. (Reproduced by permission of Defense Nuclear Agency, USA)

Figure 6.26(*a*) shows the typical electrical levels that might be experienced in the different zones. For example the external telephone lines might experience EMP or lightning both of which could cause pulses at the MV/kA level. The first barrier (including the treatment of any cables penetrating the barrier) has to reduce the pulses entering the inner space to or just below the normal background level caused by the typical mains operated equipment supposed to be present in this zone. In this case the building wall (which may need special treatment) is the first barrier. Inside the building the typical ambient is at the 100 V/1 A level. Thus an attenuation of about 100 dB (80 dB + 20 dB) is required to be produced by the first barrier. If now in the building there are circuits working at the V/mA level a further attenuation of 60 dB has to be produced by a second barrier. This results in the zone layout shown in Figure 6.26(*b*). Notice that all the conductors penetrating any barrier are shown to pass through some sort of treatment (once again indicated by the boxes with crosses) so as to prevent unwanted penetration of electrical energy along the conductors. This approach to zoning can be carried out to many different levels for noise sources both within and/or outside any nested system [12].

Finally it is necessary to consider how ground connections should be made in such zoned systems. As discussed at the beginning of this section it is not good practice to allow a ground connection to penetrate a shield before making a contact because this allows current which should be kept outside (or inside) to enter (or leave) a shielded zone. Figure 6.27(*a*) shows how topologically

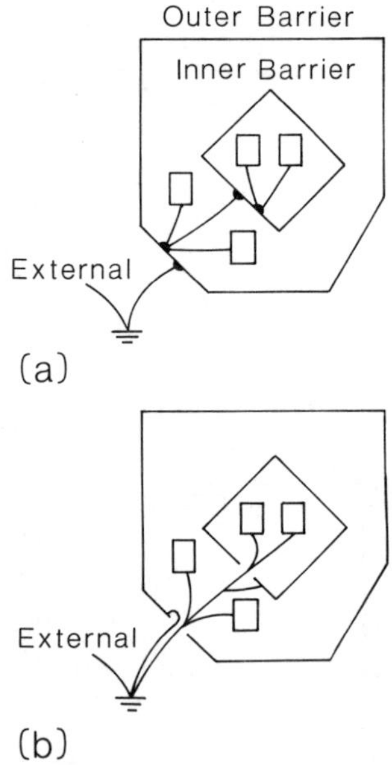

Fig. 6.27 (*a*) Topologically *correct* arrangement for grounds in a two barrier system, (*b*) topologically *incorrect* arrangement for grounds in a two barrier system. Adapted from reference [10]. (Reproduced by permission of IEEE)

correct connections should be made [11]. On the other hand a very poor arrangement of ground connections which violates both barriers is indicated in Figure 6.27(*b*).

It should be noted that these optimum ground wire arrangements may not always be essential. This is more likely to be the case when the isolation between each zone is not too high, for example when complete closed shields are not used. The general approach suggested is, however, the best and, if considered early in any design, should carry little economic penalty.

6.4 PASSIVE COMPONENTS AND FILTERS

6.4.1 Passive components

6.4.1.1 Introduction

Circuit elements are usually considered to be purely resistive, capacitive, or inductive and their connections/lead wires to have no impedance. A moment's reflection would make it apparent, however, that this can only be a very idealised picture of the real position. A solution of Maxwell's equations in any situation requires consideration of the time varying values of both the electric and magnetic fields. Such a calculation must also allow for any resistive losses. Thus resistors, capacitors, and inductors (and any other circuit element) require fairly complex models if they are to be described correctly over a large range of frequencies. These more complete models are often of importance in describing circuit behaviour when estimating EMC effects. In this section a brief description will be given of the some of the simpler models which enable this aim to be achieved. Before making any design decisions based on these models it is of course also necessary to know the power and voltage limits applicable to each component being considered.

6.4.1.2 *Conductors*

The impedance Z of round wires [3, 4, 14] consists of an inductive as well as a resistive term so that

$$Z = R + j\omega L. \tag{6.22}$$

The skin effect (see Section 2.5.6.3 for a more extensive discussion) increases the effective resistance of conductors by concentrating the current on the surface closest to the fields generating the current. The characteristic distance over which this bunching occurs is termed the skin depth. A simple high frequency formula for the resistance of a round wire of radius a is then [14]

$$R = (2\pi a\sigma\delta)^{-1} \tag{6.23}$$

where $\delta =$ the skin depth $= 1/\sqrt{\omega\mu\sigma/2}$ and $\sigma =$ the conductivity.

The inductance of a wire is made up of an internal L_i and an external component L_e so that

$$L = L_i + L_e.$$

Usually the internal term can be neglected and it is only necessary to calculate the inductance due to the external fields which is geometry dependent. For example one could use (6.17) which gives the inductance of a round wire of radius a height h above a ground plane. Rewriting this in a more usable form gives

$$L = 2 \times 10^{-9} \ln(2h/a) \text{ H cm}^{-1}. \tag{6.24}$$

It is difficult for h/a to be much less than 1 so that the typical lead inductance of most discrete components is likely to exceed 3–4 nH. From the above estimate it is clear that it is probably necessary to include the lead impedance in any circuit model of all components. Indeed in any circuit design the impedances of any connecting wires or ground planes should be estimated to see if they effect circuit behaviour or could cause EMC problems. As an example of the latter the effects of conductor impedance on grounding have already been discussed in Section 6.3.2.

6.4.1.3 *Resistors*

Circuit models of resistors depend on the resistor type being considered. One general version that includes most features is shown in Figure 6.28(*a*). Here L_L

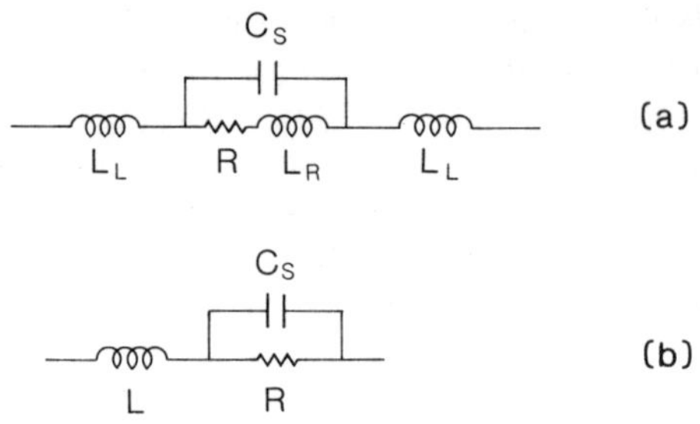

Fig. 6.28 (*a*) General equivalent circuit model for a resistor. (*b*) Simplified circuit model for resistors with low internal inductance

and L_R represent the inductance of the leads and resistive element respectively. C_S is the stray capacitance that exists between the leads and across the resistor R. For wire would resistors L_R can be appreciable but on the other hand for thin film or composition type resistors L_R can be neglected and the lead inductance tends to dominate. The stray capacitance C_S is usually in the range 0.1–0.5 pF. The simplified circuit model shown in Figure 6.28(*b*) then applies, L being calculated from (6.24). This model indicates that resistors will have appreciably different impedances than those expected as the frequency rises [2]. Resistor assemblies having high inductance generate more stray fields and are more susceptible to inductive coupling.

6.4.1.4 Capacitors

There are many different types of capacitor available but in general they all can be represented by the simplified circuit model shown in Figure 6.29. Here L represents the sum of the lead and the internal inductance. Each capacitor type has its applications and a frequency range over which it can be safely used. For example, electrolytic types give the greatest capacitance per unit volume but can only maintain their high values at low frequencies. Table 6.1 gives the operating range and internal inductance of several types of capacitor. Such parameters as these vary with construction and class of capacitor. It is therefore important to determine the detailed specification of any being proposed for a specific design so that an adequate equivalent circuit can be produced. In EMC work two common problems which highlight the need to know an accurate equivalent circuit are as follows.

(1) The occasional need to parallel large electrolytics with smaller high frequency capacitors. This is done to provide some high frequency filtering

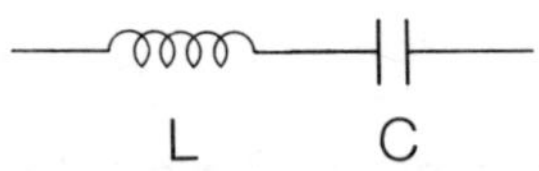

Fig. 6.29 Circuit model for a capacitor

Table 6.1. Typical values of the internal inductance of some capacitor types. Leads will increase these

Type	Approximate operating range	Internal inductance (nH)
Ceramic	10 kHz–1 GHz	1.4
Tantalum electrolytic	1–1000 Hz	20–50
Large value aluminium electrolytic	1–1000 Hz	120
Mica	10 kHz–1 GHz	45
Large value polystyrene	10 kHz–1 GHz	90

capability to the essentially low frequency smoothing function of the primary design aim.

(2) In filter design using capacitors the inductance L of the leads and the internal construction can produce resonances which can lead to poor performance at high frequencies. This is discussed further in Section 6.4.2.

6.4.1.5 *Inductors*

Inductors can be wound on a non-magnetic (air core) or more usually on a magnetic (commonly ferrite) core. Any magnetic fields generated by an inductor will spread far more widely with an air core design rather than with a magnetic core but in all cases it is worth being aware of the possible flux linkage effects with other parts of the circuit. In developing an equivalent circuit for a single inductor it is usual to ignore these secondary inductive effects. A typical low frequency circuit model is shown in Figure 6.30, L being the value of the inductance and r the DC resistance. At higher frequencies the effective resistance increases due to the skin effect and losses in the core so that the high frequency model is more correctly given by that shown in Figure 6.31. C_S is the intrinsic parallel capacitance due to interturn effects and the outside geometry. R is the high frequency resistance. Such a hybrid model can be analysed in terms of a normal parallel RLC circuit leading to the approximate equations

Fig. 6.30 Low frequency model of an inductor

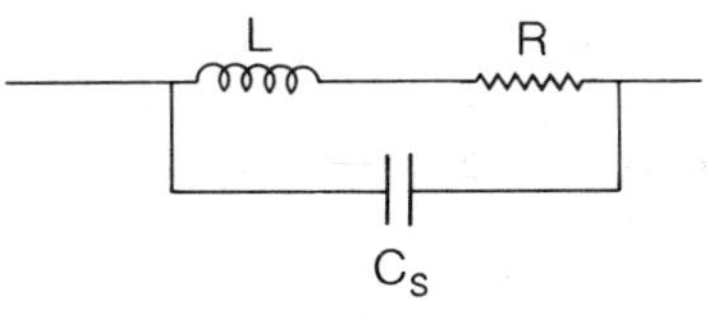

Fig. 6.31 High frequency model of an inductor

quoted below. Thus a typical ferrite core inductor will have a specification that quotes a self-resonant frequency (SRF) and a quality factor Q given by:

$$f = \mathrm{SRF} = 1/2\sqrt{LC_S} \tag{6.25}$$

and

$$Q = 2\pi fL/R. \tag{6.26}$$

When using inductors it is important to realise that resonant effects can occur in association with other stray capacitances than C_S discussed above. At high enough frequencies all inductors will behave as capacitors.

Sometimes inductance can be added to a circuit by slipping ferrite beads over selected leads. Such beads can produce 10–100 Ω impedance at 1–100 MHz but have very low impedance at low frequencies. It is thus possible to use them for some selective filtering and damping but only in low impedance circuits [2].

Many other components in electronic/electrical systems (for example all types of electrical contact) have properties which lie outside the simple operating model used to describe their first order circuit function. There is no space here to describe all such components but in any event it is necessary to consider the EMC implications of every part of such systems paying particular care over possible secondary effects not normally considered in the primary design.

6.4.2 Filters

There are two general classes of filter normally encountered in practice

(1) Mains or power supply filters. These have to operate in situations where the source and load impedance may vary depending on the frequency to be rejected.
(2) Communication filters. These are usually complex high insertion loss devices that operate with fixed load and source impedance (50–300 Ω usually).

Within these classes it is possible to employ any of the four types of filter:

(*a*) low pass;
(*b*) high pass;
(*c*) band pass;
(*d*) band elimination or notch.

Most of what follows will apply to all these filter types in general but however most of the examples will be based on low pass systems. The basic performance of a filter is described by the shape of the insertion loss frequency curve. The standard format for making insertion loss measurements is illustrated in Figure 6.32. The source impedance is Z_S. For a load impedance Z_L the load voltage is V_1 with no filter in the circuit. If now a filter is inserted as in Figure 6.32(*b*) and the source voltage is left unchanged a new load voltage V_2

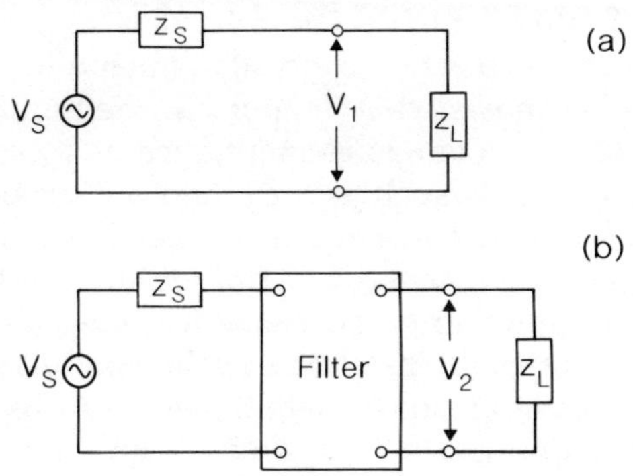

Fig. 6.32 The basis of filter insertion loss measurements

can be measured. A set of V_1 and V_2 can be measured for each frequency of interest and the insertion loss (IL) calculated from,

$$IL = 20 \log_{10}[V_1/V_2] \text{ dB}. \quad (6.27)$$

Typical normalised IL frequency curves for a low pass filter are shown in Figure 6.33. In a low pass filter one designs to have no attenuation up to some

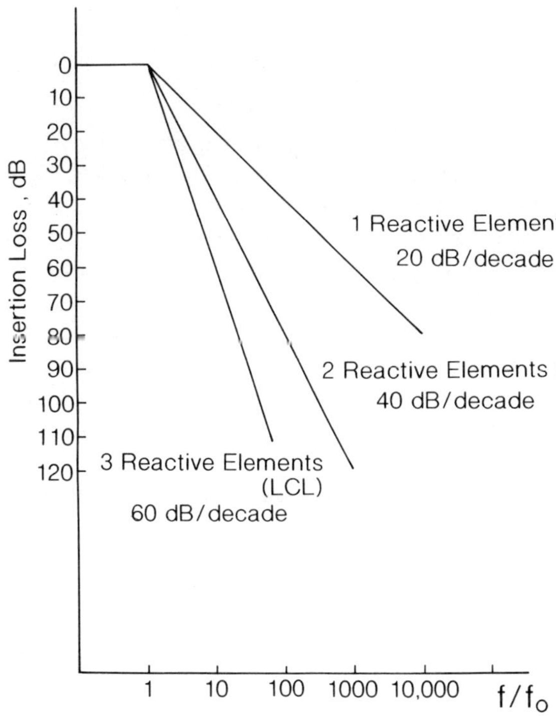

Fig. 6.33 Schematic insertion loss for a low pass filter having 1, 2 or 3 reactive elements

turnover frequency f_0 after which a fast roll off is required. The roll off rate depends on the number of reactive elements (suitably orientated L and C) in the filter [1]. Figure 6.33 shows what happens as the number is increased from one to three. In the case of a single element such as a capacitor 20 dB/decade is achieved. For two elements such as an L and a C arranged in a T network a 40 dB/decade roll off is possible. For a three element system such as an L, C and another L arranged in a T network 60 dB/decade results. In general the roll off rate increases as the number of LC elements increases [1]. This behaviour is obtained with all filter types. The choice of the filter element values together with the source and load impedances determines f_0. At high IL values (as with shield designs attempting to produce high shielding effectiveness (SE > 80–120 dB)) a number of secondary effects become important. In order to discuss these it is helpful to consider the simple case of a low pass capacitor filter as shown in Figure 6.34. In this the general model of the capacitor introduced in Section 6.4.1.4 is used. Simple algebra shows that

$$\mathrm{IL} = 20 \log_{10}\left[\frac{1 + Z_S/Z_T}{1 + Z_S/Z_L}\right] \mathrm{dB} \qquad (6.28)$$

where

$$Z_T = \frac{ZZ_L}{Z_L + Z}$$

with

$$Z = r + j\omega L + \frac{1}{j\omega C}.$$

Thus it is clear that source and load impedance effect the insertion loss. More importantly resonance effects are to be expected. Figure 6.35 shows calculated values of IL using the above equations for a number practical systems but assuming relatively low values of series inductance (lead plus internal). The figure shows that such a filter's effectiveness can be seriously degraded in important frequency ranges.

As the resonance is approached the filter performs much better but as the frequency exceeds by about a factor of 1.5 the resonant value the lead inductance starts to dominate. The only answer to these inductive resonance effects is to lower the inductance or to use as low a capacitance value as possible

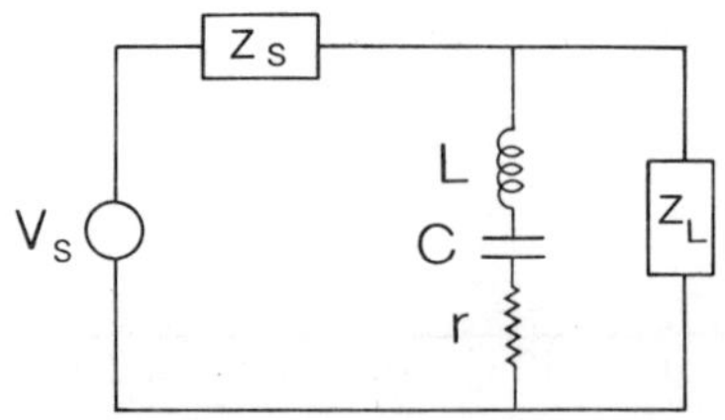

Fig. 6.34 Circuit model for a low pass capacitive filter

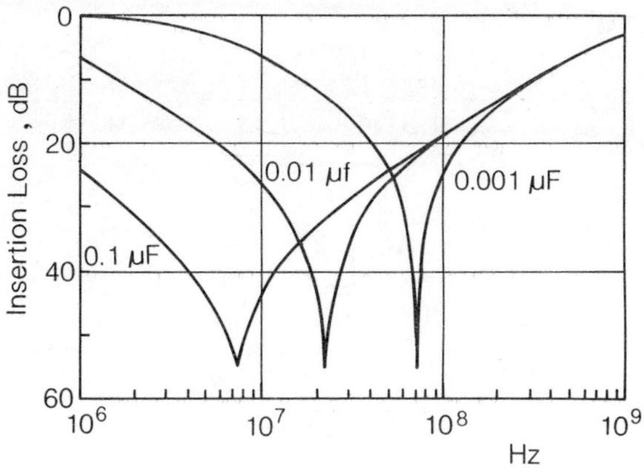

Fig. 6.35 Calculated insertion loss frequency curves (using Equation (6.28)) for $Z_s = Z_L = 50\ \Omega$, $r = 0.04\ \Omega$, with $C = 0.1$, 0.01 and $0.001\ \mu F$. In all cases the total inductance was 5 nH

(this is the recommended first approach). Some manufacturers provide capacitors that mount behind chips with lowish inductances but for really good high frequency performance it is necessary to use feed through capacitors. In this type of capacitor designed to mount on or through a metal chassis the capacitance exists between the through terminal and the shield/chassis. If any inductance remains it forms a T network filter [3]. Such capacitors (which can also be designed to incorporate absorptive lossy material [19]) can be used up into the Ghz range. Designs are also available as integral parts of multiway connectors providing some filter action on each line. It is also possible to parallel capacitors to provide a wider frequency filter action but extra resonances are possible with such arrangements and care should be taken when using this approach. The existence of the various resonant IL variations due to stray inductance (or capacitance) can occur with any filter design and as such must be considered early in any design.

The design of any real filter needs to consider many more features than have been outlined above or which could be covered in this book. Details of practical designs can be found in the literature [1, 15]. Some of the factors that need to be considered are source/load impedance, the nature of the noise to be removed, power limitations, safety requirements and the importance of line overvoltages. Finally it is worth commenting on the need for careful installation of filters. Since the filter must provide a high impedance (to achieve high insertion loss) any stray capacitance (or mutual inductance) between the input and output leads could provide a low impedance bypass negating the effect of the filter. This is illustrated in Figure 6.36(*a*). It is thus very important to keep the output well away from the input to any filter. For the filter to operate it must absorb or reflect or divert the incoming noise. In situations where a noise current is diverted to a shield as in the case of the T network shown in Figure 6.36(*b*) then the quality of the shield connection becomes crucial. As

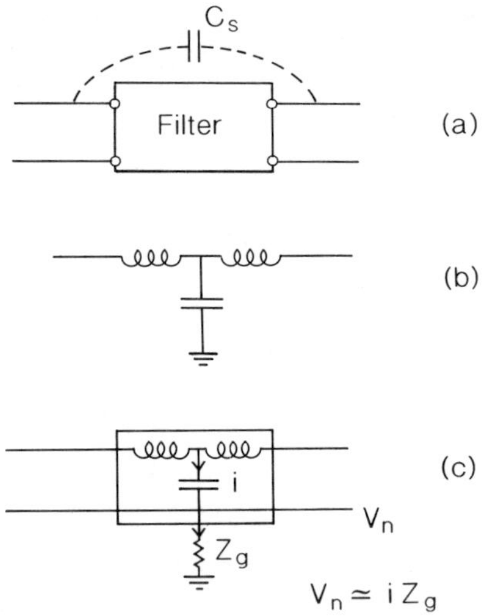

Fig. 6.36 Some common faults with filter installation

shown in Figure 6.36(*c*) any impedance Z_g will deteriorate the filters performance. The quality of this high frequency connection has to be on a par with that discussed in Chapter 4 and Sections 6.3.2 and 6.3.3. Thus high rejection filters often come in shielded containers which can be firmly mounted on a shield wall with the input leads on the outside and the output leads inside the protected space.

As was seen in Section 5.2.3, pulses with fast rise times have high frequency components which can be a source of interference. These can be removed by using filters, but an alternative technique is to use wave shaping to make a pulse which has a much smoother profile but adequate properties for the task at hand [16, 17].

Consider a pulse of amplitude A given by

$$f(t) = (A/2)(1 + \cos(2\pi t/T_P)) \tag{6.29}$$

for $|t| < T_P/2$ and zero elsewhere, see Figure 6.37. This is known as a 'raised cosine pulse', and its Fourier transform is

$$V(f) = \int_{-T_P/2}^{T_P/2} f(t)e^{-j2\pi ft}\,dt$$

$$= -\frac{A\sin(\pi f T_P)}{2\pi f(fT_P + 1)(fT_P - 1)}$$

which can be written as

$$\frac{AT_P\sin(\pi f T_P)}{\pi f T_P}\,\frac{1}{2(1-(fT_P)^2)}.$$

This is similar to that of a rectangular pulse (Equation (5.19)) but multiplied by

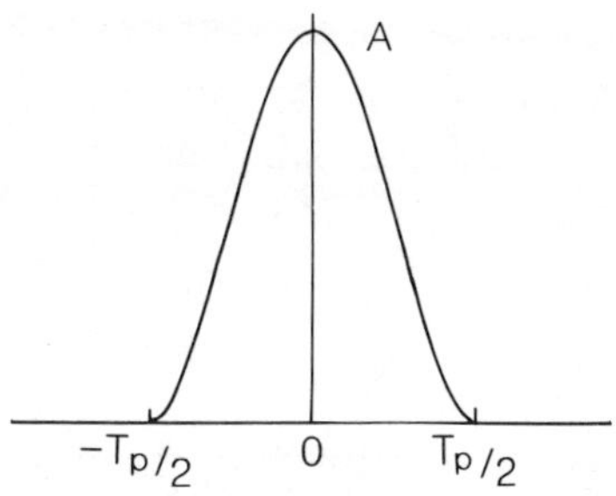

Fig. 6.37 Profile of a 'raised-cosine' pulse

an extra term $1/2(1-(fT_P)^2)$, which is very small when f is large, supressing high frequency components. This once again illustrates that it is advantageous to smooth out leading and trailing edges of pulses to remove high frequency components to reduce the effects frequency dependent coupling between circuits (see Section 2.7.1) and radiation from circuits, which depends on the square of the frequency, see (5.51).

There are on the market a number of devices which are designed to take advantage of wave shaping techniques to reduce noise problems. For example the Texas Instruments Inc. trapezoidal-waveform interface bus transceivers, such as the SN75ALS056 and SN75ALS057, are designed for two way communication along densely populated backplanes [18]. As was seen in Section 5.2.6 the high frequency components in trapezoidal-waveforms are damped, thus using them can reduce coupling between circuits. As well as using this technique, these particular chips also have internal low-pass filters to improve their noise immunity, and protection against generating unwanted transients or spikes ('glitches') during power up/down.

6.5 ISOLATION AND SUPPRESSION

6.5.1 Isolation techniques

Earlier in this chapter (Section 6.2) the problem of ground loops was discussed. Such loops can be enhance inductive pickup. Moreover because no two earth points are ever at the same potential common mode currents can be driven between the two systems. There are of course certain situations that set up specific common mode currents and some of these are illustrated in Figure 6.38. Figure 6.38(*a*) shows an idealised view of a digital system producing a differential mode current i. However the ground return impedance causes a voltage V_N which can now drive a common mode current I down the output lines. This type of common mode current and its importance in causing excessive radiation was discussed in Section 5.4.5. Ott [3] has also considered this problem and presented ways in which the effects may be reduced. Two other transient sources of common mode current are shown in Figures 6.38(*b*) and (*c*). The first of these shows the situation when lightning strikes a building

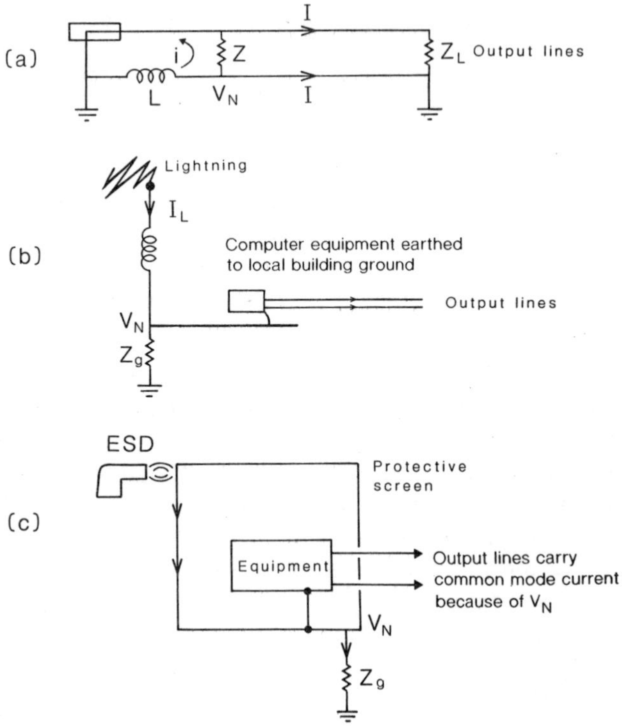

Fig. 6.38 Some sources of common mode currents

and the lightning current I_L flows down the building lightning conductors. Because of the inevitable ground impedance the whole building ground potential rises by $I_L Z_g$. In the example shown computer equipment grounded locally and through interconnect cables to computers in other buildings will cause large common mode currents to be sent between the buildings. Such events caused a fair amount of damage to improperly protected systems in the early 1980s. To stop such calamities it is essential either to use isolation (nowadays fibre optic links are commonly used) or adequate suppression. A somewhat similar situation though on a smaller scale is shown in the second example. Here an ESD has occurred to a protective shield. With good design none of the current will flow inside the shield but as the diagram shows the presence of a ground impedance generates a voltage V which causes common mode currents to be sent down the output lines which can generate problems elsewhere. In all these cases some means has to be found to prevent or negate the effect of the current flow. In this section three techniques will be discussed, namely

(*a*) balancing,
(*b*) transformers/common mode chokes,
(*c*) opto-isolators.

All these techniques have been covered fairly extensively in recent books [1, 3, 20] so that the explanations given here will be brief.

6.5.1.1 *Balanced circuits*

Balancing is a relatively cheap technique for preventing some of the major effects of common mode currents. On the other hand since it is not really an isolation system it may provide no protection to the overvoltage threats outlined in Figure 6.38(*b*) and (*c*). As defined by Ott [3] 'a balanced circuit is a two conductor circuit in which both conductors and all circuits connected to them have the same impedance with respect to the ground and to all other conductors'. This type of circuit enables the common mode currents at the load to be neutralised. A simple resistive circuit illustrating the principle is given in Figure 6.39. The two voltage sources symmetrically arranged around the ground point A generate two signal currents i_{S_1} and i_{S_2}. Both these currents flow in the same direction through the split load. Thus the load voltage V_L across XY (as would be expected for the differential current) is given by

$$V_L = (i_{S_1} + i_{S_2})(R_{L_1} + R_{L_2}).$$

If the resistors are chosen so that $R_{W_1} = R_{W_2}$, $R_{S_1} = R_{S_2}$ and $R_{L_1} = R_{L_2}$ then $i_{S_1} = i_{S_2}$.

If a voltage source V_G is now introduced between A and B common mode currents I_{G_1} and I_{G_2} now flow in the circuit. Notice that I_{G_1} and I_{G_2} oppose each other in each half of the load so that the load voltage V_L between X and Y becomes

$$V_L = (i_{S_1} + i_{S_2})(R_{L_1} + R_{L_2}) + I_G R_{L_1} - I_G R_{L_2}.$$

Now if the circuit is balanced by making:

$$R_{S_1} = R_{S_2} \qquad R_{L_1} = R_{L_2} = R_L \qquad i_{S_1} = i_{S_2} = i_S \qquad R_{W_1} = R_{W_2}$$

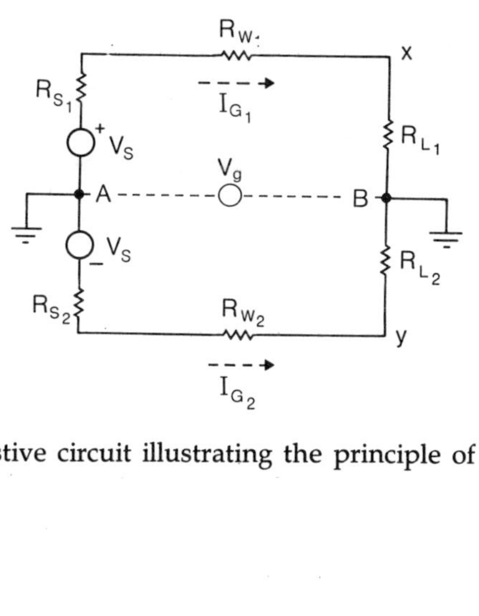

Fig. 6.39 Resistive circuit illustrating the principle of circuit balance

Then $I_{G1} = I_{G2}$ so that

$$V_L = (i_{S1} + i_{S2})(R_{L1} + R_{L2}) = 4i_S R_L.$$

Thus the common mode current term becomes zero and the voltage at the load truly reflects the differential current. It is rarely possible to balance circuits to better than 1 part in 1000–10 000 so that situations where very large common mode currents sit on top of small differential currents still present a problem. It is also difficult to balance circuits at high frequencies because all the resistive terms in the above analysis should be represented as having stray capacitances and inductances associated with each term. These are difficult to assess accurately so that impedance matching becomes almost impossible. Further details of balancing and some of the resulting limitations are to be found in Ott [3]. Finally it should be mentioned that differential amplifiers are often used in balanced systems. These amplifiers for the example discussed would amplify the difference between the voltage at X and that at Y so producing the expected result. If however the common mode current is large the voltages at X and Y can exceed the design values of the amplifier chosen so leading to the possibility of damage.

6.5.1.2 *Transformers and common mode chokes*

The most obvious way to break a ground loop is to use an isolating transformer as shown in Figure 6.40(*a*). Now any ground voltage V_g appears across the transformer interwinding capacitance. Transformers work well but they give no dc current path, necessitate a separate transformer for each signal line and the stray capacitance C_S causes their impedance to become low at high frequencies. The effect of the stray capacitance can be lessened by using a shield between the windings grounded to the load at B as shown in the diagram. Mardiguian [19] describes a 20 MHz balanced twisted pair system that uses two isolation transformers suitably grounded (at the centre tap of the side connected to the

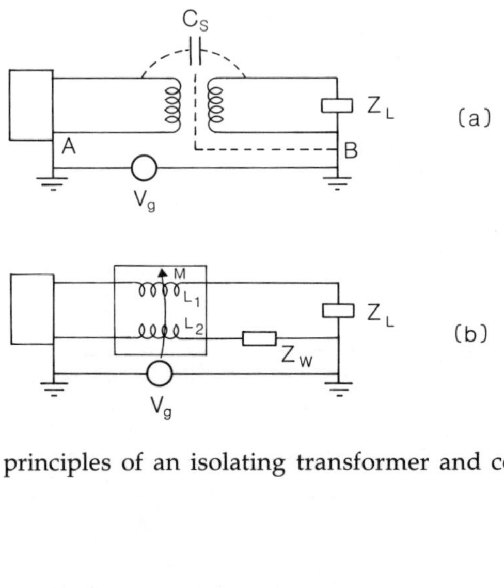

Fig. 6.40 Operational principles of an isolating transformer and common mode choke

balanced line) and screened as Baluns to enable connection of the link to unbalanced driver and load. Baluns are very useful in connecting balanced/unbalanced systems to other unbalanced/balanced systems. Many types exist [21] other than the transformer version discussed above. One common use of a balun is to connect a coaxial cable (unbalanced) to a balanced antenna (e.g. a dipole). One effect of this is to keep the radiating currents on the dipole without these currents them being returned on the outside of the coaxial line where unwanted radiation could occur.

If a transformer is connected as shown in Figure 6.40(*b*) it behaves as a common mode choke. Now there is a DC link. Ott [3] has recently presented the simple theory for both low and high frequency operation so that his results will only be quoted here.

(1) The voltage appearing across the load Z_L is due to the differential current as required provided the frequency f exceeds the limit

$$f > 5Z_w/L. \tag{6.30}$$

Note that for closely coupled coils on the same core $L_1 = L_2 = L = M$.

(2) At low frequencies the noise due to the common mode current produced by V_g is given by

$$V_L = \frac{V_g Z_w/L}{j\omega + Z_w/L}. \tag{6.31}$$

V_L is the noise voltage developed across Z_L.

(3) At high frequencies the stray capacitance across the choke dominates performance and just as in the case the single inductor discussed in Section 6.4.1.5. For low value chokes (2–20 μH) Ott [3] has shown that it is difficult to get IL values much above 10–30 dB in the 10–30 MHz region. Commercial chokes [15] of somewhat higher inductance (1–8 mH) with resonant frequencies in the range 200–1000 kHz give IL values up to 60 dB in the resonant region. Typical performance curves for two commercial

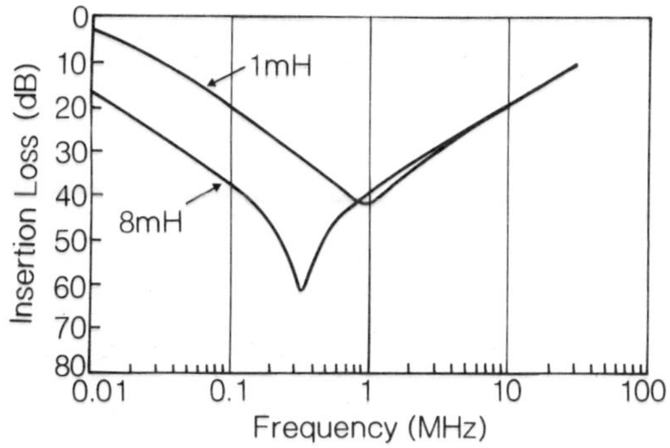

Fig. 6.41 Measured insertion loss frequency curve for two commercial common mode chokes. (Reproduced by permission of muRata Mfg. Co. Ltd.)

common mode chokes are shown in Figure 6.41. All common mode chokes behave like series capacitors at very high frequencies.

(4) Many circuits can be wound on the same core. Such chokes are good at removing low frequency noise but are restricted in their use at high frequencies.

6.5.1.3 *Opto-isolators and optical fibres*

Opto-isolators or opto-couplers are integrated devices that consist of a light source usually a photo diode (two in reverse parallel for AC input) driven by one circuit closely coupled by a transparent insulator to a receiver as shown in Figure 6.42(*a*). The optical coupling is arranged to be good but the electrical coupling weak (usually no more than 0.5–2 pF). The insulator will usually withstand 1–5 kV. The receiver drives the second isolated circuit and can have a variety of forms depending on the application [22, 23]. The commonest use is in digital circuits where very high switching speeds can be obtained. Whilst analogue coupling is possible difficulties arise for high bandwidth high linearity applications.

Larger degrees of isolation can be obtained by going to a fibre optic system as shown in Figure 6.42(*b*). Now all the cable run is replaced. Such systems are finding favour in computer networks, telephone systems and inside aircraft. In all cases the common mode currents are blocked. As well as this any extra coupling that would have occurred between the replaced cable and external fields is prevented.

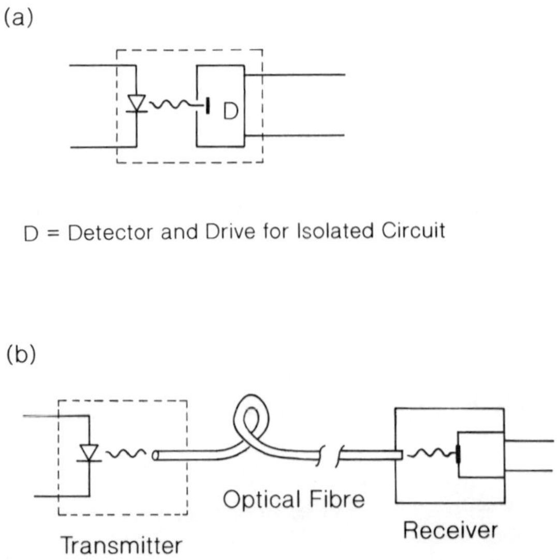

Fig. 6.42 Schematic diagram of an optoisolator and optical fibre system

6.5.2 Suppression techniques

In situations where circuits are exposed to power and voltage levels beyond their normal working range thought must be given to providing protection to prevent damage [20]. Note that any protection provided by the first three suppressor types to be discussed here will not enable the circuit to continue working normally during a transient fault event. Transient conditions of the above kind are fairly common for example in telephone systems, due to unexpected mains transients and in remotely run equipment [24]. Many of these transients (but not all) are caused by lightning. Military (and some civilian) systems also may require protection against EMP. Suppressors work by presenting a high impedance to ground until the transient overvoltage appears when very rapidly they become essentially a short circuit to ground so diverting the incoming energy. The three main types are under constant development so that for accurate performance figures it is essential to consult the latest manufacturers' data sheets. Standler [20] provides a list of current manufacturers of all types of protection equipment. The three suppressor types in ascending order of their power handling ability are as follows.

(1) *Diodes.* All types can in principle be used but for higher stand off voltages they are usually various forms of avalanche/Zener diode which work in reverse bias. The design voltage or protection voltage V_p is chosen to be about 10% above the circuit working voltage. For AC two diodes back to back are used. These devices are limited to protection levels between 2–3 to 400 V and pulsed currents between 200–1000 and 20 A for the respective voltage values. These parameters are for pulses of a few tens of microseconds. In operation diodes clamp the voltage at a value somewhat greater than V_p called the clamping voltage V_c. Foldback [25] diodes can achieve $V_c < V_p$. Avalanche diodes work well but can only deal with small pulse currents and they also have a large capacitance (500–2000 pF). There are circuits for minimising the effect of this capacitance [20] but it is still a problem in high frequency systems. Diodes have fast switching speeds and can protect low voltage circuits.

(2) *Varistors.* Varistors are non-linear resistive elements made of sintered blocks of zinc oxide (mixed with some other metallic oxides). A typical I–V curve for a varistor is shown in Figures 6.43(*a*) and (*b*). The first figure shows the symmetrical nature of the characteristic. V_p is chosen to be just to the left of the steeply rising portion indicating the onset of low impedance. The second characteristic is logarithmic and indicates that the device will pass some current at the normal operating voltage. This current can usually be arranged to be less than 0.1 mA. The clamping voltage V_c like in other devices depends on the load line of the effective source and where it will intercept the varistor characteristic. V_c will however be higher than V_p. Tin oxide varistors are available in a wide range of operating values from protection levels of 20 V at up to 2000 A to 2000 V at 70 000 A. These values as for the diodes are approximate figures for pulses some ten microsecond long. Repetitive exposure to the peak ratings lowers the performance. Like diodes varistors also have a

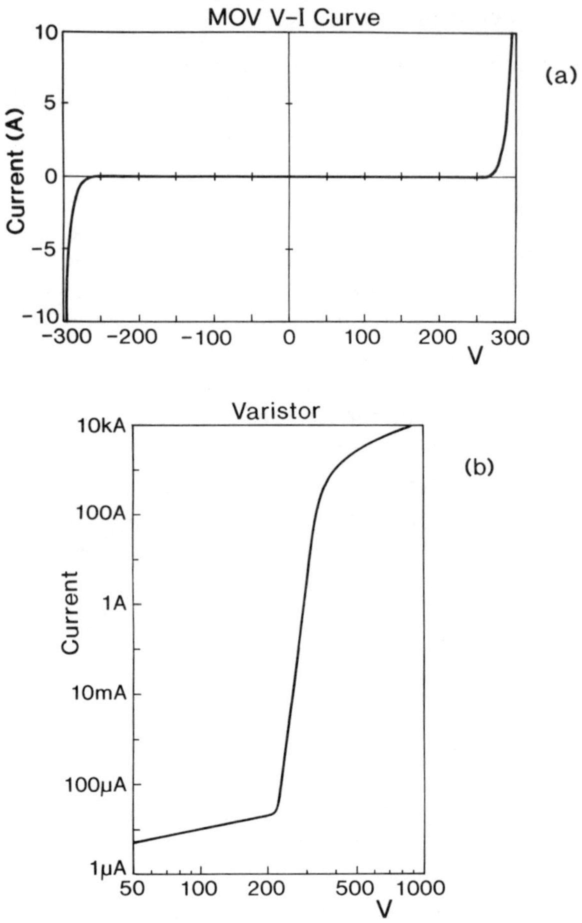

Fig. 6.43 Characteristics of typical varistors, after Standler [20]. (Reproduced by permission of Wiley, New York)

large capacitance (100–4000 pF depending on rating). The response time is fast (20–30 ns). These devices perform well in most suppressor applications. Figure 6.44 compares the performance of two commercial T network filters (three terminal capacitor with ferrite beads and varistor with ferrite beads) exposed to a 1300 V 200 ns square pulse. In Figure 6.44(*a*) the capacitor based filter saturates and a large part of the pulse is allowed through. In Figure 6.44(*b*) the capacitor is replaced by a varistor; the pulse passing the filter is now very much reduced.

(3) *Gas arresters or spark gaps*. These devices are small sealed spark gaps containing rare gases (argon, neon). Usually they have two electrodes as shown in Figure 6.45 but three electrode versions are available for certain applications [20, 25]. As a voltage pulse arrives at the arrester (above the protection level) a gas breakdown is initiated. Unfortunately this can take a little time to start and

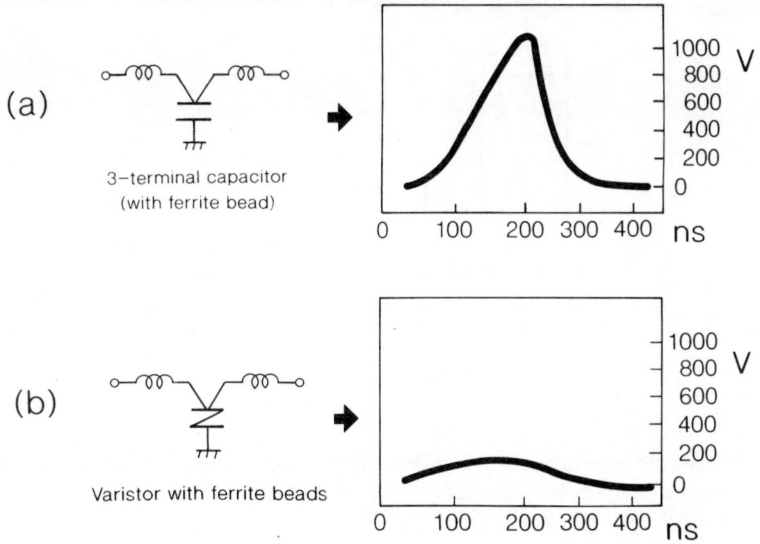

Fig. 6.44 Sketch of oscilloscope traces (50 ns/div, 200V/div) obtained of the output of two types of commercial T network filters exposed to a 1300V, 200 ns, square pulse. (*a*) Conventional EMI filter. (*b*) Type DSS710 varistor-capacitor with ferrite beads. (Reproduced by permission of muRata Mfg. Co. Inc.)

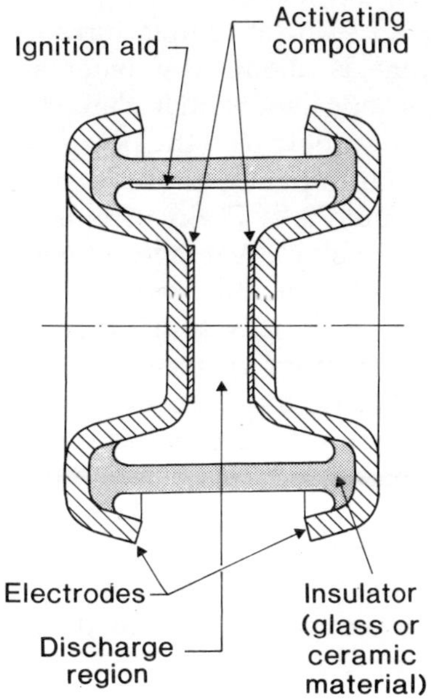

Fig. 6.45 Schematic diagram of a gas arrester. (Reproduced by permission of Siemens AG)

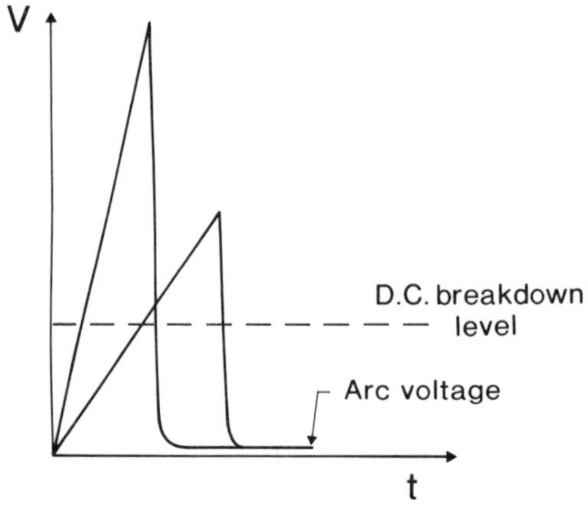

Fig. 6.46 Schematic diagram showing the effect of breakdown voltage on rate of rise of voltage

then develop a low impedance. As a consequence fast rising pulses will break down at a higher value than more slowly rising pulses. The effect of this on some linearly rising pulses is shown in Figure 6.46. The delays in breakdown mean therefore that some part of the pulse can pass this type of suppressor. Manufacturers have developed a number of techniques for minimising the delays such as the inclusion of weak radioactive sources or field enhancing structures. Once the small gap has broken down the gas discharge goes into either a glow mode ($I < 1.5$ A) or at higher currents an arc mode. In the arc mode the clamping voltage is about 20 V whereas in the glow mode it is 70–150 V. Thus this is the one device that clamps well below the protection level. This factor can present problems if the power supply can provide enough current and voltage to keep the spark gap conducting. Typical protection levels can be from 10 000 V to about 70 V. Peak currents are about the same as that for the varistors. One other major advantage for gas arresters is their very low capacitance (1–3 pF). Because all the various suppressors have optimum performance in differing regimes it is sometimes useful to combine them to provide a total system that more correctly provides the protection profile required of a specific job. For example a series connection of a spark gap and varistor from the supply line to the ground return provides a high protection voltage, a low capacitance and the capability of extinguishing the spark gap when the transient has gone. Another example is shown in Figure 6.47. A 10 kV surge is applied to a 25 V line protected first by a spark gap then a varistor and finally by a diode. The figure shows how the surge is attenuated as it passes along the combination. Other types of hybrid protection circuits are used on telecommunication systems [25].

At very high signal frequencies (several GHz) none of the above suppressor types can be used because their self capacitance is too high. One other system, however, can be used to protect such systems from overload though it is really

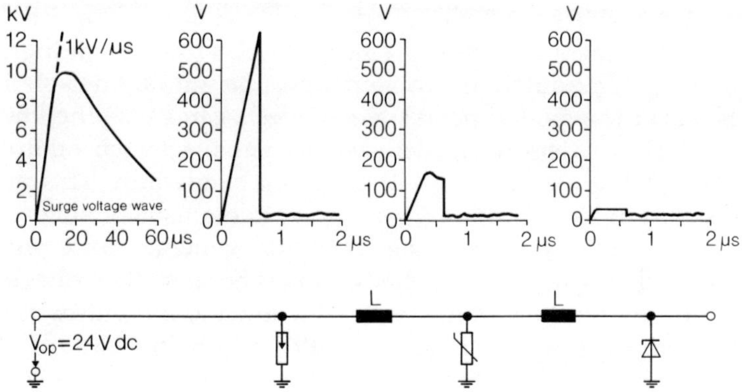

Fig. 6.47 The effect of a 10 kV pulse on three suppressors in series. (Reproduced by permission of Siemens AG)

a robust filter rather than a suppressor. A short circuited quarter wave line is mounted between the incoming signal line and ground so that the whole presents a properly matched line. The quarter wave line at the design frequency acts as an open circuit (see Section 3.5.6) leaving the incoming signal unaffected. Lower frequency noise (lightning, EMP) see the quarter wave line as short to ground and so are diverted. Such systems have been used on aircraft antenna for a long time and commercial protectors based on this principle are now available.

6.6 CIRCUIT DESIGN FOR EMC

The principles involved in circuit design which will improve EMC performance are many and have been extensively covered in recent literature [1, 3, 4, 19, 20, 26, 27]. The operating principle of the circuit and the basic circuit board layout are the building blocks of all electronic systems. It is true to say that mistakes made during the initial planning can lead to costly reworking of the system later in its development. Because of this it is crucial to get the basic circuit arrangements which are favourable for EMC built in at the earliest stages. This section therefore concentrates on some of the major factors that have to be considered.

(1) *Analog versus digital systems.* Analog circuits by and large operate with small signal levels and bandwidths. Indeed the first general point for analog circuit design is to try to use the highest signal levels practical and to choose the smallest bandwidth so limiting the possibilities of for electromagnetic disturbance. Limiting the bandwidth can sometimes be an illusion. By this is meant the fact that systems can have out of band response due to unexpected resonances. Moreover a direct interaction between any incoming high frequency noise/signal and the various non-linear devices present in any circuit

can lead to level shifts and so called audio-rectification or 'the rusty nail syndrome'. This name arose from the fact that close to certain high power HF transmitters the LF modulation can sometimes be audibly detected by nearby wire fences nailed to wooden posts). Such effects can cause the low frequency modulation of HF carriers to be detected by circuits which otherwise would appear to be working way out of their designed bandwidth. This problem is of major concern in military systems. Analog systems usually suffer disturbance by outside sources but in contrast digital circuits generate noise problems both within and outside a system using them. This is because the voltage swings for logic level shifts are typically of the order of a volt and the pulses involved have very small risetimes (e.g. 2 nS for ECL-10K, 50 nS for CMOS). The higher clocking rates being achieved with logic systems having the smallest risetimes. Because of the high clocking rates and small risetimes digital systems need large bandwidth circuits. Section 5.2.8 gives a discussion of the harmonics generated by pulse trains and by inference a measure of the bandwidth (BW) required. Thus for example it is shown that the amplitude of the harmonics reduces as the frequency f increases and for $f > 1/\pi t_r$ (t_r is the risetime) this reduction becomes 40 dB/decade. This turning point is often used as a measure of the bandwidth required for a given logic system [3, 19]. Thus this implies a bandwidth requirement of 6.4 MHz for CMOS and 160 MHz for ECL-10K. Some reduction of this may sometimes be obtained if waveshaping (as discussed in Section 6.4.2) can be used. Digital systems tend to be more immune to outside disturbance than analog circuits in spite of the larger bandwidths. The choice of logic family depends on many factors besides the noise margin and required clocking rates [3, 19] and so cannot be discussed further here. In any event first of all it is necessary to choose a logic system and the minimum clocking rate for a digital circuit or the bandwidth and signal levels for an analog circuit. After this it is then necessary to consider the arrangement of the power supplies and ground layouts on the circuit board.

(2) *Ground layouts and power rails.* The important principles governing the design of good grounding systems have already been discussed in Section 6.3. The key feature being to keep the impedance (mostly inductive) between the source and the equipment as low as possible. Thus ideally the power rail and the ground return should be closely spaced infinite copper planes. Practical requirements mean that this is only approached in multilayer board assemblies but first of all a single board will be considered.

If it is assumed that the power supply feeding a digital circuit board is located elsewhere then initially it is necessary only to consider the board contacts, supply rails and initially one dual-in-line chip (DIL) as shown in Figure 6.48(a). The main problem is the high BW requirements of the fast switching logic. Thus voltage drops across the supply lines during switching mean that the voltage at the DIL will be lowered possibly to a critical value. Moreover the current flows around the loop A_1 are a source of radiative emission. The first thing to do on a single layer circuit board is to thicken the ground return as shown in Figure 6.48(b). It is actually best if as much copper as possible is left on the board so as to produce an almost complete ground plane. Figure 6.49 shows

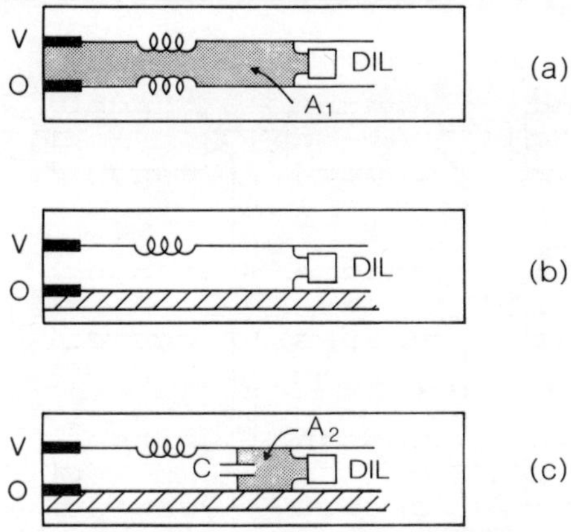

Fig. 6.48 Three methods of connecting a single Dual-in-Line chip (DIL) to a power supply through leads with inductance; (*a*) Via two simple tracks giving a loop of area A_1; (*b*) thickened ground return, (*c*) add a decoupling capacitor, which reduces the effective area of the loop to A_2

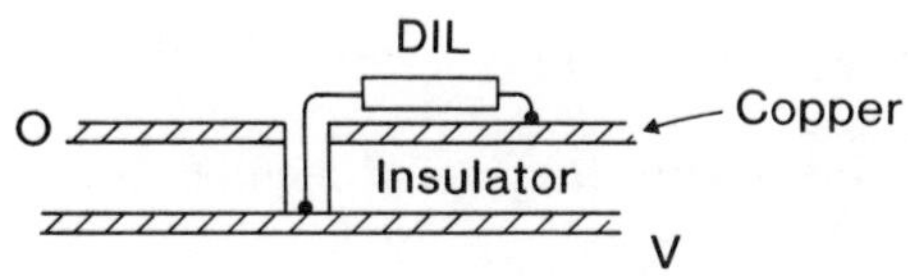

Fig. 6.49 Single circuit board with the copper left almost complete on the top face to simulate a ground plane

how this can be done on a single layer board with the copper and components one side and the circuit traces on the other side. Of course a number of compromises have always been made such as the use of wide parallel power/return lines (sometimes raised or mounted beneath the DIL) or the use of gridded ground structures to simulate a ground plane as shown in Figure 6.50. The ground plane (or the near equivalent) minimises common impedance coupling effects and the tendency to generate ground voltages which are the source of common mode currents (as discussed in Section 6.5). When a digital chip switches a fast inrush current I occurs. The size and risetime depend on the logic system employed and the exact use of the chip (fan out). This pulsed current can still produce an unacceptable voltage drop V_D at the chip terminals because of the remaining power rail inductance. This is overcome by using suitable power supply decoupling capacitors. These reduce the voltage swing and the loop area in which this high frequency current flows as shown in Figure 6.48(*c*). The reduced radiating area A_2 is clearly indicated. The placement of the decoupling capacitors for one design of a whole board is shown in Figure 6.50. It is usual to place the decoupling capacitors as close as possible to the chip it is protecting. As far as the choice of capacitor is concerned

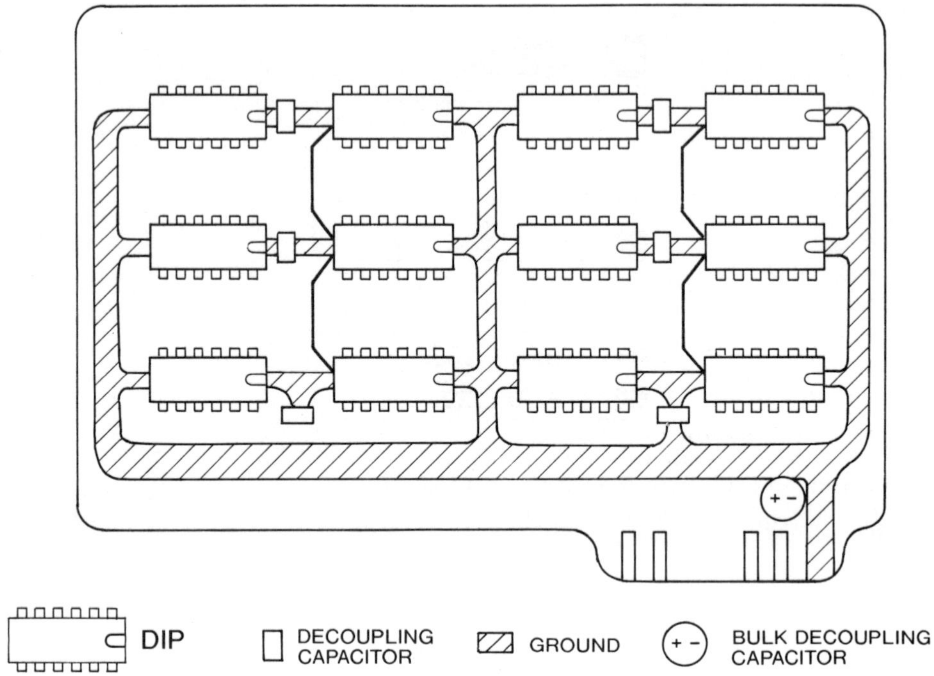

Fig. 6.50 A grid type ground plane on a printed circuit board, after Ott [3]. (Reproduced by permission of Wiley, New York)

all the strictures about lead inductance and self-resonant frequency (SRF) discussed in Section 6.4.2 still apply. The size of the capacitor must be chosen carefully. If it is too large the SRF may be too low or if it is chosen too small it may allow an unacceptable voltage swing at the chip terminals. It is easy however to estimate the value of C required assuming all the charge comes from the capacitor giving:

$$It_r = CV_D$$
$$C = It_r / V_D.$$

The voltage drop V_D allowable depends on the noise margin of the logic used and how much of this can be allocated to be associated with the voltage swing at the chip terminals [19]. The risetimes are well documented in the literature [1, 3, 19]. The current I depends on the duty of the chip [19]. Assuming $I = 5$ mA, $V_D = 0.04$ V and $t_r = 2$ ns, which might correspond to values for a typical ECL-10K chip, then it is found that $C = 250$ pF. It is important to use the above formula to calculate accurate values for C in order to ensure correct operation of the system. So far only a single chip has been considered but of course a circuit board will contain many chips as for example those indicated in Figure 6.50. The best practice is to mount a decoupling capacitor as close to the terminals of each chip as possible. The whole board must also be decoupled because eventually the charge from each local capacitor must be replenished from the main power supply with usually a longer time response so that larger capacitors can be used for this duty (with lower SRF). All this assumes that the total charge required must come from the local added decoupling capacitance but of course this can be affected by the impedance of the power rails. This will be discussed further a little later.

The circuits on a circuit board should be arranged to minimise interference between them. Thus analog circuits should be kept well away from fast digital circuits and in particular the clock. This means that they should also not share the grounding system except for a connection at the board edge. Long runs of parallel signal lines which are prone to cross talk should also be avoided. If they have to be used place a ground line between each pair. If they have to cross each other make sure they cross at right angles. Using the same strictures it pays also to place the slow speed logic, the medium speed logic and the fastest logic all in different regions of the board.

For the fastest logic systems it is more correct to analyse the power distribution/signal lines in terms of transmission line theory rather than simple circuit theory. Thus as intimated at the beginning of this section low characteristic impedance is really the requirement to be sought for a good power distribution system. The short circuit current obtainable from a charged line is V/Z_0. The characteristic impedance and other properties of transmission lines are introduced in Section 3.5.1 onwards. Appendix E gives formulae for forms of line likely to be met on circuit boards and from these formulae it is clear that closely spaced planes make the best system for voltage rails and their ground return. Such an arrangement is used in multilayer boards. An example of a four layer board is shown in Figure 6.51 together with an exploded view showing the functions of the various levels. More generally the chips are on the uppermost level followed by the ground plane on the next, lower levels will consist of interconnects and finally a voltage distribution plane. The traces that are along any plane form transmission lines with either the ground return or the voltage distribution plane. Many interconnects are made vertically. The use of such systems whilst more expensive and difficult to repair have several advantages.

(*a*) Low impedance power distribution. The large local capacitance means that not so many decoupling capacitors will be required.
(*b*) The use of transmission line geometry more extensively means that terminating the lines and keeping the pulses cleaner and freer of reflected components becomes easier.
(*c*) The ground and voltage distribution planes form good screens for the circuitry within them so that less radiation escapes from the overall system. One further advantage is that circuits either side of a conducting plane can keep the currents that flow in the plane due to each circuit separate. This reduces the common impedance coupling provided the skin depth is small compared to the plane thickness. This is most often the case for frequencies greater than 10–100 MHz.

(a) (Caption on opposite page)

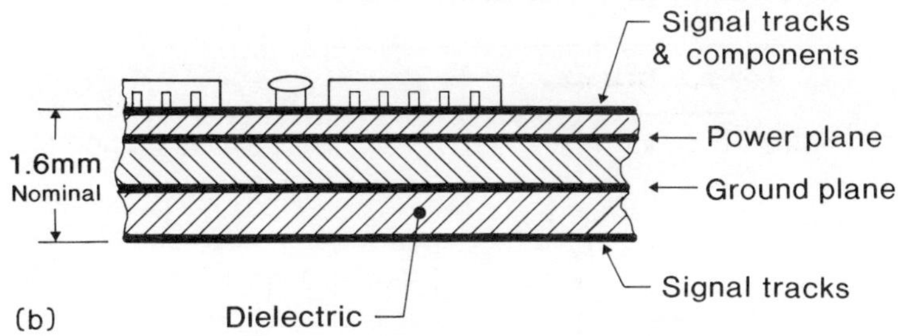

Fig. 6.51 Photograph (*a*, opposite) and cross-section diagram (*b*) of a four layer board, showing decoupling capacitors associated with each chip. (Reproduced by permission of University of Liverpool, Department of Physics)

In order to minimise radiation/susceptibility effects in all types of board it is essential to make sure that all circuit loops have the smallest possible area. Thus all signals are routed close to the ground return or better still close to the ground plane. In critical situations the use of surface mount components can lead to reduced circuit areas [19].

(3) *Power supplies and external circuitry.* So far a single circuit board has been considered but usually this will be just one of many connected by a motherboard. The general principles already discussed for the single board still apply and are often more acute since the trace runs are usually much longer. The power supply often presents a problem particularly if it is a switched mode power supply (SMPS). These can emit harmonics up to 50–100 times the switching frequency. Thus these should have good filters at the mains inlet and have a medium quality screened box around them. The DC output after filtering should always be routed with a ground return (twisted). It is also advisable to keep these power supplies away from sensitive circuits. All signal lines should be routed close to the ground return, or as twisted pairs or in coaxial form so as to minimise radiation and pickup. As for the single circuit board differential mode radiation is reduced by keeping all loops small.

The longer runs in a motherboard system and the fact that connections to the outside world may be made from it mean that common mode radiation problems need to be considered in this section though of course they can arise even on a single board. As discussed in Section 6.5.1 common mode currents arise because of ground voltages on the circuit board. Thus the first step in reducing them is to improve the grounding system for example by using a complete ground plane. Two other techniques consist of the use of common mode chokes (Section 6.5.1.2) or by filtering all output lines to a 'clean' ground [3]. The filtering scheme described by Ott [3] requires the noisy digital ground to be only connected to the clean ground by a single path. The limits of this scheme can easily be understood by considering the equivalent circuit shown in Figure 6.52. Z_C represents the impedance between the noisy ground (producing the common mode voltage V_N) and the clean ground at point 0. Z_g is the impedance between 0 and the true ground. Z_F are the impedances of the

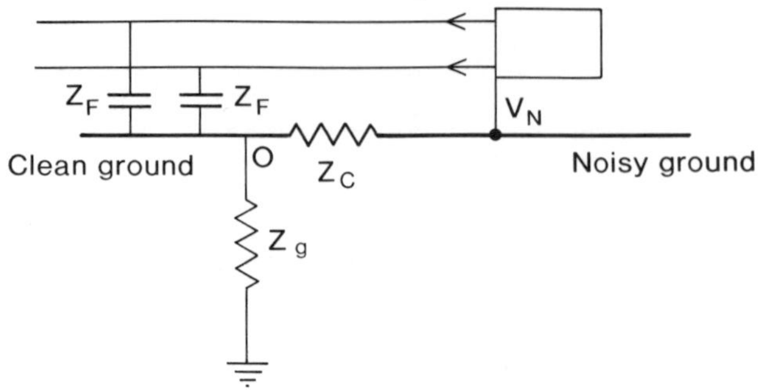

Fig. 6.52 Equivalent circuit for filtering common mode currents onto a 'clean' ground

capacitive filters. Thus for good performance it is necessary that

$$Z_g \ll Z_C.$$

Note all the impedances are small except that of the output line which is likely to be close to 50 Ω.

Ott comments that it is often necessary to use all three techniques to obtain a reasonable reduction in common mode currents. It is always good practise to measure (and then minimise) the common mode currents leaving any system since they most often dominate the emitted radiation as discussed in Section 5.4.5. This is because the effective loop areas are so much larger for situations involving common mode currents.

If the system as a whole is to be connected to the mains then it is at this time one should consider how much mains variation the system can stand. Whilst this topic lies outside the scope of this book it is essential to consider whether any sort of mains protection will be required. The various possibilities include line conditioners, ferroresonant transformers, filters, and uninterruptible power supplies. All these are discussed in recent literature [1, 20, 28, 29].

(4) *Shielding and filters.* Having designed the boards and their connections to minimise radiation and as a consequence their susceptibility to external fields it only remains to assess whether any shielding is required. To do this is not a trivial exercise since either an accurate calculation or detailed measurements will be required. Crude estimates can be made using the procedures outlined in Sections 5.4.4 and 5.4.5 noting the total number of loops and summing the results. More precise techniques using computer models of circuit board systems are being developed in industry [30–33]. Having estimated how much shielding is required then the procedures extensively discussed in Chapter 4 can be applied. If an existing cabinet has to be marginally improved then the techniques outlined by Mardiguian [19] can be helpful. In all systems it will usually be sensible to add at least 5–10 dB more shielding than estimated. Once the whole system is near completion it is worth revising the filter needs of each shield penetration. One EMC specification that can be very onerous to meet is the typical ESD requirement. Indeed some manufacturers see this type of test as revealing the overall performance of the complete system [19]. The

shielding/grounding/filtering/suppression techniques needed are fully discussed earlier in this book and elsewhere [3, 34]. Some points however that need special mention are as follows.

(a) Whenever any switch or control knob penetrates a shield it is most important that it cannot carry the pulse ESD current into the unprotected inner circuits. Thus they should be made of insulating material and some alternative path for the discharge current provided [3].
(b) Signal cables penetrating the shield should be provided with protective suppressors and filters located so as to deflect the current onto the outside of the shield, not onto the inside.
(c) Circuit boards that may be removed by non protected personnel should have an earthed trace around the whole edge of the board connected to the OV line. Any initial discharge on handling will therefore be directed to ground [3].

6.7 EMC SYSTEM DESIGN

If even simple electronic systems are considered carefully it is possible in principle to find millions of routes by which electromagnetic noise could enter or leave the system [1, 27]. Indeed it is this problem in general that makes setting up EMC standards so difficult. Fortunately this book being at an introductory level does not have to address this particular difficulty. In spite of this it helps to have some compact way of thinking about a system that is to be designed or has been built but is causing EMC problems. One way of starting is to go back to the zoning ideas of Vance *et al.* [10–13] as was suggested in Section 6.3.4. Thus one could imagine every system to be divided into N zones

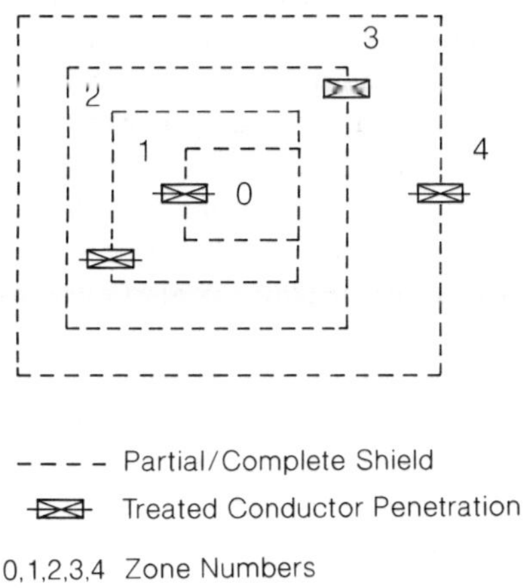

Fig. 6.53 Simple zone layout for EMC design

(hopefully N can be kept relatively small with care) as shown schematically in Figure 6.53. In this example it has been assumed that there are 4 zones with only one cable penetration in each boundary (there will usually be many). The electromagnetic field levels are considered to be of the same order of magnitude throughout a zone. The zone system in Figure 6.53 could be imagined to apply to a common circuit board problem. For example zones 0, 1, and 2 could correspond to different regions on a circuit board layout. Then zone 3 would correspond to the region inside the system cabinet and zone 4 the space outside. The problem then comes down to deciding how much each shield (partial, unintended or full) and each cable penetration treatment must attenuate any electromagnetic disturbance. This may be different for the noise entering or leaving the system. In order to meet a specific EMC code it is clear that a number of different solutions will be possible. Some simple guidelines on how to allocate protection between the various boundaries can be suggested and are given below.

(1) It is not worth attempting to get very large attenuations at one boundary. It may be easier and cheaper to spread the protection through a number of zones.
(2) The ideas presented in Section 6.3.4 suggest that it is hardly worth reducing incoming noise from one zone to another below that which is already present within the zone. This of course applies to the relevant frequency regions in each case.
(3) Any solution must be stable in time. This implies that solutions that depend on extremely good contacts (in poorly controlled situations) or the use of devices working near their limits may not be advisable.

In spite of these simplifications the problem of economic design is still very difficult. In order to sort out some of the possibilities calculations can be helpful. As a first step one could use the approximate coupling/shielding calculations scattered through this book. Nowadays there are also an increasing number of computer based packages which more accurately represent the actual situation in a number of areas [1, 31–33]. Some authors have developed expert systems [35] to help with this problem in again a rather narrow area. All these techniques require the user to have a reasonable knowledge of EMC and electromagnetic theory if they are to apply the results properly. Experience seems still to be important in reaching a balanced design.

Besides having an organised approach it is also necessary to decide now much effort should be allocated to the EMC part of any design. Most authors seem to suggest that between 10% and 20% of the total development cost should be allocated to EMC work and that this can be considered to fall under three headings.

(1) *Fundamental design.* This includes many of the points raised earlier in the various sections of this chapter. It is important at an early stage to identify possible interfering sources, coupling paths and any simple layout changes

that can render the circuit intrinsically less sensitive/noisy. Changes introduced at this stage bring the biggest cost benefits.

(2) *Preliminary testing.* Here EMC tests of the whole or parts of the system are carried out to see how close the design is to specification. The whole gamut of EMC testing may need to be carried out using the techniques briefly introduced in Chapter 1. One problem sometimes met is how to relate test results for a part of a system to those expected for the whole. On the basis of this preliminary test changes may have to be made to the initial design introducing extra filtering, shielding etc. Then the design has to be finally approved.

(3) *Commissioning tests.* These are done to get the approval of whatever EMC regulatory body is involved in laying down the specification being followed.

Some expenditure under each of these headings would seem to be inevitable. In all the work covered in this book not the least being in the overall design of systems it is clear that a good understanding of basic electromagnetism is essential if all the well known remedies to EMC problems are to be applied efficiently.

REFERENCES

[1] Violette, J. L. N., White, D. R. J. and Violette, M. F. (1987) *Electromagnetic Compatibility Handbook* Van Nostrand, Reinhold, New York.

[2] Jordan, E. C. and Balmain, K. G. (1968) *Electromagnetic Waves and Radiating Systems* 2nd edition, Prentice-Hall, New Jersey.

[3] Ott, H. W. (1988) *Noise Reduction Techniques in Electronic Systems* 2nd edition, Wiley, New York.

[4] Coekin, J. A. (1975) *High Speed Pulse Techniques* Pergamon, Oxford.

[5] *IEEE Wiring Regulations* (1981) 15th edition, The Institute of Electrical Engineers (London). (Some discussion of the various earthing arrangements allowed will be found in Linsley, T. (1989) *Advanced Electrical Installation Work* Edward Arnold, London.)

[6] B.S. 5724: Part 1 1989 {I.E.C. 601-1: 1988} and subsequent parts *Medical Electrical Equipment Part 1. General Requirements for Safety* British Standards Institution (See Appendix A for the address.)

[7] Paul, C. R. (1980) Effect of Pigtails on Crosstalk to Braid-Shield Cables *IEEE Trans on EMC* **EMC-22** 161–172.

[8] Hejase, H. A. N., Adams, A. T., Harrington, R. F. and Sarkar, T. P. (1989) Shielding Effectiveness of 'Pigtail' Connections *IEEE Trans. on EMC* **EMC-31** 63–68.

[9] Moser, J. R. (1986) Peripheral Cable-Shield Termination: The System EMC Kernel *IEEE Trans. on EMC* **EMC-28** 40–45.

[10] Graf, W. and Vance, E. F. (1982) A Topological Approach to the Unification of Electromagnetic Specifications and Standards *Proc. IEEE National Aerospace and Electronics Conf., Naecon* p 4–10. See also Vance, E. F. (1988) The role of Shielding in Interference Control *IEEE Trans. on EMC* **EMC-30** 294–297.

[11] Vance, E. F. (1980) Electromagnetic-Interference Control *IEEE Trans. on EMC* **EMC-22** 319–328.

[12] Vance, E. F., Graf, W. and Nanevicz, J. E. (1981) Unification of Electromagnetic Specifications and Standards Part 1, Evaluation of Existing Practices *Defense Nuclear Agency Report, Washington DC DNA 5433F-1.*

[13] Graf, W., Hamm, J. M. and Vance, E. F. (1983) Unification of Electromagnetic Specifications and Standards Part 2, Recommendations for Revisions of Existing Practices *Defense Nuclear Agency Report, Washington DC DNA 5433F-2.*

[14] Ramo, S., Whinnery, J. R. and Van Duzer, T. (1965) *Fields and Waves in Communication Electronics* Wiley, New York.

[15] A number of commercial companies provide excellent technical guidance for their products which often are of general relevance; see for example EMI Countermeasures (1987) *Application guidance from muRata Mfg. Co. Inc.* Murata Erie N. America Inc., Smyrna, GA, USA, and *Technical Tips* Corcom Inc., Libertyville, IL USA.

[16] *Electronic Engineers' Handbook* (1975) 1st Edition, ed. D. G. Fink, McGraw-Hill, New York.

[17] Strauss, L. L. (1970) *Wave Generation and Shaping* 2nd edition, McGraw-Hill, New York.

[18] *Interface Circuits Data Book* (1991) Texas Instruments, Dallas, Texas, USA.

[19] Mardiguian, M. (1984) *Interference Control in Computers and Microprocessor Based Equipment* Interference Control Technologies, Gainesveille, VA.

[20] Standler, R. B. (1989) *Protection of Electronic Circuits from Overvoltages* Wiley, New York.

[21] Liao, S. Y. (1988) *Engineering Applications of Electromagnetic Theory* West Publishing Company, St Paul.

[22] See commercial suppliers descriptive catalogues, for example *Optoelectronics and Image Sensor Data Book* (1987) Texas Instruments, Dallas.

[23] Chappell, A. (ed) (1976) *Optoelectronics: Theory and Practice* Texas Instruments.

[24] Woodward, J. (1988) Lightning Interference Protection of Electrical Equipment in Remote Locations *Proc. 6th Int. Conf. on EMC* York, IERE Publication 81 pp. 91–97.

[25] Tillman, R. G. and Croft, D. C. (1987) Electrical Protection of Telecomms Equipment *BR. Telecom J.* **5** 43–47.

[26] Sangwine, S. J. (1987) *Electronic Components and Technology-Engineering Applications* Van Nostrand Reinhold (UK), Wokingham.

[27] White, D. R. J., Attkinson, K. and Osburn, J. D. M. (1985) Taming EMI in microprocessor systems *IEEE Spectrum* **22** 30–37.

[28] Duell, A. W. and Roland, W. V. (1981) Power Line Disturbances, their Effect on Computer Design and Performance *Hewlett-Packard J.* **32** 28–32.

[29] Martzloff, F. (1990) Protecting Computer Systems Against Power Transients *IEEE Spectrum* **27** 37–40.

[30] Swainson, A. J. G. (1990) Radiated Emission, Susceptibility and Cross Talk on Ground Plane Printed Circuit Boards *7th Int. Conf. on EMC, University of York* IEE Conference Publication 326, pp. 37–41.

[31] Scott, K. J. (1990) An Equivalent Circuit Model for Arbitrarily-shaped Metallisation Areas on Printed Circuit Boards *7th Int. Conf. on EMC University of York* IEE Conference Publication 326, pp. 42–49.

[32] Clark, G. (1990) Observation of Electromagnetic Coupling and Power Flow in Printed Circuits *7th Int. Conf. on EMC University of York* IEE Conference Publication 326, pp. 50–57.

[33] John, W. (1990) Remarks to the Solutions of EMC Problems on Printed Circuit Boards *7th Int. Conf. on EMC, University of York* IEE Conference Publication 326, pp. 68–72.

[34] Boxleitner, W. (1989) How to Defeat Electrostatic Discharge *IEEE Spectrum* **26** 36–40.

[35] Price, M. J. S. and MacDairmaid, I. P. (1988) Developing an Expert System for EMC Design *6th Int. Conf. on EMC, University of York* IEE Conference Publication 81, pp. 331–336.

Appendix A Some Useful Contact Addresses for EMC Standards

Fax numbers are given in the form: country code—area code—local code—number.

Note. Telephone and Fax numbers change with time, but the Telex number appears to be unchanged over long periods. The date when the numbers were last checked by the authors are given.

Standard	**Address**
CENELEC	European Committee for Electrotechnical Standardisation, Rue Brederode 2, Bte 5-1000, Bruxelles, Belgium. Tel (+322) 519 68 71 Telex 26257 CENLEC B Fax 32 2 519 68 19 (Valid 22 March 1991)
CCITT,UIT	The International Telegraph and Telephon Consultative Committee, Place des Nations, CH-1211 Geneva 20, Switzerland. Tel 41 22 730 5111 Telex 421 000 UIT CH Fax 41 22 733 7256 (Valid 3 October 1990)
CNET	Centre National d'Etudes des Telecommunications, Route Tregastel, 22301 Lannion Cedex, France. Tel (96) 05 11 11 Telex 950 455 Fax 33 96 05 3216 (Valid 12 March 1991)
ETSI	European Telecommunications Standards Institute, BP 152-06561, Valbonne Cedex, France. Tel (33) 92 94 42 00 Telex 470 040 Fax 33 93 65 47 16 (Valid 12 March 1991)
FCC	Federal Communications Commission, Technical Standards Branch, Washington DC 20554, USA. Tel (202) 653 6288 Telex 710 822 0160 Fax 1 202 653 8773 (Valid 9 October 1990)

FTZ	Deutsche Bundespost TELEKOM Fernmeldetechnisches Zentralamt, Am Kavalleriesand 3, Postfach 10 00 03, D-6100 Darmstadt, Germany. Tel (0 6151) 83 (0) Telex 41 95 11 ftz d Fax 49 6151 83 4791 (Valid 20 November 1990) (Technical Guidelines on EMC in draft form only, November 1990)
IEC/CISPR	IEC, International Electrotechnical Commission, 3, Rue de Varembe, 1211 Geneva 20, Switzerland. Tel 41 22 734 01 50 Telex 28 872 ceiec ch Fax 41 22 733 38 43 (Valid 8 October 1990)
IEEE	The Institute of Electrical and Electronic Engineers, Inc., Service Centre, 445 Hoes Lane, P.O. Box 1331, Piscataway, NJ 08855-1331, USA Tel (908) 981 0060 Telex 833 233 Fax 1 908 981 0027 (Valid 16 April 1991)
JIS	Japanese Standards Organisation (JIS), c/o Standards Department Agency of Industrial Science and Technology, Ministry of International Trade and Industry, 1-3-1 Kasumigaseki, Chiyoda-Ku, Tokyo 100, Japan. Tel 81 3 501 92 95/6 Telex 02 42 42 45 Fax 81 3 580 14 18 (Valid 22 March 1991)
MIL STD	Defense Logistic Agency OASD (P&L)/DASD(PR) 5203 Leesburg Pike, Suite 1406, Falls Church, VA 22041-3466, USA. Tel (703) 274 6000/6001 (general information) Tel (703) 756 2343 (Standardisation) Fax 1 703 756 7622 (Valid 17 April 1991)
NEMA	National Electrical Manufacturers, 2101 L Street NW, Washington, DC 20037, USA. Tel (202) 457 8400 Telex 904 007 Fax 1 202 457 8411 (Valid 16 April 1991)
SEV	Schweizerischer Elektrotechnischer Verein, Seefeldstr. 301, CH-8034 Zurich, Switzerland. Tel 41 1 384 91 11 Telex 817 431 sev ch Fax 41 1 55 14 26 (Valid 3 October 1990)
UL	Underwriters Laboratory Inc, 333 Pfingsten Road, Northbrook, IL 60062-2906, USA. Tel (708) 272 8800 Telex 201227

	Fax 1 708 272 8129 (Valid 16 April 1991) (An independent, not-for-profit organisation testing for public safety)
VDE	VDE-Verlag GmbH, Bismarckstr, 33, 1000 Berlin 12, Germany. Tel 030 348 001 (0) Telex 1-81 683 Fax 49 30 3417093 (Valid 28 September 1990)
VG	Bundesamt fur Wehrtechnik und Beschaffung (BWB) Postfach 7360, 5400 Koblenz, Germany. Tel 02 61 40 01 Telex 8 62 661 Fax 49 261 4 00 67 30 (Valid 22 March 1991) (Publications available through booksellers Beuth Verlag)

United Kingdom addresses

BEAMA	The British Electrotechnical and Allied Manufacturers' Association, 8, Leicester Street, London, WC2H 7BN. Tel (071) 437 0678 Telex 263 536 Fax 44 71 437 4901 (Valid 24 May 1991)
BSI	British Standards Institution, Linford Wood, Milton Keynes, MK14 6LE. Tel (0908) 22 00 22 (Switchboard and Information) Tel (0908) 22 11 66 (Enquiries) Telex 825 777 BSIMK G Fax 44 908 320856 (Valid 22 March 1991)
DEF STAN 59.41	Ministry of Defence, Directorate of Standardisation, Kentigern House, 65 Brown Street, Glasgow G2 8EX. Tel (041) 224 2496 Telex 779 699 Fax 44 41 224 2503 (Valid 11 October 1990)
DTI	Radiocommunications Agency, Waterloo Bridge House, Waterloo Road, London SE1 8UA. Tel (071) 215 5000 (Switchboard) Tel (071) 215 2072 (24 hour Information and Library Service) Tel (081) 200 1992 (EMC Hotline) Telex 261969 DTIWBH G Fax 44 71 928 5746 (Valid 20 March 1991)
IEE/IERE	Institute of Electronic and Radio Engineers, Savoy Place, London, WC2R 0BL. Tel (071) 240 1871 Telex 261176 IEELDN G Fax 44 71 240 7735 (Valid 24 May 1991)

NAMAS National Measurement Accreditation Service,
NAMAS Executive, National Physical Laboratory, Teddington,
Middlesex, TW11 0LW.
Tel (081) 943 7140
Telex 262344 nplg
Fax 44 71 943 7134
(Valid 24 May 1991)
(Catalogue of UK calibration and testing laboratories, includes EMC)

Note: At Easter 1994 all UK telephone numbers will change to 10 digit numbers by introducing an extra number 1 after the 0, e.g. BEAMA will become (0171) 437 0678.

USA address for Department of Defense, EMC/EMI

Commander,
Space and Naval Warfare Systems,
Code SPAWAR 2243,
Washington, DC 20363-5100, USA.

Tel (703) 602-0559
Fax 1 703 602 6805/5891

(Valid 24 September 1991)
(Contact at this time: Mr. S. Caine)

Appendix B The Curl Vector and Stokes' Theorem

Of all the concepts introduced in vector analysis the curl vector seems to be the one that causes the greatest difficulties, possibly because the role it plays in describing the structure of a vector field is not understood.

Consider some fluid flowing over a flat surface, and assume that the velocity is small enough to have streamline flow, so that the velocity vectors are all parallel to each other and to the surface. Now it is conceivable that the precise distribution of velocity with depth could vary depending on the fluid and the nature of the surface, such as those sketched in Figure B.1. These velocity distributions are clearly different, hence there is a need to be able to describe *how* they are different in some coordinate independent and mathematically rigorous way; the curl vector does this.

A mental mode of what a curl vector is can be obtained by imagining a very small paddle wheel being introduced into the stream at different points, as in Figure B.2. It is clear that because the velocities on either side of the wheel are slightly different the wheel might rotate, and the rate of rotation would depend on the precise orientation of the wheel within the fluid, (being a maximum when the axis of the wheel is perpendicular to the velocity vectors), and size of the *velocity gradient* across the wheel.

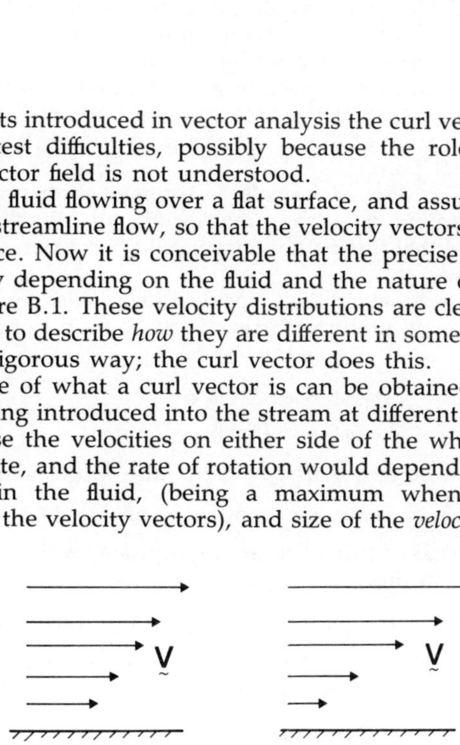

Fig. B.1 Schematic diagram of two possible velocity distributions

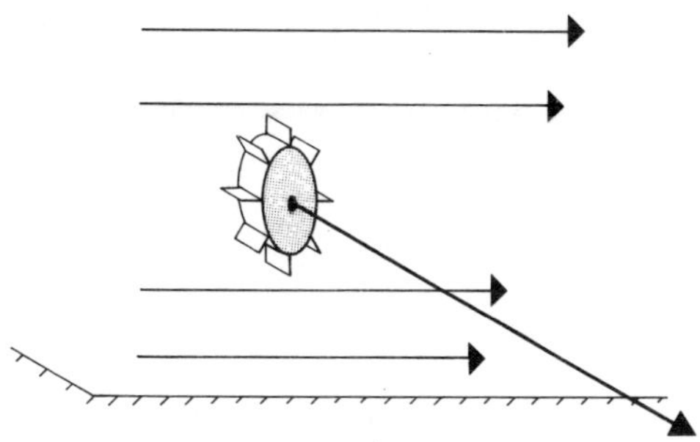

Fig. B.2 Introduction of a small paddle wheel into a flowing liquid

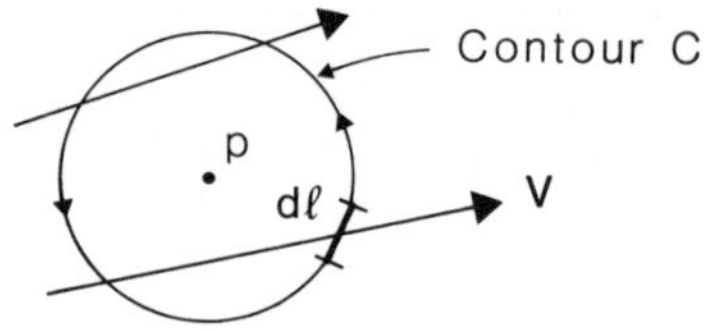

Fig. B.3 Contour C around which the line integral is evaluated

Thus at any point in the fluid another vector could be set up pointing in the same direction as the axis around which there is a maximum rotation speed, having a length proportional to the angular velocity of rotation. If this was done at a large number of points throughout the liquid, the set of vectors go generated would form a separate vector field, describing the structure of the velocity field but not its magnitude. One cannot tell from the rate of rotation of the wheel at a given point the absolute value of the velocity of the fluid at that point!

The curl vector at a point P is defined formally in terms of the path integral per unit area along a closed path around it as the area shrinks to infinitesimal size, as in Figure B.3. It must be stressed that this path integral could be zero in some cases, as the following examples will show.

The magnetic field (a vector field) at a radius r inside and outside of a straight long wire of radius R and carrying a steady uniform current i can be found using Ampère's law to be

$$\mathbf{B}_{\text{inside}} = \mu_0 i r / 2\pi R^2 \boldsymbol{\phi}$$
$$\mathbf{B}_{\text{outside}} = \mu_0 i / 2\pi r \boldsymbol{\phi}.$$

The curl of these vectors can be found in two ways, the 'professional' way and the intuitive way.

(a) For a long wire cylindrical coordinates are appropriate, with $\mathbf{z}$ along the wire, $\mathbf{r}$ radially and $\boldsymbol{\phi}$ azimuthally such that

$$\text{curl } \mathbf{B} = \{(1/r)\ \partial B_z/\partial\phi - \partial B_\phi/\partial z)\}\mathbf{r} + \{\partial B_r/\partial z - \partial B_z/\partial r\}\boldsymbol{\phi} + (1/r)\{\partial(rB_\phi)/\partial r - \partial B_r/\partial\phi\,)\mathbf{z}.$$

From symmetry the terms in $\partial/\partial z$ and $\partial/\partial\phi$ must be zero because at fixed radius the field is independent of the z and ϕ coordinates; hence

$$\text{curl } \mathbf{B} = (1/r)\{\partial(r\mathbf{B}_\phi)/\partial r\}\mathbf{z}$$

i.e. inside the wire

$$\begin{aligned}\text{curl } \mathbf{B} &= (1/r)\{\partial(r\mu_0 i r/2\pi R^2)/\partial r\}\mathbf{z}\\ &= \mu_0 i/\pi R^2 \mathbf{z}\\ &= \mu_0 \mathbf{J}\end{aligned}$$

and outside the wire

$$\text{curl } \mathbf{B} = \frac{1}{r}\frac{\varrho}{\partial r}(r\mu_0 i/2\pi r)\mathbf{z}$$

i.e.

$$\text{curl } \mathbf{B} = 0.$$

These two results are exactly what is expected from Maxwell's equation curl $\mathbf{H} = \mathbf{J} + \partial\mathbf{D}/\partial t$, where both the vectors $\mathbf{J}$ (the current density) and $\partial\mathbf{D}/\partial t$ $(=0)$ are those *at the point where the curl vector is evaluated.*

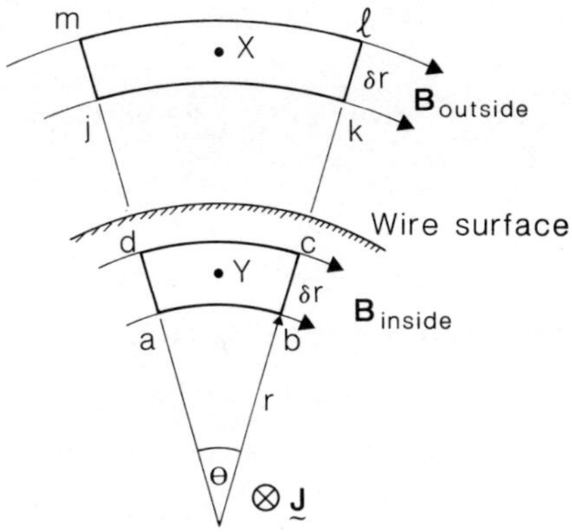

Fig. B.4 Two contours, *abcd* inside the wire, *jklm* outside the wire. The wire carries a uniform current density **J** flowing into the plane of the paper.

(*b*) For a more intuitive approach consider Figure B.4. This shows a pie-shape slice of the wire and its magnetic field, and points *X* and *Y* are where curl **B** is to be calculated. Define contours between radius *r* and $r+\partial r$ as *abcd* inside the wire and *jklm* outside the wire, around each of which the path integral $\int \mathbf{B}\cdot \mathbf{dl}$ will be evaluated.

Alongside the sides *bc, da,* kl, and *mj* the field **B** is perpendicular to the path, so that $\mathbf{B}\cdot\mathbf{dl}=0$ and there is no contribution to the path integral. Thus inside the wire

$$\begin{aligned}\int \mathbf{B}\cdot\mathbf{dl} &= B(r)\cdot \text{arc length } ab - B(r+\partial r)\cdot \text{arc length } cd\\ &= (\mu_0 ir/2\pi R^2)\cdot r\theta - (\mu_0 i(r+\partial r)/2\pi R^2)\cdot (r+\partial r)\theta\\ &= (\mu_0 i\theta/2\pi R^2)\cdot \{r^2-(r+\partial r)^2)\}\\ &= (\mu_0 i/\pi R^2)\cdot\{r\theta\cdot\partial r\}.\end{aligned}$$

Now $\{r\theta\cdot\partial r\}$ ≈ area enclosed by the contour *abcd*. Thus since

$$\text{curl } \mathbf{B} = \int \mathbf{B}\cdot\mathbf{dl} \text{ per unit area in the limit when } \partial r \text{ and } \theta\to 0$$

$$\text{curl } \mathbf{B} = \nabla\times\mathbf{B} = (\mu_0 i/\pi R^2) = \mu_0\mathbf{J}.$$

Outside the wire

$$\int \mathbf{B}\cdot\mathbf{dl} = B(r)\cdot\text{arc length } jk - B(r+\partial r)\cdot \text{arc length } lm$$

$$= (\mu_0 i/2\pi r)\cdot r\theta - (\mu_0 i/2\pi(r+\partial r))\cdot(r+\partial r)\theta = 0$$

so that

$$\nabla\times\mathbf{B}=0.$$

Careful examination of these two derivations will show why $\nabla\times\mathbf{B}$ is different in the two cases. Both path integrals depend on the difference between the line integrals along the shorter side of the contour at radius *r* and the longer side at $r+\partial r$. Inside the wire the field *increases* with radius so that the integral along the longer side of the contour becomes dominant whereas outside the wire the field *decreases* with radius and the two terms just cancel each other. Thus it is clear that the way the field changes in space, i.e. its structure, is the crucial element in determining the curl vector.

This idea can be used in the following way. The longer side of the contour (the lines *cd* or *lm* in Figure B.4) have total length of $r\theta+\partial r\cdot\theta$, while the shorter sides (*ab* or *jk*) are of length $r\theta$. Outside the wire where the Maxwell equation $\nabla\times\mathbf{H}=\mathbf{J}+\partial\mathbf{D}/\partial t=0$, the field $\mathbf{B}(r)$ must satisfy

$$\mathbf{B}(r)\cdot r\theta - B(r+\partial r)\cdot\{r\theta+\partial r\cdot\theta\} =$$
$$\{B(r+\partial r)-B(r)\}r\theta = -B(r+\partial r)\cdot\partial r\theta$$

hence

$$\{B(r+\partial r)-B(r)\}/\partial r = -B(r+\partial r)/r$$

and in the limit as $\partial r\rightarrow 0$ this becomes the differential equation

$$\mathrm{d}B(r)/\mathrm{d}r = -B(r)/r.$$

The solution to this is $B(r)=K/r$ where K is a constant, since

$$\mathrm{d}B(r)/\mathrm{d}r = -K/r^2 = -(K/r)\cdot(1/r) = -B(r)/r.$$

Thus in order to satisfy Maxwell's equations outside the long straight wire the magnetic field can only vary as $1/r$; this is an example of the uniqueness of electromagnetic fields in any given situation. Note three things.

(1) $\nabla\times\mathbf{B}$ has been found at points X and Y by examining $\mathbf{B}$ *around* these points; nowhere in the derivation was the value of $\mathbf{B}$ *at* these points needed.
(2) The curl vector is parallel to the $\mathbf{z}$ axis, i.e. points along the wire. The $\mathbf{B}$ field is in the plane perpendicular to $\nabla\times\mathbf{B}$ at any point.
(3) There is not enough information in Maxwell's equations to enable the precise form of the constant K to be found.

It is a measure of the genius of Maxwell that he recognised that electromagnetism could be formulated in terms of local variables rather than the older forms of the Coulomb and Biot–Savart laws which are formulated in terms of the spatial arrangement of charges and currents.

Once the concept of the curl of a vector field defined in terms of a path integral per unit area has been mastered, there should be little difficulty in seeing that Stokes' theorem is elegant yet basically very simple. Figure B.5 shows a surface S enclosed by a contour C situated in a vector field $\mathbf{V}$. The surface is divided into small areas ∂A with a unit vector $\mathbf{n}$ perpendicular to the surface (so that the vector area is $\mathbf{dS}=\partial A\mathbf{n}$). At each point on the surface at the centre of one of these areas ∂A the scalar product $(\nabla\times\mathbf{V})\cdot\mathbf{dS}$ is equal to the path integral of $\mathbf{V}$ taken around the edges. Doing this in the same way

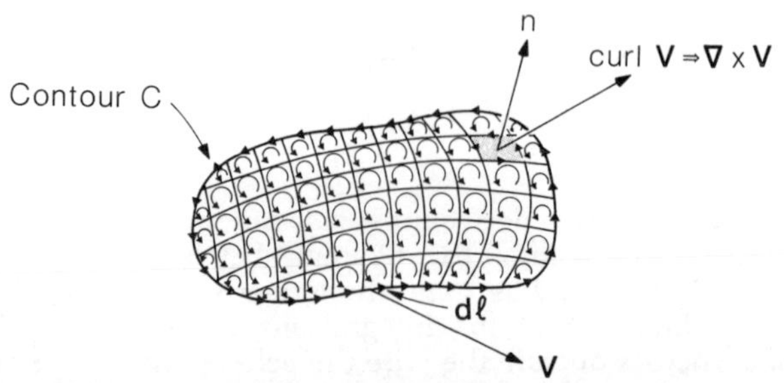

Fig. B.5 A surface defined by a contour C broken into many small elements, around each of which a line integral is performed, adapted from reference [1]. (Reproduced by permission of Chapman and Hall, Ltd.)

for all the vector areas which make up the surface and summing the result, it is clear that the line integral contributions along the common edges of the vector areas will cancel, leaving only those line integrals along the edges of the surface (the contour C) so that

$$\int_C \mathbf{V} \cdot d\mathbf{l} = \int_S (\nabla \times \mathbf{V}) \cdot d\mathbf{S}$$

This is Stokes' theorem.

REFERENCE

[1] Hague, B. (1939) *An Introduction to Vector Analysis* Wiley, New York.

Appendix C Mathematics of the Variation of Zw with Distance for a Short Wire and Magnetic Dipole Antennas

For this, the same notation as in Lorrain, Corson and Lorrain (3rd edition 1988) will be used. For the electric dipole start with their Equations (38-21) for H_ϕ and the E_θ part of Equation (38–14). For this proof it is assumed that the length of the dipole $s \ll r^3$ and $s \ll \lambda^3$

$$E_\theta = \frac{[p]}{4\pi\varepsilon_0 \bar{\lambda}^2 r}\{\bar{\lambda}^2/r^2 - 1 + j\bar{\lambda}/r\}\sin\theta$$

$$H_\phi = \frac{c[p]}{4\pi\bar{\lambda}^2 r}\{-1 + j\bar{\lambda}/r\}\sin\theta$$

where $\bar{\lambda} = \lambda/2\pi$ and $[p]$ is the time-dependent dipole moment $[p] = Q_0 s e^{j\omega t}$. Then

$$\begin{aligned} Z_W = E_\theta/H_\phi &= (1/c\varepsilon_0)\{\bar{\lambda}^2/r^2 - 1 + j\bar{\lambda}r\}/\{-1 + j\bar{\lambda}/r\} \\ &= Z_0\{(1 - \bar{\lambda}^2/r^2) - j\bar{\lambda}/r\}\{1 + j\bar{\lambda}r\}/\{1 + (\bar{\lambda}/r)^2\} \\ &= (Z_0/\{1 + (\bar{\lambda}/r)^2\})\{1 - j(\bar{\lambda}/r)^3\}. \end{aligned}$$

Thus the square of the modulus of the wave impedance is:

$$\begin{aligned} (|Z_W|)^2 &= (Z_0/\{1 + (\bar{\lambda}/r)^2\})^2\{1 - j(\bar{\lambda}/r)^3\}\{1 + j(\bar{\lambda}/r)^3\} \\ &= (Z_0/\{(1 + (\bar{\lambda}/r)^2\})^2\{1 + (\bar{\lambda}/r)^6\}. \end{aligned}$$

Hence

$$|Z_W| = Z_0\sqrt{\frac{\{1 + (\bar{\lambda}/r)^6\}}{\{1 + (\bar{\lambda}/r)^2\}}} \qquad \text{(C1)}$$

so that when $r \gg \lambda/2\pi$

$$|Z_W| \approx Z_0$$

and when $r \ll \lambda/2\pi$, $\bar{\lambda}/r \gg 1$. Hence

$$\begin{aligned} |Z_W| &\approx Z_0\surd(\bar{\lambda}/r)^6/(\bar{\lambda}/r)^2 \\ &\approx Z_0(\lambda/2\pi r). \end{aligned}$$

For the magnetic dipole case (loop of area A carrying a current $i_0 e^{j\omega t}$), start with equations (39-31) for the azimuthal electric field and the H_θ part of (39–34) for the magnetic field i.e.,

$$E_\phi = \frac{\mu_0 c\,[m]}{4\pi\lambda^2 r}\,\{1 - \mathrm{j}(\lambda/r)\}\sin\theta$$

and

$$H_\theta = \frac{[m]}{4\pi\lambda^2 r}\,\{-1 + (\lambda/r)^2 + \mathrm{j}\lambda r\}\sin\theta$$

where $[m\}$ is the time-dependent magnetic moment $Ai_0 e^{\mathrm{j}\omega t}$.

Thus

$$Z_\mathrm{W} = Z_0\{1 - \mathrm{j}(\lambda/r)\}/\{(\lambda/r)^2 - 1 + \mathrm{j}\lambda/r\}$$

which can be manipulated as in the first example to yield:

$$|\,Z_\mathrm{W}\,| = Z_0\,\frac{\sqrt{\{1 + (\lambda/r)^6\}}}{\{((\lambda/r)^2 - 1)^2 + (\lambda/r)^2\}}\,. \qquad \text{(C.2)}$$

Thus when $r \gg \lambda/2\pi$

$$|\,Z_\mathrm{W}\,| = Z_0$$

and when $\lambda/r \gg 1$

$$|\,Z_\mathrm{W}\,| = Z_0(2\pi r/\lambda).$$

Appendix D Pulse or Resonant Charging of Capacitors

In Figure 2.18 a capacitor C_1 is charged to some voltage V_0 and connected to capacitor C_s through a switch S and wires which have small inductance L and resistance R; L and R are not discrete components in the circuit. It is assumed that this is a high-Q circuit where $R/\omega L \ll 1$. At the instant the switch is closed, the current through the circuit is zero, as is the emf around it, i.e.

$$\int \mathbf{E} \cdot \mathrm{d}\mathbf{l} = 0$$

because there is no battery or changing flux $\partial\phi/\partial t$ to generate an emf. However at $t=0$ the current starts to flow from capacitor C_1 and has a non-zero value for $\partial i/\partial t$.

By inspection of the circuit it is clear that it is a resonant circuit with some damping, where the resonant angular frequency would be close to $\omega = 1/\surd(LC)$, where C is the total capacitance.

At $t=0$ the sum of all the individual voltages V_i across the components is zero, and the condition $\Sigma\, V_i = 0$ applies at all subsequent times, so that a time t later

$$V_0 - (1/C_1)\int_0^t i\,\mathrm{d}t - (1/C_s)\int_0^t i\,\mathrm{d}t - iR - L\,\mathrm{d}i/\mathrm{d}t = 0. \tag{D1}$$

Differentiating (D1) with respect to t gives

$$L\,\mathrm{d}^2 i/\mathrm{d}t^2 + R\,\mathrm{d}i/\mathrm{d}t + i/C = 0 \tag{D2}$$

where $1/C = (1/C_1) + (1/C_s)$ and C is the total capacitance.

As the circuit is a simple LRC, the solution to this equation will be oscillatory with some damping i.e.

$$i = (V_0/\omega L)\mathrm{e}^{-\alpha t}\sin(\omega t) \tag{D3}$$

where $V_0/\omega L$ is the amplitude of initial current, which comes from the fact that in a high-Q circuit at $t=0$ the impedance $\mathrm{j}\omega L$ will dominate over the very small pure resistance R. Use of the sine function ensures $i=0$ at $t=0$.

Substituting (D3) into (D2) yields, after some manipulation,

$$\omega C(2\alpha L - R)\cos(\omega t) + (CL\omega^2 - (\alpha^2 CL - \alpha CR + 1))\sin(\omega t) = 0.$$

In order to balance this equation, both terms must be zero independently, which means

$$(2\alpha L - R) = 0 \tag{D4}$$

and

$$(CL\omega^2 - (\alpha^2 CL - \alpha CR + 1)) = 0 \tag{D5}$$

so that from (D4)

$$\alpha = R/(2L) \tag{D6}$$

and, using (D6),

$$\omega^2 = (1/(LC)) - R^2/(4L^2) \tag{D7}$$

i.e.

$$\omega^2 = (1/(LC)) - \alpha^2. \tag{D8}$$

Note that at this stage although it is known that both R and L are small, their ratio is unknown. However as the resistance is small it is reasonable to expect that there is little damping, so that the damping term $\alpha \approx 0$. This will be proved shortly.

The voltage V_c across capacitor C_s at time t becomes

$$V_c = (1/C_s)\int_0^t \mathrm{d}t$$

$$= (V_0/\omega L C_s)\int_0^t \mathrm{e}^{-\alpha t}\sin(\omega t)\,\mathrm{d}t$$

$$= \frac{V_0}{L\omega C_s}\left\{\frac{\omega}{\alpha^2+\omega^2} - \mathrm{e}^{-\alpha t}\frac{(\alpha\sin(\omega t) + \omega\cos(\omega t))}{\alpha^2+\omega^2}\right\}. \tag{D9}$$

Now (D7) can be rearranged to

$$\omega^2 = (1/4L^2)(4L/C - R^2\} \tag{D10}$$

so that *when the approximation* $4L/C \gg R^2$ is valid then

$$\omega^2 \approx 1/(LC). \tag{D11}$$

Equation (D8) can be rearranged to give

$$\alpha/\omega \approx \surd(1/(LC\omega^2) - 1) \tag{D12}$$

but using (D11)

$$\alpha/\omega \approx 0 \tag{D13}$$

i.e.

$$\alpha \approx 0. \tag{D14}$$

Multiplying the top line in the second term in the brackets in (D9) by ω/ω, the voltage V_c generated across the capacitor C_s is written

$$V_c = \frac{V_0}{L\omega C_s}\left\{\frac{\omega}{\alpha^2+\omega^2} - \omega\mathrm{e}^{-\alpha t}\frac{((\alpha/\omega)\sin(\omega t) + \cos(\omega t))}{\alpha^2+\omega^2}\right\}$$

and using (D13) and (D14) this becomes

$$V_c = \frac{V_0}{L\omega^2 C_s}\{1 - \mathrm{e}^{-\alpha t}\cos(\omega t)\}$$

which, using (D11), yields

$$V_c = \frac{V_0 C}{C_s}\{1 - \mathrm{e}^{-\alpha t}\cos(\omega t)\}.$$

Now whenever αt is small (remember $\alpha \approx 0$) then $\mathrm{e}^{-\alpha t} \approx +1$ so that the maximum possible value for V_c is when $\cos(\omega t) \approx -1$, i.e.

$$V_{c\ \max} = \frac{2V_0 C}{C_s} \tag{D15}$$

where C is the total capacitance of the circuit. Writing

$$C_s = \gamma C_1$$

then

$$C = C_1(\gamma/(\gamma+1)).$$

Thus (D15) can be written

$$V_{c\ max} = \frac{2V_0}{\gamma+1} \tag{D16}$$

The maximum possible voltage is $2V_0$, and this is reached when $\gamma \approx 0$, i.e. when C_s is very small compared with C_1.

It is also interesting to look at the ratio of the energies W_1 and W_2 stored in each capacitor, where W_1 is the energy stored in C_1 before the switch was closed. It can be shown that

$$W_2/W_1 = 4\gamma/(\gamma+1)^2$$

which is equal to 1 when $\gamma = 1$, i.e. when $C_s = C_1$. In this case when the switch is closed all the energy is transferred from C_1 to C_s at the same voltage in a time scale governed by the condition that αt must be small. This interesting result is used in the design of power transmission systems to transfer large amounts of energy along a power line with little energy loss.

Appendix E Some Properties of Common Transmission Lines

The high frequency approximation for the characteristic impedance of a transmission line given by Equation (3.26) is

$$Z_0 = \sqrt{(L/C)}.$$

Thus to make lines with low impedance requires L small and C large. Capacitance can be increased by bringing the conductors closer together, making them bigger or introducing material with higher values of ε_r.

The following approximate formulae for characteristic impedances of different geometry transmission lines, shown in Figure E.1, have been collected from several sources, see the list of references. They are suitable for use in circuit board design work. For a comprehensive discussion of transmission line parameters as well as coupling coefficients between transmission lines, see Chapter 8 in reference [1]. Losses on microstrip lines, including radiation losses are discussed in [5].

(1) *Coaxial line,* neglecting the internal wire inductance, Figure E.1(a). Matick shows [1]

$$Z_0 = 60\sqrt{(\mu_r/\varepsilon_r)}\log_e(b/a)\ \Omega. \qquad \text{(E1)}$$

(2) *Open two-wire line,* wire diameter D, separation h, Figure E.1(b)

$$Z_0 = 120\sqrt{(\mu_r/\varepsilon_r)}\log_e(h/D + \sqrt{((h/D)^2 - 1)})\ \Omega. \qquad \text{(E2)}$$

(3) *Wire of diameter D a distance h over a ground plane,* Figure E.1(c).

$$Z_0 = 60\sqrt{(\mu_r/\varepsilon_r)}\log_e(2h/D + \sqrt{(2h/D)^2 - 1}))\ \Omega. \qquad \text{(E3)}$$

(4) *Wide strip line,* Figure E.1(d). Valid when the thickness is negligible, i.e. $h \gg 3t$ and $w \geqslant 8 - 10h$; see Section 8.4 and Figure 8.8 in [1] for this second justification:

$$Z_0 = 377\sqrt{(\mu_r/\varepsilon_r)}(h/w)\ \Omega. \qquad \text{(E3)}$$

For $0 < w/h < 10$ use [1]

$$Z_0 = \sqrt{(\mu_r\mu_0\varepsilon_r\varepsilon_0)}/C\ \Omega$$

where C is the capacitance per unit length given by the approximate formula

$$C = (w/h)\varepsilon_r\varepsilon_0\{1 + (h/\pi w)(1 + \log_e(2\pi w/h))\ \mathrm{F\ m^{-1}}.$$

As the ratio w/h increases this becomes $w\varepsilon_r\varepsilon_0/h$ F m^{-1}, the same as for a simple parallel plate capacitor. Thus

$$Z_0 \approx 377\sqrt{(\mu_r/\varepsilon_r)}(h/w)/\{1 + (h/\pi w)(1 + \log_e(2\pi w/h))\ \Omega. \qquad \text{(E4)}$$

(5) *General strip lines*

Case (a) A general parallel strip line with two conductors of width w_1 and w_2, of

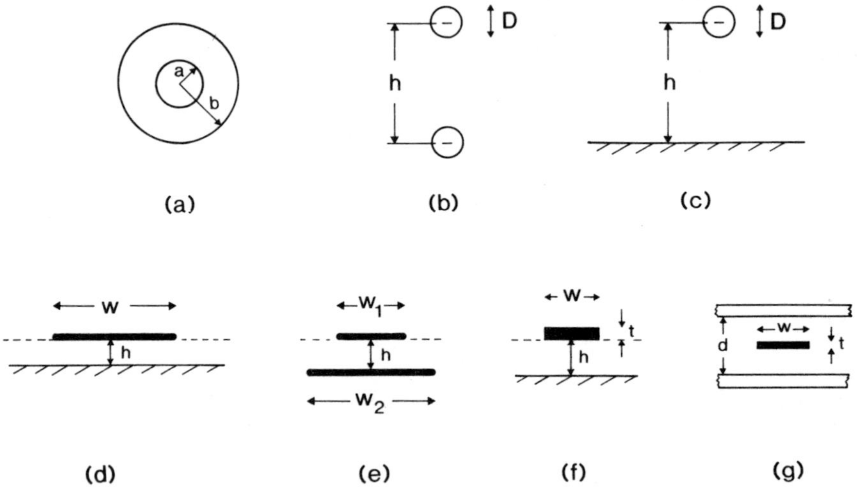

Fig. E.1 The geometrical cross-sections of several different types of transmission lines

negligible thickness, separated by a distance h, Figure E.1(e). There are no simple formulae [1], so use

$$Z_0 = \surd(\mu_r \mu_0 \varepsilon_r \varepsilon_0)/C \ \Omega \tag{E5}$$

where C is the capacitance calculated for different ratios of w_2/w_1 [2] and displayed in Figure 8-7 in reference [1].

Case (b) Microstrip line: a conductor of width w and non-negligible thickness t situated a distance h over an infinite ground plane Figure E.1(f), $\mu_r = 1$. Kaupp gives [3]

$$Z_0 = \frac{87}{\surd(\varepsilon_r + 1.41)} \log_e(5.98h/(0.8w + t)) \ \Omega. \tag{E6}$$

Liao [5] suggests $h < 0.8w$ (i.e. $w/h > 1.25$) for this to be valid.

(6) Shielded strip line of non-negligible thickness, Figure E.1(g), with $w/d > 0.35$, reference [6]

$$Z_0 = \frac{94.15}{\surd\varepsilon_r} \{wK/d + C_f/(8.854\varepsilon_r)\}^{-1} \tag{E7}$$

where

$$K = 1/(1 - t/d)$$

and the fringe capacitance per unit length is given by

$$C_f = \frac{8.854}{\pi} \varepsilon_r \{2K \log_e(K + 1) - (K - 1)\log_e(K^2 - 1)\} \text{pF m}^{-1}.$$

DISCUSSION AND SOME EXAMPLES

Figure E.2 shows the comparison of formulae (E3), (E4) and (E6) as function of (w/h), drawn setting $\mu_r = 1$ and $\varepsilon_r = 1$. Comparison with Figure 8.8 in [1] shows that both (E3) and (E4) overestimate Z_0 somewhat when $w/h \leqslant 2$ compared with the precise solution given by Black and Higgins [2]. For example they calculated $Z_0/\surd(\mu_r/\varepsilon_r) \approx 180 \ \Omega$ for

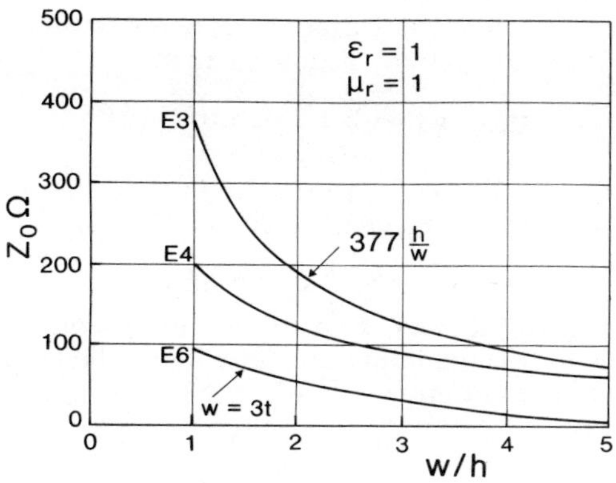

Fig. E.2 Transmission line characteristic impedance as a function of the ratio of conductor width to separation, w/h, calculated using Equations (E3), (E4) and (E6)

equal width conductors when $w/h = 1$, where (E3) gives 377 Ω and (E4) yields 200 Ω. However for $w/h > 2$, (E4) is a very good approximation. Formula (E6) applies to the limiting case when the width of one conductor goes to infinity. The impedance is lowered and (E6) is the commonly used formula. In Figure E.2 it is shown for the case when $w = 3t$. Note that all of the impedances in Figure E.2 can be reduced by using a dielectric with $\varepsilon_r > 1$.

Consider the following examples, which the reader should check and investigate which parameters are the most sensitive in changing the result.

(1) Impedance of an open wire line, Figure E.1(b). This is sold as 300 Ω twin balanced feeder cable, with the following approximate properties; $D = 0.23$ mm, $h = 9.8$ mm, $\varepsilon_r = 2.56$; thus using (E2), $Z_0 = 296\ \Omega$.
(2) Wire of the same diameter as in example 1, at the same distance over an infinite ground plane, Figure E.1(c), i.e. $D = 0.23$ mm, $h = 9.8$ mm, $\varepsilon_r = 2.56$; thus using (E3) $Z_0 = 172\ \Omega$. Because of the increased capacitance the impedance is lower than in example 1.
(3) Wides strip line, Figure E.1(d). If h is to be kept at 9.8 mm, then the width $w \geqslant 10h$ for (E3) to be used, so with $h = 9.8$ mm, $w = 98$ mm, $\varepsilon_r = 2.56$; then $Z_0 = 23.56\ \Omega$. Increasing the width of the conductor from 0.23 mm to 98 mm compared to example 2 has further increased the capacitance, so reducing the impedance.
(4) Microstrip line, Figure E.1(f). With $h = 0.3$ mm, $w = 1.0$ mm, $t = w/3 = 0.33$ mm, $\varepsilon_r = 2.56$ then using (E6) $Z_0 = 20\ \Omega$. Thus one can construct a line with about the same impedance as example 3, but having much smaller dimensions.

ATTENUATION CONSTANT OF A COAXIAL TRANSMISSION LINE

The high frequency approximation for the attenuation constant of transmission line, Equation (3.27), can be written

$$\alpha = R/(2Z_0) + GZ_0/2 \text{ nepers m}^{-1}. \tag{E8}$$

For a continuous dielectric of conductivity σ_d then [7]

$$G = 2\pi\sigma_d/\log_e(b/a)\ \Omega^{-1}\ \text{m}^{-1} \tag{E9}$$

and if the current is concentrated into a surface layer of order one skin depth on the inner and outer conductors with conductivity σ_a and σ_b then [7]

$$R = \sqrt{(f/10^7)}\left\{\frac{1}{a\sqrt{\sigma_a}} + \frac{1}{b\sqrt{\sigma_b}}\right\} \Omega \text{ m}^{-1} \tag{E10}$$

where f is the frequency. Substituting (E1), (E9) and (E10) into (E8), and differentiating, one finds that the attenuation is a minimum when [7]

$$\log_e(b/a) = 1 + \sqrt{(\sigma_a/\sigma_b)}/(b/a). \tag{E11}$$

When the inner and outer conductors are made from the same material this is satisfied when $b/a = 3.59$. Thus from (E1) the impedance of a coaxial line (with $\mu_r = 1$ and ignoring dielectric losses) for minimum attenuation is

$$Z_0 \approx 77/\sqrt{\varepsilon_r}\ \Omega.$$

For dielectrics such as polystyrene where $\varepsilon_r = 2.56$ this gives $Z_0 = 47.92\ \Omega$, nominal 50 Ω cable. This is the reason why 50 Ω cable is commonly used.

IMPEDANCE OF TRANSMISSION LINES WITH FINITE CONDUCTIVITIES OR COMPLEX GEOMETRIES

Reference [8] contains further exact formulae for the impedance of other line configurations as well as the impedances of finitely conducting cylindrical conductors.

There is an interesting analogue technique for finding the impedance of a transmission line with arbitrary geometry by drawing its cross-section with conducting paint on paper having a surface resistance of 376.7 $\Omega\,\text{m}^{-2}$ ('space paper') and measuring Z_0 directly with an ohmmeter. See Section 8.9 in reference [1] and Section 10-5 in reference [9].

REFERENCES

[1] Matrick, R. E. (1969) *Transmission Lines for Digital and Communications Networks* McGraw-Hill, New York.
[2] Black, K. G. and Higgins, T. J. (1955) Rigorous Determination of the Parameters of Microstrip Transmission Lines *IRE Trans. Microwave Theory Techn.* **MTT-3** 93.
[3] Kaupp, H. R. (1967) Characteristics of microstrip transmission lines *IEEE Trans. on EMC* **EMC-16** 185–193.
[4] Assodourian, F. and Rimol, E. (1952) *Simplified Theory of Microwave Transmission Systems, Proc. IRE* **40** 1651–1657.
[5] Liao, S. Y. (1988) *Engineering Applications of Electromagnetic Theory* West Publishing, St Paul.
[6] Cohn, S. (1954) Characteristic Impedance of the Shielded-strip Transmission Line *IRE Trans. Microwave Theory and Techniques* **MTT-2** 52.
[7] Jackson, W. (1945) *High Frequency Transmission Lines* Wiley, New York.
[8] Vance, E. F. (1978) *Coupling to Shielding Cables* Wiley, New York.
[9] Kraus, J. D. (1984) *Electromagnetics* 3rd Edition, McGraw-Hill, New York.

Index

ABB Research Centre, Switzerland 209
absorbers 87–92
  hypothetical case study 88–90
  perfect 51
  performance 90
  real 91
  size 92
absorbtion 137, 140
AF *see* antenna factor
aliasing in FFT 178
Amnesty International v
Ampère's Law 61, 63
amplifiers, dangers to 250
anechoic room 18, 91, 206
antenna
  broadband 200–202
  energy flow around 38
  factor 19, 190, 192
  general properties 186–188
  log-periodic 201
  magnetic dipole *see* small loop
  reasons for study in EMC 185
  short wire 34, 38–40, 139
  slot 188–189
  small loop 34, 40, 85, 139, 143
  Yagi–Uda 201
anti-aliasing filter 177, 178
aperture gain 150, 189
aperture, effects on currents 152
aperture, effects on screened rooms 151
apertures, list of possible treatments 235
attenuation constant in material 73, 82
attenuation constant in waveguides 111–112
audio-rectification 258
Austin, B. xiii

background radiation, trend 12
balancing, to defeat common mode currents 249–250
Balmain, K. G. 185
Baluns 251
bandwidth 257
  fast switching logic 258
  IEEE definition 184
  interference 184
  of detectors 167, 184
  of Fourier series 171
BCI 21
beats 3
Bedrosian, G. 135
Benenson, P. v
Booker, H. G. 189
braided cables, transfer impedance 161–162
Bridges, J. E. 128, 132, 133, 134, 135
British Aerospace PLC xiii
BSI (British Standards Institute) 16, 17
Bulk current injection *see* BCI
BW *see* bandwidth

capacitance, between two wires 66
capacitance, to describe shields 124–128
capacitors, equivalent circuit 240–241
Casey, K. F. 151
cavity resonance 114–116
CENELEC 16
Chatterton, P. A. ii
Cheyne, F. (Ms) xiii
CISPR 16, 17
CMOS 258
coated glass 85
coaxial cable, 50 ohm 100
Coekin, J. A. 221
Collins, S. (Ms) xiii
common mode chokes 250–252
  IL at high frequency 251
common mode current 247
  protection from 248–252
common mode radiation, from cables 198–200
compatibility level 7
compatibility margin 7
conductivity 30
  complex 84–85
conductor, definition 52
conductor, wavelength in 54
conductors, inductance of 239
Cooley, J. W. 177
Cooley, W. W. 134

copper, impedance 57, 77
copper, skin depth 54–56, 77
corrosion, of joints 155
coupling 33
  between cables with pigtails 231–232
  capacitive 65–66
  coefficient between transmission lines 105
  common impedance 222, 224
  frequency dependent 179–180
  inductive 65–66, 179–180
  inductive between planar loops 218–222
  modes 3
  plane wave to loop 214–215
  to part-shielded cable 216–217
Crawford cell 111, 116, 206–208
Crawford, M. L. 206
curl of a vector field 30–32, 273–277
cut-off frequency in waveguides 112

Dainton, J. xiii
data security 5
DATASTOP, metal layer window 156
decibel 82
decibels 191–192
decoupling capacitors 259–263
diamagnetic 27
dielectric, definition 52
dielectric, impedance 43–44
dielectrics, quarter wave 50
diffusion 58–60
diffusion time 59, 110
directional couplers 101–102
directivity 187
dispersion 97
distortion, harmonic 168
distortion, intermodulation 168
disturbance 1
  on mains supply 5, 13–14
dominant mode, in waveguide 112, 113
ducting, of magnetic fields 119, 121, 123, 124, 144
duty cycle 178, 180, 197, 198

ECCOSORB AN75, AN77, AN79 90–91
ECL 258, 260
eddy currents, in shields 124
eddy currents, time constant 135
Edwards, D. N. xiii
EEC xiii, 16, 17, 18
effective aperture, of antenna 188
electric field, in conductors 62
electric polarisation 26, 27
Electro-Metrics Inc. 201, 202
electromotive force 61
electrostatic discharge *see* ESD
EM disturbance 1
EM fields 25
EM interference 1
EMC
  accident example 4
  analogue vs digital circuits 257–258
  coupling estimates 214–223
  equivalent standards 17
  formal definition 6
  frequency dependence 3
  general design guidelines 257–267
  general problems 213–214
  general solutions 15–16
  handbooks 36, 153, 155
  ignition and detonation hazards 5
  problems, examples 4, 5
  regulations 16–17
  shielding 47
  standards, useful addresses 269–272
  testing 16
  tests 201, 210
    sensitivity 21
Emerson & Cuming (UK) Ltd. 90
Emerson & Cuming Microwave Products-Grace Electronic Materials 91
emf *see* electromotive force
EMI 1
emission tests 18
EMP 10, 19, 21, 29, 59, 191, 208, 237, 253, 257
  diffusion into cavities 135
  test pulse 12
  transmission of pulses 59–60
  use of lightning for tests 19
EN *see* Euronorm
end-driven wire model 199–200
energy flow near antenna 38
erf *see* error function
error function (erf) 58–59
ESD 5, 9, 248, 264, 265
  test waveform 10
ETSI 16
Euro-norm (EN) 16
EUT 18, 185

far field 32, 35–37, 65
Faraday's Law 61, 63, 68
fast fourier transform *see* FFT
FCC 16, 191, 197
  class B limits 198, 200
ferrite beads 163, 242, 255
ferromagnetic 28–29, 60
Feynman, R. P. 111, 114

FFT 176, 177–178
fields, reactive 40
filters 235, 242–247, 264–265
  anti-aliasing 177, 178
  IL curves 243, 245
  IL formulae 244
  insertion loss (IL) 242–245
  installation faults 246
  installation guidelines 245–246
  T network 245
flux linkage 62
Fourier coefficients 171
  approximation 181–184
Fourier components, measured 204–205
Fourier series, pulse train 172–175
Fourier series, trapezoidal wave 178–179
Fourier theory 167, 169–175
Fourier transform 175–176
  of aperture 189
  of raised cosine pulse 246
  of single pulse 175
Franceschetti, G. 135
Franks, A.-M. (Ms) xiv
frequency allocations 14
frequency angular 2
frequency dependent coupling to coil 179–180
frequency domain, 170, 176
Fresnel equations 47

gain, of antenna 187
gain, of apertures 150, 189
gas arrestors 254–257
gaskets 155
General Electric Co. 14, 67, 68
gigahertz transverse electromagnetic cell *see* GTEM
Graf, W. 157, 159, 161, 233
ground, loops 225, 230, 247
grounding 223–238
  at high frequency 228–233
  choice of shield end 230
  effects of pigtails 230–233
  for nested shields 234–238
  geometry of connection 230–233
  in zoned systems 237–238
  layout on PCB 258–263
  medical equipment 227
  poor design 227
  regulations 227
  rule of thumb 229
  safety 225–227
  screened rooms 233–235
group velocity 56, 59
GTEM cell 116, 208–210

harmonics 3, 168, 170, 174
Harrison, C. W. 130, 134, 146
Harrison, D. 158
Hejase, H. A. N. 232
high impedance field 38–40
high temperature superconductors 124
Hoefer, W. J. R. 100
Hope, Bob vi
Hubing, T. H. 199

IEC 16
IEE 17, 227
IEEE 6, 17
IL, of common mode chokes 251
IL, of filters 242–245
immunity 7
  level 7
  limit 7
  margin 7
impedance
  arbitrary medium 56
  free space 44
  ground 9
  importance of 46, 51
  in front of boundary 47, 74
  near field 44–45
  of conductors 53–58
inductive coupling, to loop 63–64
inductors, equivalent circuit 241–242
interference 1
ionosphere 85
isolators 235
isolators, opto- 252

Jeans, Sir James 37
joint designs 153–155
Jordan, E. C. 185

Kaufman, J. F. 199
King, L. V. 134

Laplace's equation 33, 121
Lee, K. S. H. 135
Lenz's law 61
Liao, S. Y. 93, 111, 188
light, speed/velocity 2
lightning 8, 223–224, 248, 257
  unsuitable for EMP tests 19
LISN 21
loop antenna 34, 40
Lorentz force 26
low impedance field 38–40
lumped circuit approximation 101

magnetic shielding 85
  using non-magnetic conductors 145–146
magnetisation 26, 27, 29
magnetostatic shielding 121–124
  of large radius cylinder 123
mains impedance 21
mains variation 13
Mardiguian, M. 193, 250, 264
Marvin, A. C. 92, 195
MATHCAD 75, 188
Matick, R. E. 101, 102, 103, 104
Maxwell's equations, detailed discussion 30–32
Maxwell, J. C. vi
mean free path for electrons 85
Meissner effect 124
method of moments *see* MOM
MIL SPEC 59-41 (1986) 19
MIL-STD-285, 148, 149, 156
military systems 5, 258
MININEC 200
mode, common 4
mode, conductive 3, 4
mode, differential 4
modes, cavity resonance 115
modes, radiative 3, 4
MOM codes 199, 200
MSC/EMAS 200
multi-layered media, reflection coefficient 86–87
multiple shields 139
  see also nested shields
muRata Mfg. Co. Inc. 251, 255
mutual inductance 103, 109

NAMAS 17
near field 32, 35–37, 65
  wave impedance of 85
NEC 200
NEMP *see* EMP
neper 73, 82
nested shields 123
  design guide 236–237
  earthing arrangements 233–238
niobium 124
noise 8
  bandwidth 15, 184
  between coupled transmission lines 101–104
  broadband 184
  carried by signal wire 235
  classification 15
  induced in coil 64, 179–180
  model for capacitive coupling 66
  narrowband 184
  reduction by waveshaping 246–247
  solar 9
  sources 7
  via capacitive coupling 66
nomograms 184
non-linear devices 168, 257
Nyquist critical frequency 178

open field site, general layout 20
optical fibre 252
Ott, H. W. 21, 66, 113, 150, 151, 152, 153, 221, 230, 247, 250, 251, 260, 263, 264
out of band response 257
oxidation, of joints 155

Papas, C. H. 130, 134, 146
paramagnetic 27
Paul, C. R. 231
permeability 27, 143
permittivity 27
phase constant 73
phase velocity 43, 56
Pilkington Glass Ltd. 156
Poisson's equation 33
power pattern, of antenna 186
power rails, layout on PCB 258–263
Poynting theorem 37
Poynting vector 37, 139, 186, 190
propagation constant 51–53, 87
pulse charging 67, 281, 283

quantum mechanics 29
Quine, J. P. 113

radiation resistance, of antenna 187–188
radiation
  allow for reflections 193, 196, 197
  common mode 185–186
  computer prediction 200
  differential mode 185–186
  end-driven wire model 199–200
  from cables 198–200
  from isolated loops 192–198
  reduction with decoupling capacitors 259
  man-made 9, 14
radio transmitters 13
radomes 50
Rainford Racks Ltd. xiii
raised cosine pulse 246–247
Ray Proof Ltd. 21, 147, 209
re-reflection 108, 137, 140
  between boundaries 79–81

reflection coefficient 47
  magnetic, from metal 85
reflection loss 137, 145
refractive index 43
resistors, equivalent circuit 240
resonance
  in cavities 114–116
  in filters 244–245
  in screened rooms 149
reverberation chamber 116, 200, 210
RFI 1, 2
rise time 170

saturation, magnetic 29, 123, 143
Schelkunoff, S. A. 109
screened room 82, 147
  effects of apertures 154
  grounding methods 233–236
  mesh walls 150
SE (shielding effectiveness)
  approximate formula for thin films 84
  by current return paths 221
  design for nested shields 236–237
  for electric fields 120
  for magnetic fields 120
  list of optimum treatments 235
  low frequency magnetic formulae 135
  of apertures 112–113, 150
  of cable with pigtail 232–233
  of electrostatic shields 130
  of partial cable shield 220–221
  of slots 153
  of spherical shields around a source 139–145
  of thin shields 138
  of thin walled tube 109
  of windows 156
  relation to neper 82
self inductance 103
self resonant frequency 260
SGEMP 11
Shenfeld, S. 134
shield penetration, general guidelines 265
shield penetration, optimum treatments 235
shield
  arrangement of entry cables 235–236
  between transformer coils 250
  braided 109–110
  low frequency magnetic 132–135
  nesting 123
  practical 145–149
  superconducting 124
shielded cables, coupling to signal wires 108–110
shielding
  ducting model 119
  effects of bonds and seams 136
  electrostatic 124–128
  equivalent circuit model 119, 128–136
  estimating the effects of defects 159–161
  extended plane wave model 139–145
  importance of shield shape 60, 135
  magnetic, using non-magnetic conductors 145–146
  of low frequency H waves 143
  plane wave model 120, 136–139
  quasi-static 121–136
  relating theory to practical shields 145–149
  transmission line model 120, 136–139
  using thin films 143
shielding effectiveness *see* SE
short wire antenna 34, 38–40
Siemens AG 255, 257
Siemens Plessey Assessment Services Ltd. 20, 203, 204, 205
signalling, mains 5
skin depth 53, 88, 109, 129, 130, 132, 136, 261
  increase resistance 55
  water 54
skin effect 96, 100, 129, 133, 222, 239
slots and apertures 150–155
slots, effects on SE of screened rooms 154
Smith, N. A. xiii
SMPS 263
space paper 288
spectrum analysers 176–178
spectrum utilisation 5
SRF *see* self resonant frequency
Standler, R. B. 21, 253, 254
Starfish, nuclear test 10
Stealth 92
Stokes' theorem, 60, 277
structure of fields 32
superscreen cables 163
suppressors 235
  diodes 253–254
  spark gaps 254–257
susceptibility
  electric 27
  electromagnetic 6
  magnetic 27
  tests 18, 203
switch mode power supplies 263
systems, military 5, 258

T network filter 245
Table
  Antennas and associated equipment 201
  Cut-off frequencies in waveguides 112
  Dimensions of Crawford cell 208
  Equivalent EMC standards 17
  FCC Part 15, Subpart J limits 198
  Formulae for magnetic shielding 135
  Frequencies and wavelengths 2
  Frequencies output from non-linear device 169
  Internal inductance of capacitors 241
  Near and far field estimates 37
  Properties of hypothetical absorber 89
  Relative permittivity 28
  Resonant frequencies in cavity 115
  Selected UK frequency allocations 14
  Shielding effectiveness dB 82
  Skin depth in copper 55
  Some properties of magnetic materials 123
  Window shielding effectiveness 156
table-top products, radiation from 199–200
TE wave, in waveguide 112, 113
TEM wave 42–44
  in conductor 54
  on transmission line 99, 105
  transmission through dielectric slab 78–81
  velocity on transmission line 97
Texas Instruments Inc. 247
thin films
  conducting 83, 143
  methods of manufacture 157
  realistic 85
Thomas, A. K. 122, 123
time domain 167–168, 176
TM wave, in waveguide 112
transfer admittance 159
transfer impedance 109, 158–163, 222
  measured 162
  thin walled tube 109–110
transformers 250–251
transients, switching 66–70
transients, very early 69–70
transmission coefficient 47, 78–81
  optical frequencies 84–85
  effective 80
  electric 74
  magnetic 74, 81, 137
  thin film approximation 84
transmission line 56, 69
  admittance 107
  analogy with em fields 93, 100
  attenuation constant, minimum of 287–288
  capacitance per unit length 96
  complex geometries 288
  coupling between two 101–105
  couping of em fields 105–108
  general properties 285–288
  impedance 94–95, 107
  lossy 95–99
  propagation constant 96
  short 99–100
  theory 92–100
  use on PCBs 261
  voltage reflection coefficient 99
  with distributed voltage source 106–108
transmission, of low impedance waves 85
transmission, through conductors 83–85
trapezoidal waveform 178, 182, 196
  chips 247
triangular waveform 171
Tukey, J. W. 177

uniqueness theorem 31
Upton, M. E. G. 195

Vance, E. F. 106, 107, 108, 109, 148, 157, 158, 159, 161, 233, 236, 265
varistors 254
VDE 16, 17
Violette, J. L. N. 140
VLF magnetic field 124

water, refractive index 43
wave equation 42
wave impedance 48
  as function of distance 45
  at low frequencies 140
  in front of conductor 78
  in front of dielectric 76
  inside dielectric 77
  near antennae 279–280
  near loops 143
wave shaping 246–247
waveguides 56, 110–112
  cut-off frequency 111, 112
  frequencies below cut-off 113
  modes 112
  reduction of cut-off frequency 113
  standard 112
wavelength 2
  in conductor 54, 56
waves
  in dielectric 41

non-plane 47
sinusoidal 2
speed in conductors 56
TEM 42
Wheeler, H. A. 133
Wilkinson, L. xiii
windows, conducting 155–157
windows, dielectric 49
windows, half wave dielectric 77, 81

zoning 265–266
design guide 236–237